IWNMS 2004

IWNMS 2004

Proceedings of the International Workshop on Nanomaterials, Magnetic Ions and Magnetic Semiconductors Studied Mostly by Hyperfine Interactions (IWNMS 2004) held in Baroda, India, 10–14 February 2004

Edited by

D. R. S. SOMAYAJULU

M. S. University of Baroda, Vadodara, Gujarat, India

and

K.-P. LIEB

Georg-August University, Göttingen, Germany

Reprinted from *Hyperfine Interactions*
Volume 160, Nos. 1–4 (2005)

Springer

A C.I.P. Catalogue record for this book is available from the Library of Congress.

ISBN 3-642-42724-3
ISBN 978-3-642-42724-4

Published by Springer
P.O. Box 990, 3300 AZ Dordrecht, The Netherlands

Sold and distributed in North, Central and South America
by Springer
101 Philip Drive, Norwell, MA 02061, U.S.A.

In all other countries, sold and distributed
by Springer
P.O. Box 990, 3300 AZ Dordrecht, The Netherlands

Printed on acid-free paper

Hyperfine Interactions (2005)
DOI 10.1007/s10751-005-9142-z

Sponsors

1. Oil and Natural Gas Corporation Ltd. (ONGC)
2. Department of Science and Technology (DST)
3. Council of Scientific and Industrial Research (CSIR)
4. University Grant Commission (UGC)
5. Gas Authority of India Ltd. (GAIL)
6. Indian Petrochemical Corporation Ltd. (IPCL)
7. Hind Hivac Co. (P) Ltd.(HHV)
8. Indian National Science Academy (INSA)

Hyperfine Interactions (2005)
DOI 10.1007/s10751-005-9139-7

Table of Contents

Preface 1

Invited Talks

Y. MANZHUR, J. M. PRANDOLINI, K. POTZGER, A. WEBER, W.-D. ZEITZ H. H. BERTSCHAT and M. DIETRICH / Surface & Interface Magnetism Using Radioactive Probes 3–15

V. SAMOKHVALOV, M. DIETRICH, F. SCHNEIDER, S. UNTERRIC KER and THE ISOLDE COLLABORATION / The Ferromagnetic Semiconductor $HgCr_2Se_4$ as Investigated with Different Nuclear Probes by the PAC Method 17–26

T. BUTZ, D. SPEMANN, K.-H. HAN, R. HÖHNE, A. SETZER and P. ESQUINAZI / The Role of Nuclear Nanoprobes in Inducing Magnetic Ordering in Graphite 27–37

K. P. LIEB, K. ZHANG, G. A. MÜLLER, R. GUPTA and P. SCHAAF / Heavy Ion Irradiated Ferromagnetic Films: The Cases of Cobalt and Iron 39–56

K. SATO, P. H. DEDERICHS and H. KATAYAMA-YOSHIDA / Exchange Interactions and Curie Temperatures in Dilute Magnetic Semiconductors 57–65

SUGATA RAY and D. D. SARMA / Sr_2FeMoO_6: A Prototype to Understand a New Class of Magnetic Materials 67–79

M. BANGAL, S. ASHTAPUTRE, S. MARATHE, A. ETHIRA J, N. HEBALKAR, S. W. GOSAVI, J. URBAN and S. K. KULKARNI / Semiconductor Nanoparticles 81–94

D. K. AVASTHI / Nanostructuring by Energetic Ion Beams 95–106

RATNESH GUPTA, K. P. LIEB, G. A. MÜLLER, P. SCHAAF and K. ZHANG / Xenon-ion Induced Magnetic and Structural Modifications of Ferromagnetic Alloys 107–121

AJAY GUPTA / Depth Resolved Structural Studies in Multilayers Using X-ray Standing Waves 123–142

RAVI KUMAR, S. K. SHARMA, ANJANA DOGRA, V. V. SIVA KUMAR, S. N. DOLIA, A. GUPTA, M. KNOBEL and M. SINGH / Magnetic Study of Nanocrystalline Ferrites and the Effect of Swift Heavy Ion Irradiation 143-156

Poster Papers

S. KAVITA, V. RAGHAVENDRA REDDY and AJAY GUPTA / Preparation of Fe/Pt films with Perpendicular Magnetic Anisotropy 157–163

DILEEP KUMAR and AJAY GUPTA / Effects of Interface Roughness on Interlayer Coupling in Fe/Cr/Fe Structure 165–172

P. B. JOSHI, G. R. MARATHE, ARUN PRATAP and VINOD KURUP / Effect of Addition of Process Control Agent (PCA) on the Nanocrystalline Behavior of Elemental Silver during High Energy Milling 173–180

K. ASOKAN, J. C. JAN, K. V. R. RAO, J. W. CHIOU, H. M. TSAI and W. F. PONG / Electron- and Hole-Doping Effects in Manganites Studied by X-ray Absorption Spectroscopy 181–187

C. N. MURTHY / Nanoencapsulation of Fullerenes in Organic Structures with Nonpolar Cavities 189–192

D. S. RANA, C. M. THAKER, K. R. MAVANI, J. H. MARKNA, R. N. PARMAR, N. A. SHAH, D. G. KUBERKAR and S. K. MALIK / Transport and Magnetic Properties of Eu and Sr Doped Manganite Compound $La_{0.7}Ca_{0.3}MnO_3$ 193–197

TEJASHREE M. BHAVE, C. BALASUBRAMANIAN, HARSHADA NAGAR, SHAILAJA KULKARNI, RENU PASRICHA, P. P. BAKARE, S. K. DATE and S. V. BHORASKAR / Oriented Growth of Nanocrystalline Gamma Ferric Oxide in Ectrophoretically Deposited Films 199–209

KINNARI PAREKH, R. V. UPADHYAY and R. V. MEHTA / Magnetic and Rheological Characterization of Fe_3O_4 Ferrofluid: Particle Size Effects 211–217

S. N. DOLIA and S. K. JAIN / Magnetic Behaviour of Nano-Particles of $Ni_{0.8}Cu_{0.2}Fe_2O_4$ 219–225

N. LAKSHMI, RAM KRIPAL SHARMA and K. VENUGOPALAN / Evidence of Clustering in Heusler like Ferromagnetic Alloys 227–233

S. G. PARMAR, RAVI KUMAR and R. G. KULKARNI / Influence of 50 MeV Lithium Ion Irradiation on Hyperfine Interactions of $Cr_{0.5}Li_{0.5}Fe_2O_4$ 235–239

D. R. S. SOMAYAJULU, NARENDRA PATEL, MUKESH CHAWDA, MITESH SARKAR and K. C. SEBASTIAN / Concentration Dependence of Room Temperature Magnetic Ordering in Dilute Fe: $Ge_{1-x}Te_x$ Alloy 241–246

NARENDRA PATEL, MUKESH CHAWDA, MITESH SARKAR, K. C. SEBASTIAN, D. R. S. SOMAYAJULU and AJAY GUPTA / Magnetic Ordering in $Fe_{0.008}Ge_{1-x}D_x$ (D = As, Bi) 247–252

T. GOVERDHAN REDDY, P. YADAGIRI REDDY, V. RAGHAVENDRA REDDY, AJAY GUPTA and K. RAMA REDDY / ^{151}Eu Mossbauer Spectroscopy in $La_{0.48}Eu_{0.29}Ca_{0.33}MnO_3$ 253–260

Author Index 261–262

Hyperfine Interactions (2005) 160:1
DOI 10.1007/s10751-005-9141-0

Preface

The International Workshop on "Nanomaterials, Magnetic Ions and Magnetic Semiconductors Studied mostly by Hyperfine Interactions" was held during February 10–14, 2004 at the Physics Department, M. S. University of Baroda, Baroda, India. It resulted from the need felt by many hyperfine physicists to strengthen the interaction of young research students with scientists active in these newly developing areas. With 94 participants from seven different countries, the workshop discussed a wide variety of topics in these specific and some allied research fields.

There were 24 invited talks, 29 poster presentations and two tutorials. Large fraction of the invited talks and poster presentations is included in this dedicated issue of Hyperfine Interactions. We are grateful to the authors and referees for their combined efforts in ensuring a good scientific quality of the papers and to meet the high standards of the journal. Thanks are also due to Mukesh Chawda, Narendra Patel, Mitesh Sarkar, K.C. Sebastian, Sangita Parikh (Baroda) and Lucie Hamdi (Goettingen) for help and assistance in the editorial work.

D.R.S. Somayajulu and Klaus-Peter Lieb
Guest Editors

Hyperfine Interactions (2005) 160:3–15
DOI 10.1007/s10751-005-9143-y

Surface and Interface Magnetism Using Radioactive Probes

Y. MANZHUR[1], M. J. PRANDOLINI[1,*], K. POTZGER[1], A. WEBER[1], W.-D. ZEITZ[1], H. H. BERTSCHAT[1] and M. DIETRICH[2,3]
[1]*Bereich Strukturforschung, Hahn–Meitner-Institut Berlin GmbH, D-14091, Berlin, Germany*
[2]*Technische Physik, Universität des Saarlandes, D-66041 Saarbrücken, Germany*
[3]*ISOLDE - Collaboration, CERN, CH-1211, Genève 23, Switzerland*

Abstract. Magnetic properties of impurities at ferromagnetic surfaces and interfaces have been investigated performing Perturbed Angular Correlation (PAC) measurements in the ultra-high vacuum chamber ASPIC (Apparatus for Surface Physics and Interfaces at CERN) using different PAC probes. We present the measurements of magnetic hyperfine fields (B_{hf}) at ^{111}Cd probe atoms (i) in Pd covered by Ni, (ii) on Pd-decorated Ni surfaces, (iii) and on pure Ni surfaces at a variety of local structures like terraces, steps, kinks. The results yield a deep insight into the interplay of structural surface roughness and magnetic roughness on the atomic scale. Correlating the experimental B_{hf} values with the number of their nearest Ni and Pd neighbours, the coordination number, nonlinear dependences were found. These findings are compared with recent theoretical studies which were prompted by the experiments.

Key Words: nuclear methods, PAC, surface and interface magnetism.

1. Introduction

Many *overall* properties (important for possible technical applications) of a thin-layer system sensitively depend on the *local properties on the atomic scale* at the interface or at the surface. These properties are not accessible to most conventional experimental methods, especially at buried layers and interfaces. Experimental techniques using radioactive atoms as local probes are expected to contribute substantially in gaining further knowledge. A variety of nuclear methods can be applied for investigations on the atomic scale, most suitable are methods which measure hyperfine interactions, e.g., Mößbauer spectroscopy [1, 2].

Electric and magnetic hyperfine interactions of radioactive nuclei provide insight into local properties; the electric quadrupole moment of the probe nucleus interacts with electric field gradients (EFG) when there is a noncubic geometry of the adjacent atoms, the magnetic dipole moment interacts with magnetic hyperfine fields (B_{hf}) caused by polarised electrons. The short range of the hyperfine

* Author for correspondence.

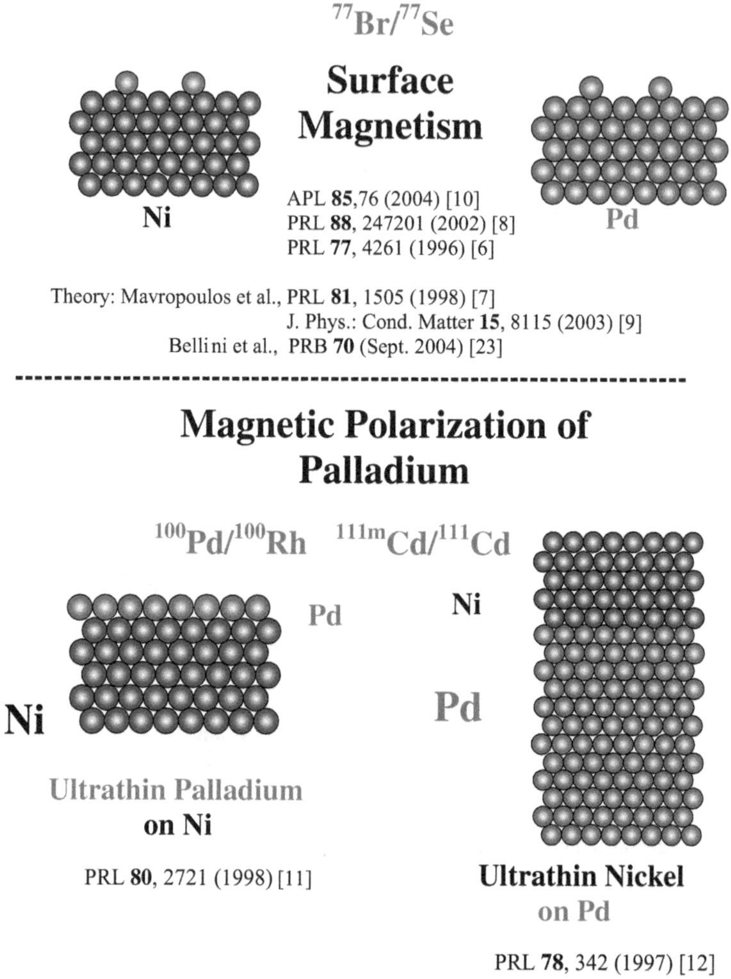

Figure 1. Icons for three categories of experiments. *Top*: surface studies; *Bottom*: ultra-thin layers of Pd on Ni bulk; ultra-thin Ni layers on Pd bulk; radioactive probe atoms in *red* colour, each position is measured at a time.

interaction allows for monolayer resolution of the measurement in layered systems or even for atomic resolution, e.g., on surfaces. The long penetration range of the nuclear radiation (in most cases gamma-radiation) enables measurements in any depth of the sample, in particular at buried interfaces.

With the application of isolated (radioactive) atoms in basic research of ultra-thin and ultra-small structures, i.e., with the determination of the properties of single atoms *in* their immediate atomic environment, the general strive for miniaturisation of electronic circuits (going from larger to smaller and smaller units in size) can be regarded from a very different point of view: We *start* with the smallest possible units, isolated atoms embedded in picostructures. Out of the many fields of ultra-thin layers where nuclear methods can be applied, the present contribution describes our selection: investigation of surface and interface magnetism, which by itself is already a broad field [3–5].

The ultimate aim in magnetic storage (as an example for applications) is the use of single magnetic atoms (e.g., rare earth atoms) carrying information by the occupation of different magnetic and/or chemical and/or structural states. In order to prevent "cross-talk", these atoms (or small clusters of these atoms) have to be separated from each other by embedding them into two-dimensional structures of nonmagnetic atoms. Such magnetic/nonmagnetic structures are in the range of picometers and could be the basis for tera-bit memories. Our basic research may serve for such goals.

2. Three Categories of Experiments

Figure 1 shows the three categories of experiments performed by our group in the first half of the nineties of the last century, inserted are the references [6–23]. The upper part shows the icon representing radioactive probes (^{77}Se, ^{111}Cd) on a ferromagnetic Ni surface. Here, the nonmagnetic atomic probes experience induced magnetic hyperfine interactions because of the ferromagnetic substrate and simultaneously they experience electric quadrupole hyperfine interactions because of the noncubic environment. For a better distinction of these interactions, the same probes were also measured on isoelectronic palladium where only electric hyperfine interactions are present, second icon. The lower left icon shows probe atoms (^{100}Rh, ^{111}Cd) embedded in one atomic monolayer of Pd grown on a Ni single crystal. The probes in Pd show ferromagnetic-like behaviour of Pd. The third type of experiments is found in the lower right icon of Figure 1. On a single crystal of Pd, radioactive probes are positioned and some additional Pd layers are grown (four or seven, as shown, each at a time) with the result, that the position of the probes is well defined in depth. Finally, a few atomic layers of Ni are grown on top allowing for the observation of magnetic interactions exactly at the probe position induced by the ferromagnetic coverage.

This contribution presents the studies *continuing* on the above described three basic experiments.

3. Ultra-thin Ni on Pd

One of our recent studies is based on the preceding investigation of induced magnetic interactions in a single crystal of Pd, when Pd is covered by thin Ni layers [12]. In ref. [12], induced long-range magnetic interactions in Pd single crystals covered with ferromagnetic Ni were observed in perturbed angular correlation (PAC) measurements applying monolayer-resolved sample preparation, see Figure 2. The ^{100}Pd/^{100}Rh probes sensed magnetic interactions even seven atomic layers away from the Ni–Pd interface. The results were explained assum-

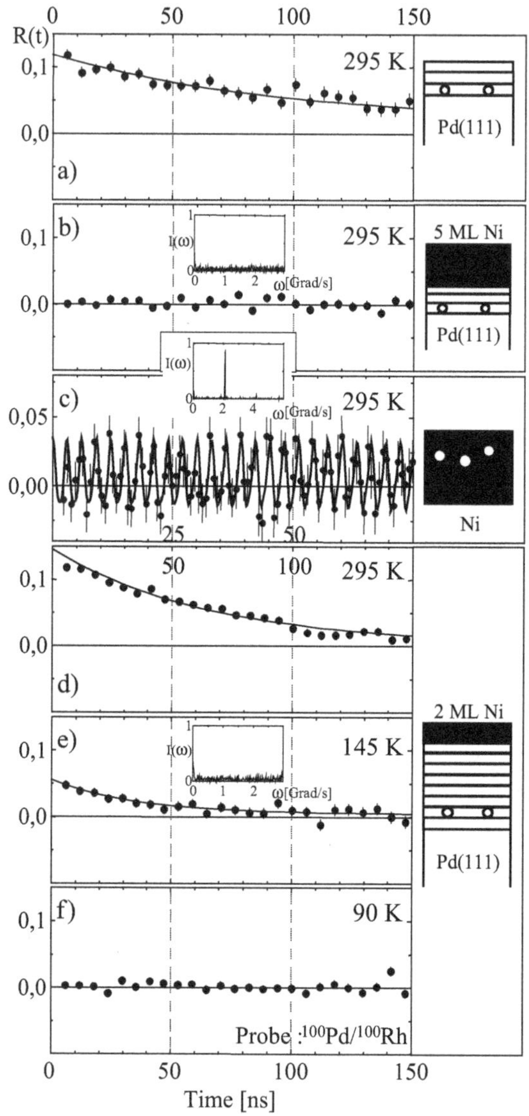

Figure 2. PAC-time spectra of $^{100}Pd/^{100}Rh$, positioned in Ni/Pd-layer systems as indicated on the right hand side of each spectrum. *Solid lines* in (a) and (d) are exponential fits accounting for broad distributions of weak EFGs. In order to keep the Pd probe atoms in position no annealing was performed. In the data of (b), (e) and (f) no oscillations could be detected. For comparison, the PAC spectrum with magnetic oscillations for ^{100}Rh in the bulk of Ni is shown in (c) (*different scales*) corresponding to the known value of B_{hf} = 20.2 T [14]. (Taken from Ref. [12].)

ing fluctuations of magnetic moments strongly interacting with the ^{100}Rh nuclei. The fluctuations are comparable to the fluctuations in Ni above T_C [13], where fluctuating spin clusters become larger and the fluctuations slower, the closer T_C is approached. In the Ni–Pd system, the spin clusters might even be con-

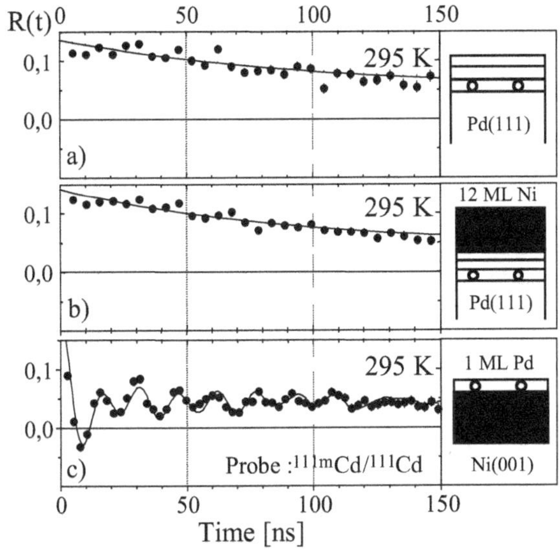

Figure 3. PAC-time spectra of ^{111m}Cd/^{111}Cd in Pd as indicated on the *right*. No static magnetic hyperfine fields causing oscillations are observed at the *sp*-element Cd, when a single crystal of Pd is covered by 12 ml of Ni. For comparison, a PAC spectrum of static magnetic (and electric) hyperfine interactions of ^{111}Cd positioned in 1 ml Pd on a Ni single crystal in zero external field is shown in (c). (Taken from Ref. [12].)

siderably larger, which seems plausible in a layer system for geometrical reasons. In a cross-check experiment, PAC measurements with ^{111}Cd were performed. ^{111}In (decaying to ^{111}Cd) was positioned three monolayers away from the Ni/Pd interface, but no manifestation for a stable hyperfine field could be found, see Figure 3, in contrast to ^{111}Cd in ultra-thin Pd on a single crystal of Ni [11].

These observations raised the question about the magnetic behaviour of the nonmagnetic Cd impurity directly at the Pd/Ni interface, when the ^{111}Cd probe atom is positioned in the topmost layer of the Pd single crystal and has also contact with Ni atoms. The series of PAC experiments is shown in Figure 4. First, the probe atoms in the uncovered topmost Pd layer were recorded were only the EFG was measured in accordance with earlier experiments [15], this time with a considerably larger ratio-function amplitude due an improved sample preparation. In the next step, the surface which contained the ^{111}In/^{111}Cd probes was covered with one monolayer of Ni (partially, as shown in Figure 4). One monolayer of Ni is paramagnetic at room temperature and only the EFG for the probes at the interface can be measured. Finally, four monolayer of Ni were grown in addition at room temperature and the PAC measurement was performed at 83 K. Five monolayer of Ni are already ferromagnetic at RT. As can be observed from Figure 4, in particular at the Fourier transforms, no static magnetic hyperfine fields could be detected although the probes are in contact with

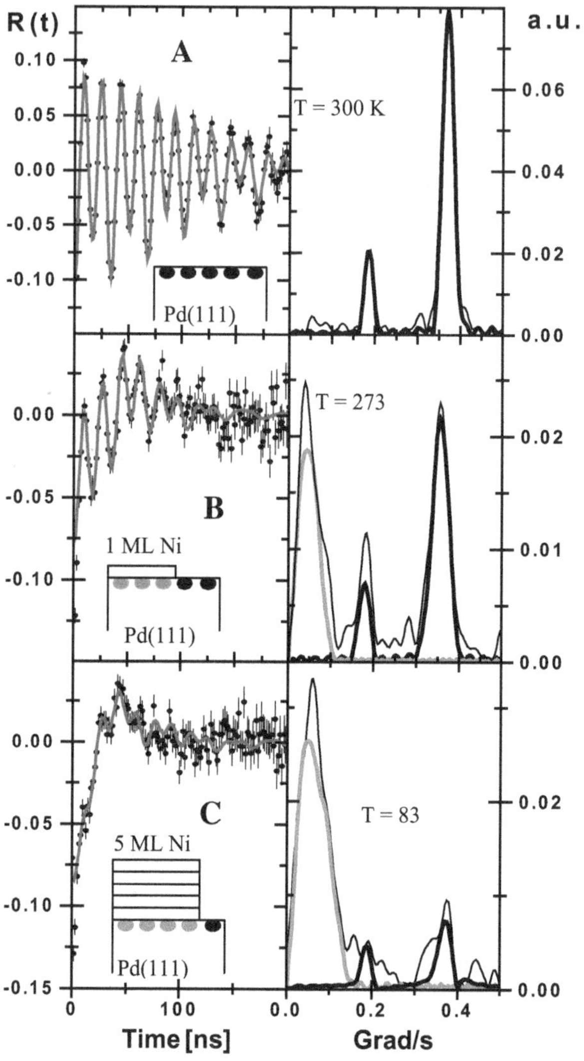

Figure 4. PAC-time spectra of $^{111}In/^{111}Cd$ in Pd in contact with Ni at the positions, shown schematically in the insets. The colours of the frequencies in the Fourier transforms correspond to the colours of the probe atoms. In the *middle* and *lower part* the uncovered (*blue*) fraction (experimentally arranged and controlled by Auger spectroscopy) serves for monitoring.

ferromagnetic Ni. This result is an indication for *fluctuating* fields in Pd even at the immediate interface.

4. Ultra-thin Pd on a Ni Single Crystal

Measurements of the magnetic behaviour of nonmagnetic Cd probes in ultra-thin Pd on a single crystal of Ni [11] is in contrast to the result of the preceding

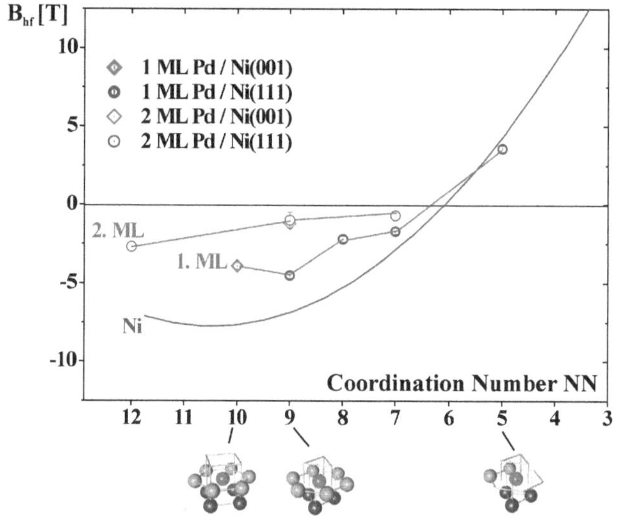

Figure 5. Magnetic hyperfine fields of ^{111}Cd in spin polarised Pd *versus* mixed coordination numbers. The *blue line* anticipates the data of Figure 7. The *green lines* correspond to measurements, were the probes are incorporated in the first or second monolayer of Pd on Ni. Some icons for visualisation of the probe sites are shown on the *x*-axis, Ni atoms in *blue*, Pd atoms in *green*, probes in *red* colour. Note that in this system even NN = 10 can be realised, since Pd grows on Ni(001) with an (111) orientation as on Ni(111).

section. In ultra-thin Pd magnetic moments are induced by the ferromagnetic Ni substrate. These moments cannot be measured directly by hyperfine interactions; they were calculated to be 0.2 μ_B [16]. For ^{111}Cd in one (or two) monolayers of Pd on Ni(001) [11] as well as on Ni(111) [17], *static* magnetic hyperfine fields were measured confirming a ferromagnetic order in Pd induced by the ferromagnetic Ni substrate [1, 18]. In detail, the experiments were performed as follows. Proceeding systematically by growing 0.5, 1, 1.5, etc. monolaycrs of Pd on Ni, with (001) and (111) orientation, the neighbouring atoms next to the Cd probe consisted of a varying number of Pd and Ni neighbours. Thus, a variety of different mixed coordination numbers (number of nearest neighbours NN) for the probe impurity Cd could be achieved, even pure Pd NN. The respective hyperfine fields were measured. The results are collected in Figure 5. It can be seen that the B_{hf} values become smaller with increasing number of Pd neighbours at Cd. This is expected because of the considerably smaller (induced) magnetic moment at Pd as compared to the Ni bulk moment of 0.6 μ_B.

A rather remarkable observation emerged during the course of this series of experiments [11]. The different lattice parameters of Ni and Pd cause an incommensurable growth of Pd resulting in (16 × 2)Pd unit cells on Ni(001). Observing the Pd atoms with respect to the Ni atoms of the substrate, we note that the atomic Ni/Pd geometry varies from Pd atom to Pd atom. Consequently,

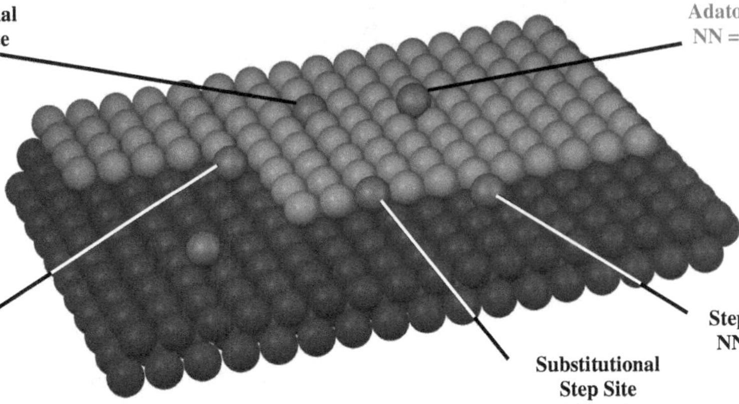

Figure 6. 3D visualisation of an idealised surface of a Ni single crystal with (001) orientation and a step, indicated by the different blue colours. All possible regular lattice sites for impurity atoms as local probes are indicated in red. The probe atoms can be distinguished by the number of their immediate Ni neighbours, the coordination number NN, e.g., the adatom has the coordination number NN = 4. In order to obtain NN = 3, the adatom has to be positioned on Ni(111).

the 3*d*–4*d*-electron hybridisation between Ni and Pd and the induced magnetic moments in Pd vary from neighbour to neighbour. However, for the Cd impurity in PAC measurements, only *one* discrete frequency for the combined interaction was observed [see Figure 3, part c)] suggesting that only selected sites are occupied by the impurity within the unit cells.

A similar experiment with Pd on Ni(111) where Pd forms (13×13)Pd unit cells leads to a similar result. Only *one* magnetic hyperfine field is detectable for Cd incorporated in the monolayer Pd on Ni(111) [17].

Taking these results as a consequence of the coordination-number dependence into consideration, the following conclusions are drawn. (i) The Cd atoms, embedded in an atomic Pd layer on a two-dimensional Ni substrate, occupy only selected sites. (ii) One might get an ordered structure of impurities in the monolayer of Pd when a suitable amount of stable Cd atoms is added in the growth process.

In the following section, the *magnetic* nature of the embedded impurities shall be studied closer.

5. Pure Ni Surfaces

In order to study the magnetic hyperfine fields at Cd impurities on pure Ni surfaces, the different surface orientations may be regarded first. In Figure 6, all regular sites for impurities on the Ni(001) surface are shown, of course, a similar presentation of impurities on Ni(111) looks a little different.

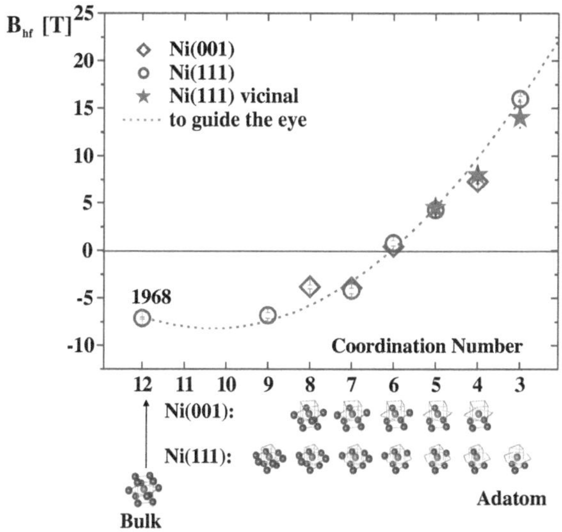

Figure 7. Magnetic hyperfine fields B_{hf} at ^{111}Cd *versus* decreasing coordination number NN of neighbouring Ni atoms. The coordination numbers for the different surface orientations are depicted in the various icons on the *x*-axis. The values for NN = 4 and 6 of Ni(001) are taken from Ref. [21]. The bulk measurement is included for which the only sign assignment exists [19]. For NN = 3 and 4, positive signs were predicted [20]. The red stars refer to a measurement on a stepped Ni(111) crystal where these three sites were measured simultaneously [10].

By a carefully chosen sample preparation, mainly by the choice of temperature, each of these surface sites can predominantly be occupied by the probe atoms. Depositing the radioactive probes at lower temperatures, $T < 90$ K, essentially only adatom sites are occupied on Ni(001) because of the restricted surface motion for Cd. For Ni(111), we found stable adatom sites for $T \leq 36$ K [8]. Increasing the sample temperature after deposition, the probes first move to steps, then to kinks, and at further increased temperatures the probes are found to be incorporated in the first row of the steps. At room temperature, the probes usually are found to be incorporated in terraces. The occupation of terrace sites can be optimised by annealing up to 400 K. The achievable fraction of different surface sites does not only depend on the mobility of the impurity but also on the temperature-dependent mobility of the host atoms on the surface. A further successful manipulation is achieved, when the probes are deposited at a certain temperature and then a submonolyer of Ni is grown by molecular beam epitaxy (MBE). Thus, by a rather complex sample preparation procedure, selected sites can be occupied in order to obtain PAC spectra for an unambiguous interpretation.

Since each site of the Cd impurity is characterised by the coordination number of nearest Ni neighbours (NN), the measured $|B_{hf}|$ values can be ordered *versus* the (decreasing) coordination number, Figure 7. The bulk value is included with $B_{hf} = -6.9(1)$ T [19], the only measurement so far, where also the sign was deter-

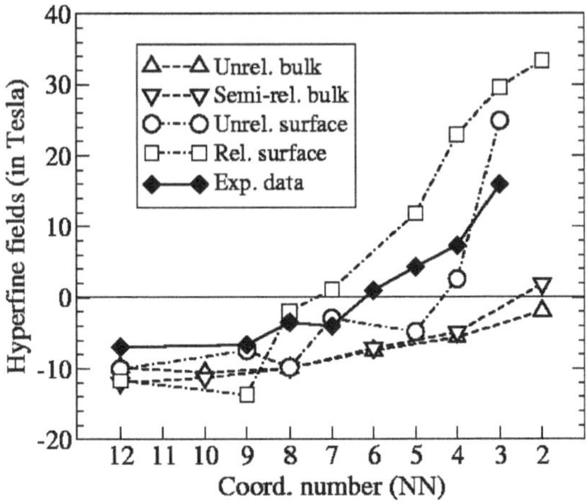

Figure 8. Calculated coordination-number dependence of the Cd hyperfine field in and on Ni [23]. *Circles* show the results for an unrelaxed surface and are very close to the results of Mavropoulos [22]. *Open squares* represent the calculations taking into account surface relaxations around the impurity. The experimental data are from Figure 7 [8]. (*Triangles* refer to simulations with bulk cells [23].) (By courtesy of S. Cottenier and V. Bellini.)

mined experimentally. Calculations for the magnetic hyperfine fields at Cd as adatoms on Ni(001) and Ni(111) predict positive fields [20]; therefore, starting with a negative field for NN = 12 (bulk) and ending with positive fields for NN = 4 and NN = 3 on Ni(001) and Ni(111), respectively, the nonlinear curve (Figure 7) for increasing magnetic hyperfine fields *versus* decreasing coordination numbers was found [8], two experimental results of Voigt *et al.* [21] are included.

The behaviour of magnetic hyperfine fields *versus* the coordination number is the subject of several recent calculations with different approaches. The publication on the experimental coordination-number dependence [8] caused a theoretical study by Ph. Mavropoulos [22] extending the work of Ref. [7] where the 4*sp* elements were treated. This time, the calculations of the magnetic hyperfine fields were applied to the 5*sp* impurities for most of the coordination numbers. The result shows that also a strong coordination dependence is observed for calculated hyperfine fields, and, in addition, the coordination-number dependence is different and characteristic for each impurity. For Cd, the experimental trend was confirmed: increasing B_{hf} *versus* decreasing NN.

Meanwhile, a second theoretical study was performed [23]. In this study, lattice relaxation around the impurity at the surface was taken into account. Figure 8 shows the results along with the experimental values taken from Figure 7.

6. Conclusions

In parallel to our publication on the coordination-number dependence and shortly after, a series of studies in theory and experiment was published where the

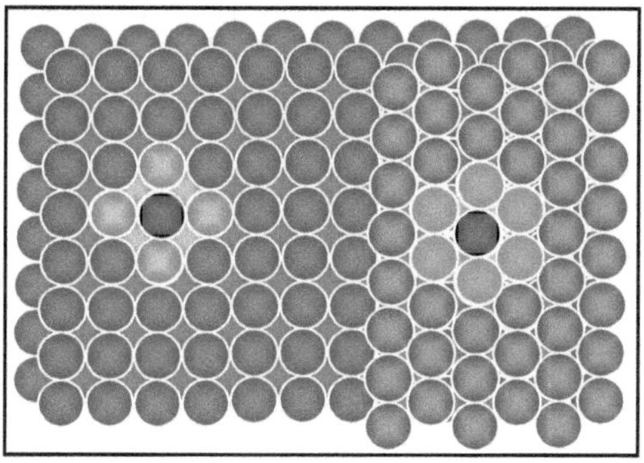

Figure 9. Impurities (*red* colour) in a Ni(001) surface (*blue* colour) and in one monolayer Pd(111) (*green* colour) on Ni(001), both at terrace sites. The nearest neighbours (NN) of the impurities are displayed by lighter colours. The impurity in Ni has eight NN of Ni; the impurity in Pd has four NN of Ni and six NN of Pd.

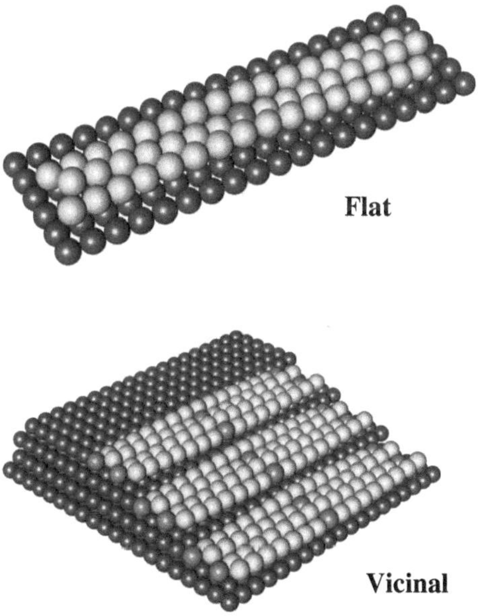

Figure 10. Artist's views. *Top*: An ordered structure of isolated impurities (e.g., rare earth atoms) embedded in a monolayer (e.g., Pd) on a ferromagnetic surface (e.g., Ni). Pd forms 10^{13} unit cells/cm^2 on Ni of which one is shown. The impurities occupy selected sites within the unit cell. *Bottom*: Ordered structure of ultra-small clusters of impurities on a regular Pd-covered stepped surface.

coordination number plays an important role on the magnetic properties. These studies treated small metallic clusters attached to surfaces. Adatoms and small clusters on Ag and Au surfaces where treated theoretically by Cabria *et al.* [24]; Lazarovits *et al.* [25] studied small Fe, Co and Ni clusters on a Ag surface. The magnetism of Co clusters in a Cu surface was considered by Klautau *et al.* [26].

Experiments on small Fe clusters on ultra-thin Ni were published by Lau *et al.* [27] and on small Co clusters on Pt surfaces by Gambardella *et al.* [28]. Already in earlier publications much attention was paid to magnetic systems where single transition-metal atoms were attached to metallic surfaces and their enhanced *d*-moments and spin-orbit coupling were calculated [29–31]. However, much less attention was paid to the change of the properties, in particular, the magnetic properties of the substrate atoms which are in contact with the clusters or adatoms. If we compare all these studies with our experimentally found coordination-number dependence for nonmagnetic impurities, we might conclude that the nonmagnetic impurities on a ferromagnetic substrate form cluster-like magnetic units together with their nearest neighbours: *impurity-induced magnetic units*. The nearest neighbours with their localised *d*-band are distinguished from all other atoms of the surface by the *sd* hybridisation with the impurity leading to the induced valence polarisation at the impurity [7].

Summarising, the studies on the atomic scale of the behaviour of single nonmagnetic atoms at ferromagnetic surfaces and interfaces result in two observations:

(i) The impurities induce cluster-like magnetic units by the inclusion of their nearest neighbours mediated by the *sd* hybridisation, illustrated in Figure 9.
(ii) Within the unit cell of an epitaxily grown Pd layer on Ni, the impurities select only one type of sites (out of many) hinting at the opportunity for ordered structures, see an artist's view in Figure 10.

Acknowledgements

M. J. P. wishes to thank the Deutsche Forschungsgemeinschaft (DFG) for generous support. M. D. acknowledges the financial support by the BMBF (05 KK1TSA/7). This work was supported by the European Commission (HPRI-CT-1998-00018).

References

1. Bland J. A. C. and Heinrich B. (eds.), *Ultrathin Magnetic Structures*, Springer-Verlag, Berlin, 1994.
2. Shinjo T., *Surf. Sci. Rep.* **12** (1991), 49 and Refs. therein. Gradmann, U. In: K. H. J. Buschow (ed.), *Handbook of Magnetic Materials*, Vol. 7, Elsevier S.P., Amsterdam, 1993, p 1.
3. O'Handley Robert C., *Modern Magnetic Materials*, John Wiley & Sons, New York, 2000, p. 432.
4. Nalwa H. S. (ed.), *Magnetic Nanostructures*, American Scientific Publishers, California, 2002.
5. *Magnetismus von Festkörpern und Grenzflächen*, Vorlesungsmanuskripte Forschungszentrum Jülich, 1993, ISBN 3-89336-110-3. *Magnetische Schichtsysteme in Forschung und Anwendung*, Vorlesungsmanuskripte Forschungszentrum Jülich, 1998, ISBN 3-89336-235-5.

6. Granzer H., Bertschat H. H., Haas H., Zeitz W.-D., Lohmüller J. and Schatz G., *Phys. Rev. Lett.* **77** (1996), 4261.
7. Mavropoulos P., Stefanou N., Nonas B., Zeller R. and Dederichs P. H., *Phys. Rev. Lett.* **81** (1998), 1505; Mavropoulos P., Stefanou N., Nonas B., Zeller R. and Dederichs P. H., *Phil. Magazine B* **78** (1998), 435.
8. Potzger K., Weber A., Bertschat H. H., Zeitz W.-D. and Dietrich M., *Phys. Rev. Lett.* **88** (2002), 247201.
9. Mavropoulos P., *J. Phys., Condens. Matter* **15** (2003), 8115.
10. Prandolini M. J., Manzhur Y., Weber A., Potzger K., Bertschat H. H., Ueno H., Miyoshi H. and Dietrich M., *Appl. Phys. Lett.* **85** (2004), 76.
11. Bertschat H. H., Blaschek H.-H., Granzer H., Potzger K., Seeger S., Zeitz W.-D., Niehus H., Burchard A. and Forkel-Wirth D., *Phys. Rev. Lett.* **80** (1998), 2721.
12. Bertschat H. H., Granzer H., Haas H., Kowallik R., Seeger S. and Zeitz W.-D., *Phys. Rev. Lett.* **78** (1997), 342.
13. Gottlieb A. M. and Hohenemser C., *Phys. Rev. Lett.* **31** (1973), 1222. Böni P., Mook H. A., Martinez J. L. and Shirane G., *Phys. Rev. B* **47** (1993), 3171.
14. Koiĉki S., Koster T. A., Pollak R., Quitmann D. and Shirley D. A., *Phys. Rev. B* **32** (1970), 351.
15. Hunger E. and Haas H., *Surf. Sci.* **234** (1990), 273.
16. Blügel S., *Europhys. Lett.* **7** (1988), 743.
17. Potzger K., Weber A., Zeitz W.-D., Bertschat H. H. and Dietrich M., *Phys. Rev. B* **72** (2005), 054435.
18. Gradmann U. and Bergholz R., *Phys. Rev. Lett.* **52** (1984), 771. Gradmann U. In: Buschow K. H. J. (ed.), *Handbook of Magnetic Materials*, Vol. 7, Elsevier S.P., Amsterdam, 1993, p. 1.
19. Shirley D. A., Rosenblum S. S. and Matthias E., *Phys. Rev.* **170** (1968), 363.
20. Lindgren B. and Ghandour A., *Hyperfine Interact.* **78** (1993), 291. B. Lindgren, *Z. Naturforsch.* **57a** (2002), 544.
21. Voigt J., Ph.D. thesis, Universität Konstanz, 1990, unpublished. Voigt J., Ding X. L., Fink R., Krausch G., Luckscheiter B., Platzer R., Wöhrmann U. and Schatz G., *Phys. Rev. Lett.* **66** (1991), 3199.
22. Mavropoulos P., *J. Phys.: Condens. Matter* **15** (2003), 8115.
23. Bellini V., Cottenier S., Çakmak M., Manghi F. and Rots M., to appear in *Phys. Rev. B* **70**, (Sept. 2004).
24. Cabria I., Nonas B., Zeller R. and Dederichs P. H., *Phys. Rev. B* **65** (2002), 0544414.
25. Lazarovits B., Szunyogh L. and Weinberger P., *Phys. Rev. B* **65** (2002), 104441.
26. Klautau A. B. and Frota-Pessôa S., *Surf. Sci.* **497** (2002), 385.
27. Lau J. T., Föhlisch A., Nietubyć R., Reif M. and Wurth W., *Phys. Rev. Lett.* **89** (2002), 057201.
28. Gambardella P., Rusponi S., Veronese M., Dhesi S. S., Grazioli C., Dallmeyer A., Cabria I., Zeller R., Dederichs P. H., Kern K., Carbone C. and Brune H., *Science* **300** (2003), 1130.
29. Lang P., Stepanyuk V. S., Wildberger K., Zeller R. and Dederichs P. H., *Solid State Commun.* **92** (1994), 755.
30. Stepanyuk V. S., Hergert W., Wildberger K., Zeller R. and Dederichs P. H., *Phys. Rev. B* **53** (1996), 2121.
31. Nonas B., Cabria I., Zeller R., Dederichs P. H., Huhne T. and Ebert H., *Phys. Rev. Lett.* **86** (2001), 2146.

Hyperfine Interactions (2005) 160:17–26
DOI 10.1007/s10751-005-9146-8

The Ferromagnetic Semiconductor $HgCr_2Se_4$ as Investigated with Different Nuclear Probes by the PAC Method

V. SAMOKHVALOV[1,a], M. DIETRICH[2,b], F. SCHNEIDER[1]
S. UNTERRICKER[1,*] and THE ISOLDE COLLABORATION[3]
[1]*Institut für Angewandte Physik, TU Bergakademie Freiberg, D-09596 Freiberg (Sachsen), Germany; e-mail: unterr@physik.tu-freiberg.de*
[2]*Technische Physik, Universität des Saarlandes, D-66041 Saarbrücken, Germany*
[3]*CERN, CH-1211 Geneva 23, Switzerland*

Abstract. The hyperfine interactions of six different PAC (perturbed angular correlations) probes in the ferromagnetic spinel semiconductor $HgCr_2Se_4$ have been investigated. Thereby the site occupation of the probes was determined and opposed to that in other comparable substances. All of the three lattice sites could be tested. Theoretically calculated hyperfine fields (WIEN97) were compared with the experimental values.

Key Words: doping behavior, hyperfine fields, implanted radioisotopes, magnetic semiconductors, PAC-method.

1. Introduction

Spin electronics has the potential to extend the technological and scientific applications of microelectronics by using injection, transport and controlling of spins for processing information in electronic components. Thereby ferromagnetic semiconductors have good chances to be used as basic materials in spin electronics. The ferromagnetic spinel semiconductors (FMSS) of type AB_2C_4 with A = Cd, Hg, Cu, B = Cr and C = Se, S belong to the oldest magnetic semiconductors known. They behave like real semiconductors, possess relatively simple ferromagnetic properties and Curie temperatures up to 430 K (in the case of $CuCr_2Se_4$). Recently the epitaxial growth of $CdCr_2Se_4$ films on both GaAs and GaP has been carried out [1] and the possibility of injection of polarized spins from a n-$HgCr_2Se_4$ layer into n-InSb was demonstrated [2]. That's why the FMSS can fulfill some requirements on materials for spin electronics. Never-

* Author for correspondence.
[a] Present Address: AMD Saxony, Dresden, Germany.
[b] Present Address: Deutsche Solar AG, Freiberg, Germany.

theless the chances of the Mn substituted III–V semiconductors to be used in this new field should be better but the FMSS are surely interesting systems to study semiconducting and ferromagnetic properties together.

Within the scope of our investigation of ferromagnetic spinel semiconductors we used the hyperfine interaction of implanted nuclear probes to get a microscopic insight into the properties of these substances. By the PAC (= perturbed angular correlations) method the substances $CdCr_2Se_4$, $CdCr_2S_4$, $HgCr_2Se_4$ and for comparison the metallic $CuCr_2Se_4$ were investigated. In this paper only the results for $HgCr_2Se_4$ with a Curie temperature of T_C = 106 K will be presented and discussed. In our experiments we used a total of six radioactive PAC probes: $^{111}In(^{111}Cd)$, ^{111m}Cd, $^{111}Ag(^{111}Cd)$, $^{117}Cd(^{117}In)$, ^{199m}Hg, and $^{77}Br(^{77}Se)$. Thereby In, Ag, and Br are essential doping atoms in $HgCr_2Se_4$. They can be used to adjust properties like conductivity, magnetic and optical properties in this semiconductor. The implantation of the probes was made by help of the ISOLDE on-line separator at CERN-Geneva (Switzerland). The ISOLDE separator delivers these probes with high isotopic purity and intensity. Therefore also short-lived radioisotopes like ^{111m}Cd (49 min) and ^{199m}Hg (43 min) can be used for PAC measurements without any problems.

By the PAC method the electric and magnetic hyperfine interaction of the radioactive probe nuclei in their environment can be analyzed. Results of the PAC method are besides the hyperfine interaction parameters magnetic hyperfine field (B_{hf}) and electric field gradient (efg) also the lattice site determination of the probe atoms, the behavior of defects, the annealing of implantation damage and more generally a microscopic view on the environment of the probes. The method works at any temperature.

Results of the investigations of $CdCr_2Se_4$, $CdCr_2S_4$ and $CuCr_2Se_4$ which were treated in the same project can be found in [3, 4]. All the substances show similarities but also interesting differences in the behavior of their hyperfine interaction parameters. By comparison of the results a comprehensive picture of the FMSS is created.

The measured hyperfine fields can be used to test the quality of theoretical charge and spin density distributions determined by modern ab-initio density functional methods. In this way we can obtain a deep microscopic understanding of the substances investigated. We applied the WIEN97 program [5], which is based on the FLAPW-method (full potential linear augmented plane waves) to calculate the full potential from a self consistent charge density distribution. The efg are determined by the second derivatives of the potential and the magnetic hyperfine fields by the distribution of polarized spins.

2. Methodical and experimental details

In a semiclassical picture the hyperfine interaction of probe nuclei possessing magnetic moments and electric quadrupole moments in environments characterized by magnetic hyperfine fields (B_{hf}) and electric field gradients (efg) can be described by the precession of nuclear spins. The emission characteristics of atomic nuclei is anisotropic regarding their spin direction. If we have an ensemble of polarized spins we will observe such an anisotropic distribution of emitted γ rays. The PAC probes decay by a γ–γ cascade. Thereby the first γ-quantum is registered in a start detector and the second one in the stop detector. In this way the coincidence spectrum reflects the anisotropic angular correlation characteristics. By the precession of the probe spins the angular correlation rotates (perturbation of the angular correlation) and modulates the coincidence rate $N(\Theta, t)$. Θ is the angle between two detectors and t is the decay time of the intermediate level. From detector combinations which measure under different angles Θ a coincidence ratio $R(t)$ is calculated, the periodicity of which reflects the nuclear spin precession. By a Fourier transformation of $R(t)$ the interaction frequencies can be determined from the power spectrum $F(\omega)$ and with the more or less precisely known nuclear moments the hyperfine fields B_{hf} and efg are evaluated from the interaction frequencies. Very often the probe environments are not uniform. This results in a damping of the $R(t)$ pattern and a line broadening of the Fourier peaks. In the case of cubic probe environments the quadrupole interaction vanishes and without a magnetic field the $R(t)$ spectrum shows the time independent unperturbed anisotropy. A pure magnetic interaction $R(t)$ is reflected by a periodicity of twice the Larmor frequency ω_L, if the sample is magnetically polarized perpendicular to the detector plane. For a pure electric quadrupole interaction depending on the spin I of the intermediate level more than one interaction frequency can occur. In the case of an axially symmetric efg with I = 5/2, for instance, three interaction frequencies ω_0, $2\omega_0$ and $3\omega_0$ are observed. From ω_0 the quadrupole coupling constant ν_Q is calculated ($\nu_Q = eQV_{zz}/h$, Q: quadrupole moment, V_{zz}: main component of the efg).

The $HgCr_2Se_4$ single crystals were grown by closed-tube vapour transport. They have an octahedral shape. All of the $A^{2+}B_2^{3+}C_4^{2-}$ FMSS crystallize in the normal spinel structure. The point symmetry at the A-site is cubic $(\overline{4}3m)$ and at the B-site $(\overline{3}m)$ as well as the C-site (3 m) axially symmetric efg exist. At all three sites magnetic hyperfine fields are expected in the ferromagnetic state.

The following six probes were used for the investigations (spins, half lives and energies of the essential intermediate level in brackets): ^{111}In(^{111}Cd 5/2+, 84 ns, 245 keV), ^{111m}Cd(5/2+, 84 ns, 245 keV), ^{111}Ag(^{111}Cd 5/2+, 84 ns, 245 keV), ^{117}Cd(^{117}In 3/2+, 54 ns, 660 keV), ^{199m}Hg(5/2−, 2.5 ns, 158 keV) and ^{77}Br(^{77}Se 5/2−, 9.7 ns, 250 keV). They all were implanted into the single crystals at the ISOLDE-GPS on-line separator with an energy of 60 keV. The implantation damage could be annealed by a thermal treatment for 10 min at 500°C in evacuated silica ampoules.

The PAC apparatus consisted of four detectors with BaF_2 scintillators (time resolution <1 ns) and fast–slow coincidence electronics. Normally the four detectors were symmetrically arranged at 90 degrees separation and with the sam-

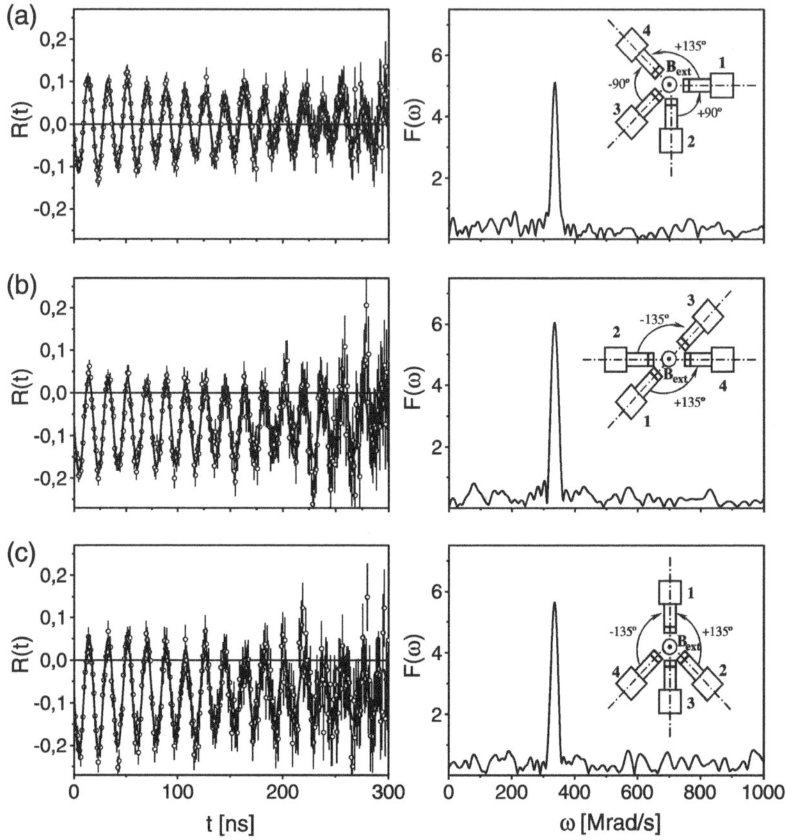

Figure 1. Sign-sensitive measurements of ^{111m}Cd in $CdCr_2Se_4$ in the FM state at 77 K with an external magnetic polarizing field of 0.5 T perpendicular to the detector plane (field direction up). *Left*: time-dependent coincidence ratio *R*(*t*); *Right*: corresponding Fourier power spectrum *F*(ω). The three different detector arrangements discussed in the text are inserted in the Fourier box. Only the optimal geometry of **a** gives an unshifted sine dependence.

ple in the centre. Up to 12 subspectra N(Θ, *t*) (eight 90° and four 180° spectra) were measured simultaneously. During the measurements the sample temperature could be chosen between 10 K and R. T. with a cryostat and by a heating device up to 1000 K. Additionally a small permanent magnet arrangement with two FeNdB magnets was used in the ferromagnetic state to polarize the samples perpendicular to the detector plane.

The sign determination of the magnetic hyperfine fields with the normal 90°/ 180° detector combination is not possible. This holds also for magnetically polarized samples. To determine the sign a ±135° detector arrangement is suitable (the two signs are determined relative to the direction of the perpendicular polarizing field and the geometrical sequence of the start and stop detectors). With a four detector arrangement only eight subspectra can be used at most (four

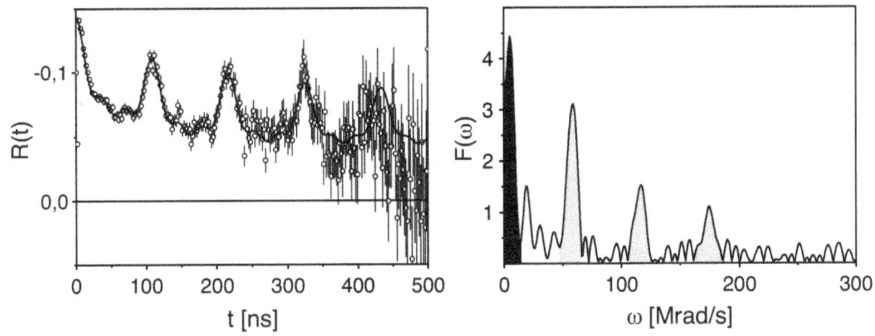

Figure 2. Measurement of $^{111}In(^{111}Cd)$ in $HgCr_2Se_4$, PM state at R. T. Left R(t), right F(ω). The In probes substitute the cubic (A-)site as shown by the practically time independent part of R(t) (there is a small influence of randomly distributed distant defects) but also the noncubic (B-)site with an axially symmetric efg. In the Fourier spectrum the peak of the probes at the (A-)site is rendered *dark* and the three peaks for the B-site *gray*. The implantation damage was annealed by a thermal treatment 600°C/15 min in vacuum.

+135° and four −135°, thereby 225° and 45° arrangements are equivalent to 135° arrangements if the change of signs is considered). In principle three of such four detector assemblies are possible which give four +135° and four −135° pairs. But only one of them is optimal because one has also to take into account the different detector efficiencies and energy window adjustments in practice. Due to the high symmetry of the four detector 90°/180° arrangement these influences are cancelled by forming the ratio R(t). In the case of the ±135° assembly this is granted only for the optimal geometry. The other two arrangements cause a shift of the R(t) zero line, normally. In Figure 1 an example is shown from a measurement with ^{111m}Cd in $CdCr_2Se_4$ and a pure magnetic interaction in the ferromagnetic state.

The corresponding three different detector arrangements are inserted into the Fourier spectra. Only the optimal arrangement of Figure 1a gives the correct sine dependence symmetrically to the zero line. From the phase at $t = 0$ the sign of the magnetic hyperfine field and from the frequency its magnitude can be determined (in this case $B_{hf} = + 11.3(1)$ T). Further details to these considerations can be found in [6].

3. Results and discussion

The probe $^{111}In(^{111}Cd)$: In the paramagnetic (PM) state two different environments are observed. About 60% of the probes are at cubic sites with only little influence of distant defects. The remainder of 40% is influenced by an axially symmetric quadrupole interaction (Figure 2).

By quenching the sample and also by measuring at higher temperatures the fraction of probes at cubic sites increases. The quadrupole coupling constant ν_Q of the noncubic environments increases with rising temperature of mea-

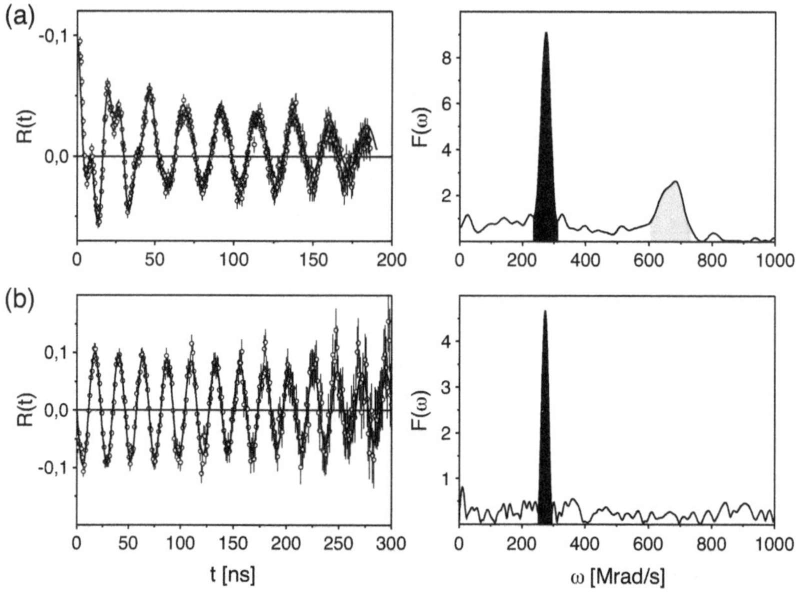

Figure 3. (a) The same sample as in Figure 2 but now in the FM state (77 K). 90°/180° detector arrangement with an external polarizing field of 0.5 T perpendicular to the detector plane. The two rendered peaks show again the A-site probes (*dark*) and the B-site probes (*gray*) with the larger field. (b) A sign sensitive measurement with ^{111m}Cd in $HgCr_2Se_4$ (FM state, 77 K, ±135° arrangement, external field perpendicular). Cd as a divalent atom substitutes the A-site. The frequency and the magnetic hyperfine field are the same as for ^{111}In(^{111}Cd) at the A-site. The interpretation is so simple because for both measurements the hyperfine interaction of Cd(5/2+) is applied.

surement. Obviously In probes can be substituted at cubic A-sites and also at non cubic B-sites. This can be proved by measurements in the ferromagnetic (FM) state (Figure 3a). Here, as visible in the Fourier spectrum, two Larmor frequencies are observed. The smaller one is exactly the frequency which was also measured for ^{111m}Cd in FM $HgCr_2Se_4$ (Figure 3b). The two frequencies correspond to the magnetic hyperfine fields $|B_{hfA}| = 9.4(1)$ T and $|B_{hfB}| = 23.3(2)$ T at 77 K.

The probe ^{111m}*Cd:* Obviously ^{111m}Cd substitutes exclusively the Hg position (A-sites). Both Hg and Cd are divalent. Figure 3b shows a sign sensitive measurement (see Section 2). The pure magnetic interaction confirms that all ^{111m}Cd probes are substituted at the A-site. The corresponding magnetic hyperfine field is $B_{hfA}(\mathrm{Cd}) = +9.4(1)$ T at 77 K.

Figure 4 shows an additional sign sensitive measurement with the ^{111}In(^{111}Cd) probe. Again the two different sites are visible. The Fourier peak with the higher interaction frequency is influenced by a combined magnetic dipole and electric quadrupole interaction taking place at the B-site. The strength of the quadrupole interaction ($\omega_0 = 51$ Mrad/s) has to be compared with the much larger $2\omega_L = 682$ Mrad/s of the magnetic interaction, both at 77 K.

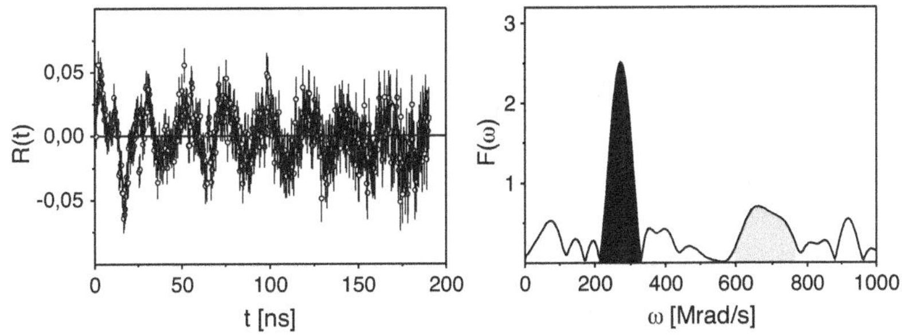

Figure 4. The $HgCr_2Se_4$ specimen with implanted and annealed $^{111}In(^{111}Cd)$ is measured in the sign sensitive ±135° geometry (polarizing field perpendicular to the detector plane), 77 K, FM state. The two marked peaks in the Fourier spectrum correspond to A-site occupation (*dark*) and B-site occupation (*gray*). The zero phase in the R(*t*) picture is the same for the A- and B-site. Therefore both fields have a *positive sign*.

Therefore the quadrupole interaction is only a small perturbation (broadening of the Fourier peak). The sign at the A-site is again positive as discussed for ^{111m}Cd. At the B-site the sign is positive too. This is interesting, because NMR measurements with the host probe ^{53}Cr result in negative fields at the B-site for all of the FMSS. In ferromagnetic metals like Fe and Ni a characteristic dependence of the magnetic fields with sign changes is known in dependence on the atomic number of the probe. It is discussed by the sensitive balance between the contributions of polarized valence and core s-electrons to the magnetic hyperfine fields. Thus the complete results for the magnetic hyperfine fields of Cd probes at A- and B-sites in $HgCr_2Se_4$ are $B_{hfA}(Cd) = +9.4(1)$ T and $B_{hfB}(Cd) = +23.2(5)$ T, both at 77 K.

The probe $^{111}Ag(^{111}Cd)$: Ag probes substitute solely the A-sites in $HgCr_2Se_4$. The magnetic field is the same as measured with ^{111}In (A-site) and ^{111m}Cd.

We observed with the different probes ^{111}In, ^{111m}Cd, and ^{111}Ag the environment of the implants In, Cd, and Ag. These atoms show different doping behavior. But, after the corresponding decay, we have measured the hyperfine fields with the same probe ^{111}Cd(5/2+, 84 ns, 245 keV) and therefore the results are very simply and directly to compare and to interpret.

The probe $^{117}Cd(^{117}In)$: The ^{117}Cd probe is again positioned at the A-site. But now the hyperfine field is measured with the ^{117}In nucleus. The differing electronic structures of In and Cd are the reason for the little increase of the field $B_{hfA}(In) = 9.9(2)$ T in comparison with the Cd field.

The probe ^{199m}Hg: The host probe ^{199m}Hg allows us to compare the magnetic hyperfine fields with that of NMR measurements [7]. A special feature of this probe is its very short half-life of 2.5 ns in the intermediate state. That allows us to use only a time range of about 20 ns. Figure 5 shows a measurement in the ferromagnetic state at 77 K. The interaction frequency is very high and the hyperfine fields are $B_{hfA}(Hg) = (+)31.6(6)$ T at 77 K and $B_{hfA}(Hg) = (+)41.6(9)$ T

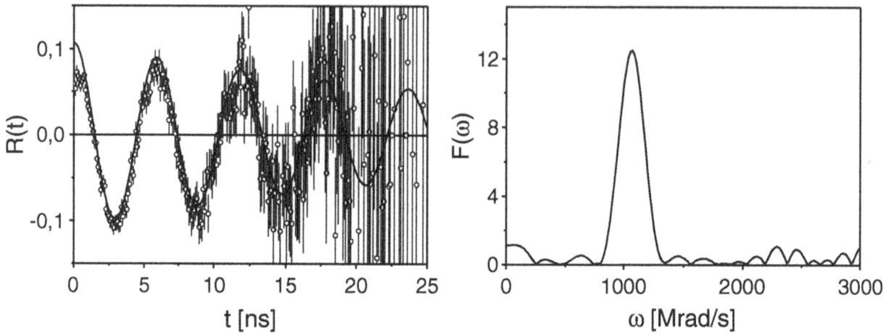

Figure 5. The measurement shows the case of ^{199m}Hg in $HgCr_2Se_4$, FM at 77 K. After implantation the sample was annealed (500°C/15 min in vacuum). The detector arrangement is 90°/180° with a perpendicular polarizing field. The Fourier peak shows the high interaction frequency of $2\omega_L$ = 1068(6) Mrad/s for the Hg site.

at 10 K. The NMR value is +44.6 at 4.2 K. The high magnetic fields are a consequence of the large atomic number of Hg.

The probe ^{77}*Br(*^{77}*Se):* ^{77}Br is the only probe to be substituted at the anion (C-)site. Unfortunately for this probe a time range of only 50 ns can be used. Because of the complicated combined electric quadrupole and magnetic dipole interaction at the C-site the interpretation of the spectra is not straightforward and unequivocal. This should be considered in the interpretation of the data. The quadrupole frequency was determined to be ν_Q = 100(10) MHz and the magnetic field to B_{hfC}(Se) = −5.2(5) T at 77 K.

4. Theory of hyperfine fields

Today first principles calculations using the density functional theory (DFT) in the local spin density approximation (LSDA) are reliable tools in solid state physics. Many of solid state properties, including the hyperfine fields, can be calculated with it. In the ferromagnetic state spin polarized calculations are necessary. We applied the WIEN97 code [5] to calculate the efg and B_{hf} in the FMSS. The results will now be compared with the experiments. Thereby in the present state we can only speak about the results of pure systems, where the probes are not impurities.

In a first step the anion parameter u was calculated by minimizing the total energy and by zero crossing of the forces at the nuclei. Such theoretical u-parameters should be more precise than the experimental ones. The result is u_{theo} = 0.2661 which has to be compared with u_{exp} = 0.2650 in the case of $HgCr_2Se_4$.

Experimental and theoretical (WIEN97) results for $HgCr_2Se_4$ are summarized in Table I (efg) and Table II (B_{hf}). A NMR measurement for the efg with ^{53}Cr (B-site) [7] is in accordance with our calculation. The PAC efg for the ^{111}Cd

Table I. Electric field gradients V_{zz} in 10^{21} V/m^2

	V_{zzA}(Hg)	V_{zzB}(Cr)	V_{zzB}(Cd)		V_{zzC}(Se)
	PAC	NMR	PAC	corr.	PAC
Experiment	0	1.1 [7]	2.7	1.3 [8]	5.4(15)
Theory	0	−1.31			−11.5

Details see text.

Table II. Magnetic hyperfine fields in *T*

	B_{hfA}(Hg)		B_{hfB}(Cr)	B_{hfB}(Cd)	B_{hfC}(Se)	
	NMR	PAC (10 K)	NMR	PAC (77 K)	NMR	PAC (77 K)
Experiment	+44.6	+41.6(9)	−17.9	+23.2(2)	−9.17	−5.2 (5)
Theory	+44.3		−3.4		−2.5	

The NMR results (4.2 K) are from [7].

probe (B-site) can not directly be compared with the theory, because Cd is an impurity. Therefore we used a Sternheimer correction with factors $(1-\gamma_\infty) = 12$ for Cr^{3+} and 25.4 for Cd^{2+} [8], respectively. The corrected (corr.) value meets the theoretical one, but this is not a first principles calculation. The theoretical anion-site efg is about two times larger than the experimental one. Moreover the great uncertainty of the latter one should be considered.

The magnetic hyperfine fields of Table II show a good agreement between theory and experiment only at the A-site. The theoretical B- and C-site values are much too small but the calculated signs are correct.

5. Comparison with the other FMSS and conclusions

In the FMSS the probes Cd, Ag, and Hg occupy the A-site. As to In probes, the investigations unambiguously have shown that in $HgCr_2Se_4$ as well as in $CdCr_2Se_4$ and $CdCr_2S_4$, these probes can substitute both the A- and the B-site. The ratio of A- to B-substitution can be enlarged by quenching the samples or by measurements at higher temperatures. The A-site substitution of In increases in the sequence $CdCr_2Se_4$, $CdCr_2S_4$, $HgCr_2Se_4$. This can qualitatively be explained by the ratio of tetrahedral and octahedral bond lengths, which increases from $CdCr_2Se_4$ to $HgCr_2Se_4$. Large tetrahedral bond lengths prefer the A-site occupation. The Br probes substitute C-sites as expected.

By doping with atoms like In, Ag, and Br the electrical and magnetic properties of these substances can be influenced considerably. Therefore the investigation of the site occupation of such atoms is essential for applications of FMSS.

In the case of $CuCr_2Se_4$ we observed a different behavior. In probes substitute exclusively the A-site as already observed for Cd and Ag [3]. A basic rule for normal spinels states that the larger of the two cations substitute the A-site. This is met in $CuCr_2Se_4$ but for the Cd and Hg compounds the A-sites are occupied by large atoms which is related to large u-parameters. Therefore now the valency of In(3+) gets importance as well. There are many normal spinels with In at the B-site.

Our WIEN97 calculations of hyperfine fields delivered good results for the efg. The same is valid for the other FMSS. But the theoretical internal magnetic fields are realistic only at the A-sites for all of the FMSS whereas at the B- and C-sites the fields are essentially too small. Only the signs are correct in every case. The reason is not clear at the moment.

Future theoretical investigations should include so-called impurity cases, where the probes are non-host atoms or where they do not occupy the normal site. In this case a basis at least of 56 atoms has to be used. Such systems require much computational expense but we have already obtained good results for the B-site efg in the non-magnetic spinels $CdIn_2S_4$, $HgIn_2S_4$ and $ZnAl_2S_4$. In this case it is very essential to consider the relaxation of the atoms in the neighbourhood of the probes [6].

Acknowledgements

We thank V. Tezlevan and I. M. Tiginyanu, Academy of Sciences of Moldova, Chisinau for the $HgCr_2Se_4$ crystals and J. Kortus, MPI für Festkörperforschung Stuttgart for his stimulating remarks to the WIEN97 calculations. The financial support of the German Ministry of Education and Research BMBF-contracts 05 KK1OFA/5 and 05 KK1TSA/7 is kindly appreciated.

References

1. Park Y. D., Hanbicki A. T., Mattson J. E. and Jonker B. T., *Appl. Phys. Lett.* **81N8** (2002), 1471.
2. Osipov V. V., Viglin N. A. and Samokhvalov A. A., *Phys. Lett.* **A247** (1998), 353.
3. Unterricker S., Samokhvalov V., Burlakov I., Degering D., Dietrich M., Deicher M. and The ISOLDE collaboration, *Hyperfine Interact.* **136/137** (2001), 275.
4. Samokhvalov V., Unterricker S., Burlakov I., Schneider F., Dietrich M., Tsurkan V., Tiginyanu I. M. and The ISOLDE collaboration, *J. Phys. Chem. Solids* **64** (2003), 2069.
5. Blaha P., Schwarz K. and Luitz J., WIEN97. ISBN 3-9501031-0-4, 1999.
6. Samohvalov V., PAC investigations of ferromagnetic spinel semiconductors. Thesis, TU Bergakademie Freiberg, Germany, 25.07.2003.
7. Stauss G. H., *Phys. Rev.* **181** (1969), 636.
8. Gusev A. A., Resnik I. M. and Tsitrin V. A., *J. Phys., Condens. Matter.* **7** (1995), 4855.

Hyperfine Interactions (2005) 160:27–37
DOI 10.1007/s10751-005-9147-7

The Role of Nuclear Nanoprobes in Inducing Magnetic Ordering in Graphite

T. BUTZ*, D. SPEMANN, K.-H. HAN, R. HÖHNE, A. SETZER and P. ESQUINAZI
Institut für Experimentelle Physik II, Fakultät für Physik und Geowissenschaften, Universität Leipzig, Leipzig, Germany; e-mail: butz@physik.uni-leipzig.de

Abstract. In this article several aspects concerning the induction of magnetic ordering in graphite via proton bombardment using nuclear nanoprobes are addressed such as proton range and lateral straggling, defect densities, heat load, and simultaneous impurity analysis via particle induced X-ray emission.

1. Introduction

Graphite consists of honeycomb layers of carbon atoms with sp^2-hybridized bonds, the interaction between layers being of the van der Waals type. Hence, it can be considered quasi-two-dimensional. On the contrary, diamond has a face centred cubic lattice with two carbon atoms in the primitive base with sp^3-hybridized bonds. Both polymorphs can be obtained with impurity levels in the ppm range or below, very likely with the exception of hydrogen in graphite. Whereas defect densities in diamond can be rather low, this is generally not the case for graphite. Highly oriented pyrolitic graphite (HOPG) is probably the best material as far as purity is concerned, but high purity single crystals do not seem to be available. Graphite is an electrical conductor with a low charge carrier density whereas diamond is a large gap insulator. Both polymorphs are diamagnetic. It is interesting to note that other allotropes like fullerenes, nanotubes, and nanofoams exist.

There is a theoretical prediction [1] that a carbon-based material with a mixture of sp^2- and sp^3-bonds can lead to ferromagnetic ordering of localized π-electrons with a Curie temperature above room temperature. Experimental evidence is accumulating [2] that ferromagnetic or ferrimagnetic ordering at room temperature in carbon-based materials can be achieved in a variety of ways, the most fascinating being the use of a proton nanobeam. It enables to

* Author for correspondence.

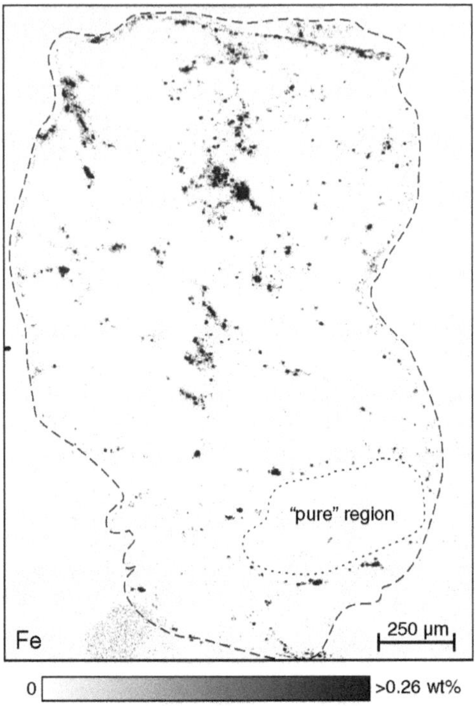

Figure 1. Distribution of the main ferromagnetic impurity Fe found in a C_{60} polymer sample. As can be seen, Fe is located in *small spots* at very high concentrations surrounded by *large areas* of almost pure carbon. The *dashed line* indicates the outer surface of the sample. The *dotted line* encircles a "pure" region.

write magnetic structures with micrometer and sub-micrometer dimensions while – at the same time – allowing for impurity analysis via particle induced X-ray emission (PIXE) [3]. It is believed that the defects induced by proton bombardment lead to a mixture of sp^2/sp^3-bonds and that hydrogen atoms may be relevant.

In this article, the basic properties of the interaction of high energy protons with graphite will be elucidated. At the present stage, most of the pertinent questions are still unanswered and several experiments are herewith proposed.

2. Impurity analysis via PIXE

In this paragraph we briefly summarize the relevant experimental details of the Leipzig high-energy nanoprobe LIPSION. The accelerator is a single ended 3 MV SINGLETRON™ with an RF-ion source for protons and alpha particles. The focusing system can deliver proton beams with diameters as low as 40 nm, however at very low currents of the order of 0.1 fA. For the present purpose, we worked at 2.25 MeV and currents of 500 pA with a beam diameter of 1 micron. The X-ray detector is an Ortec HPGe IGLET-X and subtends 187 msrad solid

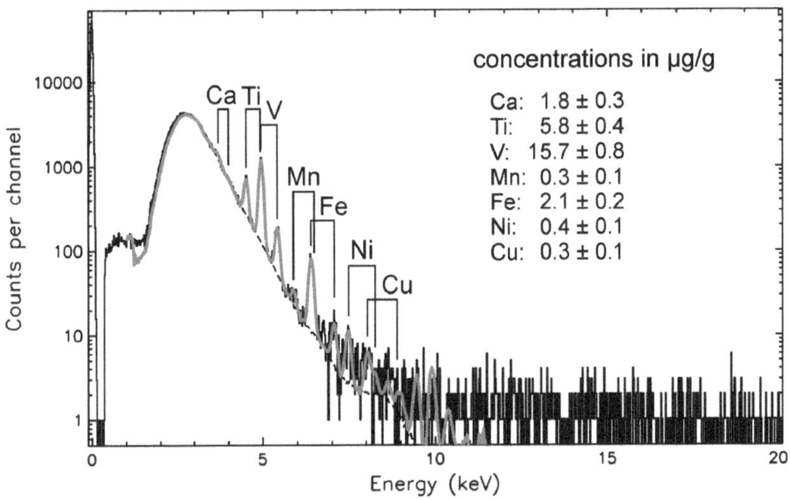

Figure 2. Typical broad beam PIXE spectrum from a HOPG sample. The main impurities are Ti, V, and Fe. The minimum detection limit for other elements heavier than Si is ~0.3 µg/g.

angle. A total charge of 0.5 µC suffices to obtain a minimum detection limit for Fe impurities below 1 µg/g. On the contrary, with our broad-beam PIXE facility we work with 2 MeV protons and currents of 150 nA and a beam diameter of 0.8 mm. Here, the X-ray detector subtends a solid angle of 0.81 msrad and about 20 µC are required to reach a minimum detection limit of 1 µg/g for Fe. Thus the fluences are in the range of 1 µC/mm^2 and 50 µC/mm^2 for the nanobeam and the broad-beam, respectively. The nanobeam value is 5–7 orders of magnitude lower than the fluences actually applied for the induction of magnetic ordering in graphite, as will be discussed below. There is another advantage of the nanobeam compared to the broad-beam: PIXE-maps for all relevant impurity elements are obtained contrary to the integral value for the broad-beam method, a rather important issue considering the grossly inhomogeneous distribution of impurities like Fe which we often encountered (see Figure 1). A typical broad beam PIXE spectrum for a HOPG sample is shown in Figure 2. It shows the presence of quite a number of impurities, the Fe content being (2.1 ± 0.2) µg/g.

We want to stress that it is crucial to measure the impurity content of the starting material as well as after each preparation and measurement step. It makes no sense to start with a high purity material and to cut it with a stainless steel knife afterwards.

Since we suspect that highly oriented pyrolitic graphite (HOPG) could contain more hydrogen than we actually implant, hydrogen content measurements should be carried out as soon as possible. The classical method is the resonant nuclear reaction $^{15}N + {}^{1}H \rightarrow {}^{12}C + {}^{4}He + \gamma$ (4.43 MeV). However, the best achievable minimum detection limits are around 200 ppm only, which may not yet be sufficient for the present purpose. The technique has been further developed by the accelerator laboratory in Helsinki and 10 ppm H in silicon were quoted as

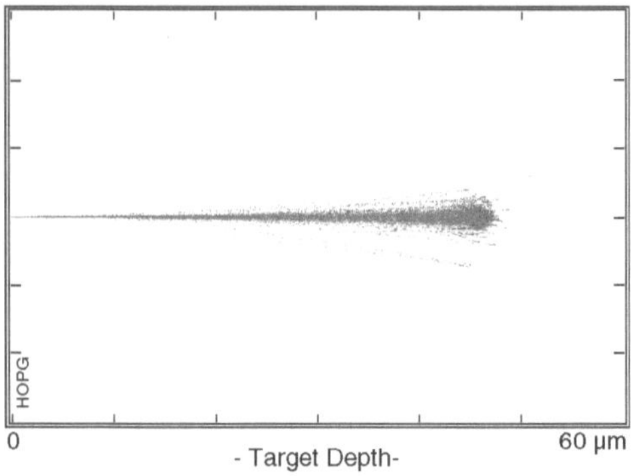

Figure 3. SRIM2003 simulation for 2.25 MeV protons on HOPG. The range of 46 µm is clearly visible and the lateral straggling has a FWHM of 2.5 µm. Abscissa and ordinate have the same scale.

minimum detection limit [4]. The elastic recoil detection of high energy protons, detected in coincidence with the projectile with a high-efficiency segmented particle detector, appears to be the best choice which offers sensitivity in the ppm range [5].

3. The interaction of MeV-protons with graphite

Most of the results discussed in this paragraph are based on the simulation code SRIM2003 [6]. Figure 3 shows the result of 1000 2.25 MeV-protons impinging onto graphite. The range of about 46 µm is clearly visible. This means that all protons will be stopped in a sample which is substantially thicker than the range whereas for samples substantially thinner essentially all protons traverse the sample. In the latter case, the only chance to stop protons in the sample comes from those rare events in which protons are Rutherford backscattered from light atoms, e.g. hydrogen atoms, and loose enough energy in the scattering process such that their range is less than the path out of the sample. The extreme case of backscattering under 180° at a hydrogen atom would actually lead to Rutherford forward scattering with the knocked-on hydrogen atom most probably leaving the sample in the beam direction. Hence, the hydrogen atom content remains unaffected. The cross-section for 180°-backscattering, however, is low. For backscattering angles of 90° and below with rapidly rising cross-sections there is a finite chance for further scattering processes. However, note that the product of at least two cross-sections is required which yields a low probability for such processes. A simulation with SRIM2003 for a graphite sample of 20 µm thickness with 50,000 2.25 MeV protons yielded four protons which came to rest in the sample. For graphite with 1% H this number increased to six. Both numbers have to be taken with great precaution because of low statistics and, above all, because non-Rutherford cross-section for protons on carbon atoms are important and not implemented in SRIM2003. In any case, for 20 µm thick

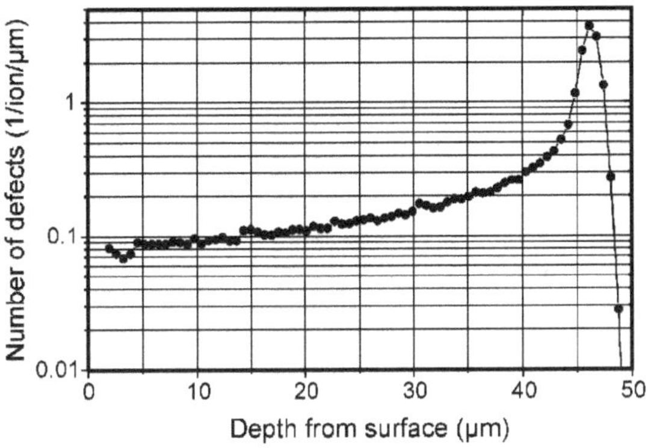

Figure 4. The number of defects per ion and 1 µm depth interval for 2.25 MeV protons in HOPG obtained from SRIM2003 Monte Carlo simulations.

samples the implantation of protons is undetectably small. Strikingly, enhancement of ferromagnetism has also been observed in disordered carbon thin films (deposited on Si substrates) after proton irradiation [16]. Therefore, experiments with unsupported samples of about 20 µm thickness are highly desirable.

There is another aspect which has to be considered: the heat load due to the beam. We typically worked with currents of 500 pA in the nanoprobe. For a spot of 2 µm × 2 µm this amounts to roughly 300 W/mm^2 for thick samples. Most of the energy will be deposited near the end of the ion track. Depending on the lateral dimensions and the thickness of the sample as well as the target holder, heat is transported away from the beam spot. The details are difficult to calculate, but the measurement and control of the temperature is of prime importance. For thin samples, this is relatively unimportant. We have evidence that at higher currents the magnetic response in MFM-measurements is strongest at the rim of the spot suggesting annealing effects in the centre. Since the stopping power or linear energy transfer (LET) is considered to be very important, we propose to use other ions like alpha particles with the same LET as for protons for comparison.

Figure 3 also shows the lateral straggling of the proton beam versus depth. This means, that the ion track diameter increases with depth, especially towards the end of the track. This has consequences for the estimated defect densities. A typical profile for the number of defects created in HOPG per 2.25 MeV-proton and per 1 µm interval is shown in Figure 4. It is clear that most of the defects are created near the end of the track when the protons are already slowed down sufficiently such that they pass over the maximum of the electronic stopping power and enter the nuclear stopping regime. For fluences between 0.001–75 nC/µm^2 we get in the near surface region between $4.7 \cdot 10^{-6}$ and 0.35 displacements per carbon atom, i.e., complete amorphization for the highest fluence. However, these numbers should be taken with great precautions because several assumptions are made which are not well justified. First, a displacement energy

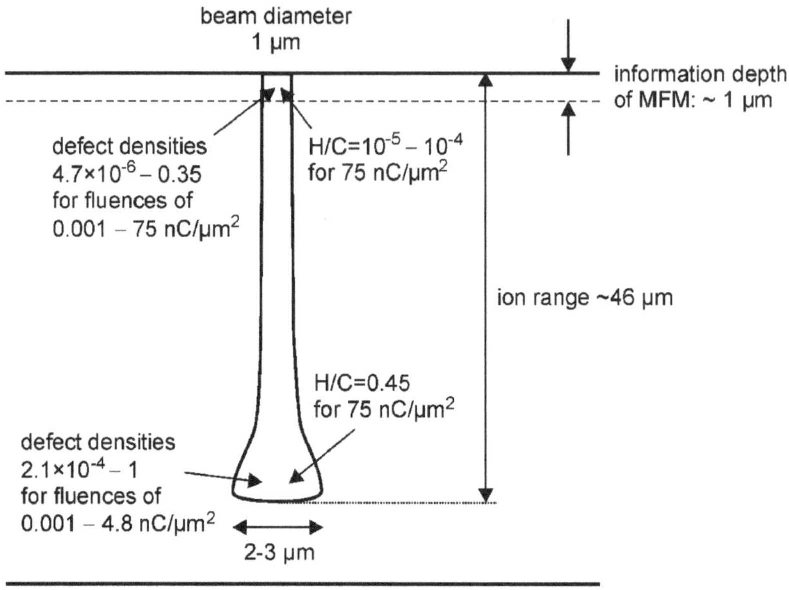

Figure 5. Sketch of the modified area in graphite due to 2.25 MeV proton bombardment.

of 35 eV for Frenkel pairs in HOPG is assumed in the SRIM2003-simulation. Changing this number to 28 eV yields defect densities which are about 30% higher. Second, annealing of defects at room temperature (and above!) is neglected. Hence, these numbers represent at best upper limits.

According to what is known from heavy ion tracks with high stopping powers there is a threshold for the stopping power above which a continuous ion track is formed with a diameter of typically 5–10 nm for a single ion. Within the track, temperatures above the melting point are achieved. In our case, however, we do neither expect continuous tracks for each proton nor temperatures above the melting point of graphite but rather the production of uncorrelated defects of the Frenkel pair type. The van der Waals gap is the first choice for an interstitial position of a carbon atom. The radius of the cylinder in which defects are created is more difficult to estimate. In the first part of the track essentially the so-called δ-electrons can lead to displacements of carbon atoms and their range might be of the order of several tens of nanometers [7]. However, these play little role because of the mass ratio between protons and electrons. On the other hand, the cross-sections for nuclear stopping are rather small at 2.25 MeV proton energy. Towards the end of the track, the lateral beam straggling becomes important and the damaged area can be several microns wide (see Figure 3). For fluences of 75 nC/μm^2 we have 5×10^{11} protons/μm^2, i.e., the regions where defects are created by each individual proton heavily overlap. Figure 5 summarizes what is known about the area in graphite which is modified by 2.25 Me proton bombardment.

Most of what happens after the defect formation is still unclear. A huge wealth of data exist on neutron irradiation damage in graphite motivated by its use as moderator in nuclear reactors [8]. Irradiation by electrons have also been studied extensively. However, the damage induced by protons is expected to be different, especially in the near surface region where defects are likely to be isolated. First, the Frenkel pair could annihilate provided the interstitial carbon atom – probably in the van der Waals gap – is mobile enough. We believe that this process occurs frequently under our conditions and is, of course, of no use for the induction of magnetic ordering. The formation of C_2 or higher aggregates has been reported [8]. Yet another possibility is that an interstitial carbon atom in the van der Waals gap attracts another interstitial carbon atom in a neighbouring gallery thereby lowering the elastic deformation energy. Self-diffusion coefficients were reported in the range of 10^{-13} to 10^{-14} cm^2/s with activation energies of about 7 eV [8]. This is somewhat surprising in the light of the enormous variety of graphite intercalation compounds. However, these data might not be relevant for the formation of an interstitial pair across a gallery. A dangling bond could attract a hydrogen atom – not necessarily from the proton implantation but rather already present as impurity. Interestingly enough, there is little information on residual hydrogen from pyrolysis of hydrocarbons [8], possibly because detection methods for contents of 100 ppm or below were not readily available and such quantities were of little relevance for applications as moderators in reactors. Hydrogen atoms in the van der Waals gap or within the graphite layers should be highly mobile, contrary to hydrogen atoms trapped at defects. Investigations with deuterium [9] indicate that deuterium does not readily diffuse out of HOPG but is rather chemically bound up to D/C ratios of about 0.45, probably at lattice defects. A broad range of binding energies up to about 2 eV is reported, i.e., even 3000–3500 K are not sufficient to eliminate all incorporated hydrogen. For the highest fluence we estimate a H/C ratio of 0.45 around the end of the track for our thick samples. Therefore, we expect all implanted protons to be trapped in the graphite host in case of thick samples. This implies that hydrogen implanted at a depth of about 45 μm cannot readily diffuse to the near-surface region. Another possibility is the formation of two sp^3-bonds across the layers – such as suggested in [1] – possibly mediated by the strain field of distant defects. This mechanism is stimulated by graphite intercalation simulations [10] in which long-range strain fields play an important role. Mobile vacancies which could form a divacancy in adjacent layers with associated sp^3-bond formation have been suggested [11]. Many more suggestions could be made; however, at present it suffices to mention that the defect formation process by high energy protons is a non-equilibrium a-thermal process and it appears rather unlikely that ordered arrays of defects are formed by migration of interstitial carbon atoms or vacancies, maybe with the exception of the di-interstitial across the gallery. Nevertheless, the layer structure should remain essentially intact, at least for low fluences.

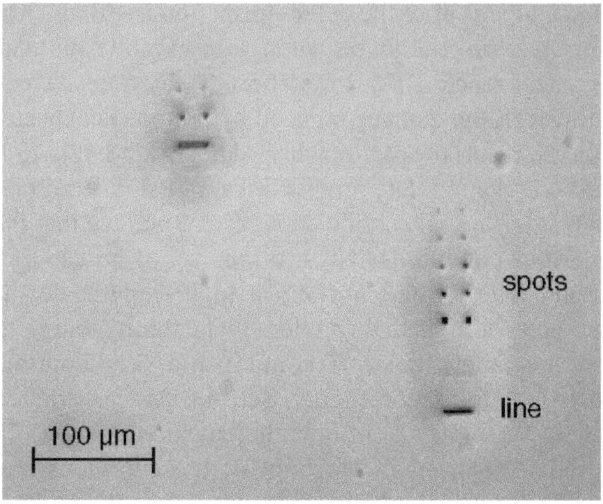

Figure 6. Optical micrograph of the HOPG surface after a proton microbeam irradiation. In general, two spots were irradiated with the same ion fluence and several ion fluences were used. Left: 2×2 μm^2 spots, right: 1×1 μm^2. Additionally, 20 μm lines were irradiated with both beam spot sizes.

For thick targets and sufficiently high fluences, the target actually swells which can be easily measured by an AFM or is even visible under an optical microscope. This effect should be absent for thin enough targets.

4. Magnetic spots

Figure 6 lower right corner shows a sample of ZYA grade HOPG from Advanced Ceramics Co. with two columns of 10 spots each, 1 μm in diameter and separated by 20 μm; the fluences on both sites are nominally identical and range from about 0.1 μC/μm^2 for the topmost spots to about 75 μC/μm^2 for the lowest spots. The topmost five spots in each column are invisible in the optical microscope. At the bottom, two line scans were performed with about 15 μC/μm^2. These spots were written with 2.25 MeV protons and a current of 500 pA. Figure 7 shows a traverse through a spot with 0.12 nC/μm^2 measured with Atomic Force Microscopy (AFM) and Magnetic Force Microscopy (MFM) in the tapping/lift mode. The tip for the MFM was made of a batch fabricated silicon cantilever with pyramidal tip coated with a magnetic CoCr film alloy magnetized normal to the sample surface (i.e., z direction). As can be seen on the left panel, the swelling of the sample in the area of this spot is about 5 nm. For fluences of 75 μC/μm^2 we observed a swelling of about 300 nm. The right panel shows the phase shift in degrees as observed with the MFM. Very sharp steps at and near the irradiated spot show up which are not correlated to the topography. This is best illustrated by subjecting the sample to magnetic fields of ~1 kOe

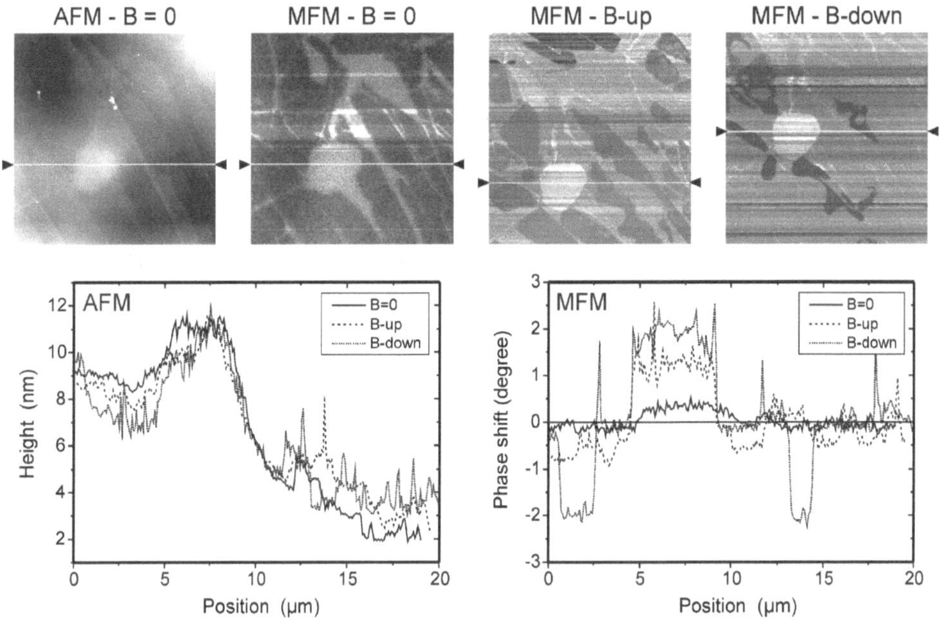

Figure 7. AFM (*top left*) and MFM images (scan size: 20 μm × 20 μm) for a 2 × 2 μm^2 spot irradiated with 2.25 MeV protons at a fluence of 7.5 × 10^{16} cm^{-2}. The AFM and MFM line scans shown below were extracted from the images as indicated by the *black triangles*. Since the sample had to be removed from the MFM for the exposure to an external magnetic field, the subsequent repositioning of the spot in the MFM was accurate to ~2 μm only.

perpendicular to the layers (up and down) outside the MFM and revisiting the spot afterwards. The bright (dark) areas correspond to magnetic domains which change after the temporal application of a magnetic field.

5. Discussion

Whatever defect or defect complex carries the localized moment and whatever coupling leads to magnetic ordering is entirely unknown at present. Two experiments are worth mentioning in this context: (i) Ruffieux *et al.* [12] showed that a single hydrogen atom interacting with a graphite surface modifies the electronic structure over a distance of 20 to 25 lattice constants; (ii) muon spin rotation/relaxation experiments [13] indicate that a μ^+ – a light hydrogen “isotope” – in graphite triggers a local magnetic moment around it. The Fermi surface topology of graphite as well as the rather low charge carrier density might provide a clue how unpaired localized moments can couple ferro/ferrimagnetically. Exchange and superexchange models appear less suitable, contrary to band models. The fact that the magnetic moments are most likely not ordered on a sort of superlattice should play no role. It is worth considering whether theories developed for dilute magnetic semiconductors [14–16] could be applied to magnetic carbon, too.

The essential question is still open: “where is the ferro/ferrimagnetically ordered region?”. The sharp edges in the phase shift of the MFM measurements suggest a location near the surface, say 1 μm deep, where – as stated above – the defect densities are low, at least for low fluences. The ion beam considerations would rather suggest a depth around 40–50 μm where there are high defect densities and where the implanted hydrogen is highly concentrated. Since magnetic domains are visible over distances as large as 20 μm, a deeper ordered region is also conceivable for the MFM data. Maybe, both near-surface regions as well as deeper regions exist. Again, in order to untangle these two possibilities, experiments with unsupported thin graphite samples are essential.

Micro-Raman experiments yielded ratios of the disorder mode at 1360 cm^{-1} to the E_{2g2}-mode at 1580 cm^{-1} as high as 0.6 and XPS-measurements yielded a sp^3/sp^2-ratio of up to 0.3 for the highest fluences [17]. It is less clear why for low fluences and low defect densities there is still a significant sp^3/sp^2-ratio. One could speculate that a single essential defect might trigger the formation of several sp^3-bonds in a sort of collapse like a house of cards. It would be interesting to perform IR-experiments with an information depth much greater than the implantation range, even with much lower lateral resolution.

Although many pertinent questions are still open, it is clear that nuclear microprobes offer the unique possibility to write magnetic structures with micrometer dimensions into graphite with wide applications in material and life sciences.

References

1. Ovchinnikov A. A. and Shamovsky I. L., The structure of the ferromagnetic phase of carbon, *J. Mol. Struct., Theochem* **251** (1991), 133.
2. Esquinazi P., Han K.-H., Höhne R., Spemann D., Setzer A. and Butz T., International Symposium on Structure and Dynamics of Heterogeneous Systems, SDHS’03, November 20–21 (2003), University of Duisburg-Essen, will be published in a special issue of “Phase Transitions”, and references therein.
3. Johansson S. A. E., Campbell J. L. and Malmqvist K. G. (eds.), *Particle Induced X-Ray Emission Spectrometry (PIXE)*. John Wiley & Sons, New York, Chichester Brisbane Toronto Singapore, 1995.
4. Torri P., Keinonen J. and Nordlund K., *Nucl. Instrum. Methods* **B84** (1994), 105.
5. Reichart P., Dollinger G., Bergmaier A., Datzmann G., Hauptner A. and Körner H.-J., *Nucl. Instrum. Methods* **B197** (2002), 134.
6. Ziegler J. F. (ed.), *The Stopping and Range of Ions in Matter*, Vol. 2–6, Pergamon Press, New York, 1977–1985.
7. van Kan J. A., Sum T. C., Osipowicz T. and Watt F., *Nucl. Instrum. Methods* **B161–163** (2000), 366.
8. Kelly B. T., Physics of Graphite, Applied Science Publ., Barking (UK) 1981, and refs. therein.

9. Siegele R., Roth J., Scherzer B. M. U. and Pennycook S. Z., *J. Appl. Phys.* **73**(5) (1993), 2225.
10. Kirczenow G., *Phys. Rev. Lett.* **49** (1982), 1853; **52** (1984), 437; **55** (1985), 2810.
11. Telling R. H., Ewels C. P., El-Barbary A. A. and Heggie M. I., *Nat. Mater.* **2** (2003), 333.
12. Ruffieux R., Gröning O., Schwaller P. and Schlappbach L., *et al.*, *Phys. Rev. Lett.* **84** (2000), 4910.
13. Chakhalian J. A., Kiefl R. F., Dunsinger S. R. and McFarlane W. A., *et al.*, *Phys. Rev.* **B66** (2002), 155107.
14. Dietl T., Ohno H., Matsukura F., Cibert J. and Ferrand D., *Science* **287** (2000), 1019.
15. Akai H., *Phys. Rev. Lett.* **81** (1998), 3002.
16. Sato K., Katayama-Yoshida H. and Dederichs P. H., *J. Supercond.* **16**(1) (2003), 31.
17. Höhne R., Esquinazi P., Han K.-H., Spemann D., Setzer A., Schaufuss U., Riede V., Butz T., Streubel P. and Hesse R., Ferromagnetic structures in graphite and amorphous carbon films produced by high energy proton irradiation, In: Raabe D. (ed.), *Proceedings of the 16th International Conference on Soft Magnetic Materials 16 (SMM16)*, 2004, p. 185, ISBN 3-514-00711-X.

Hyperfine Interactions (2005) 160:39–56
DOI 10.1007/s10751-005-9148-6

Heavy Ion Irradiated Ferromagnetic Films: The Cases of Cobalt and Iron

K. P. LIEB*, K. ZHANG, G. A. MÜLLER, R. GUPTA† and P. SCHAAF
II. Physikalisches Institut and SFB 602, Universität Göttingen, Tammannstr. 1, D-37077 Göttingen, Germany; e-mail: plieb@gwdg.de

Abstract. Polycrystalline, e-gun deposited Co, Fe and Co/Fe films, tens of nanometers thick, have been irradiated with Ne, Kr, Xe and/or Fe ions to fluences of up to 5×10^{16} ions/cm^2. Changes in the magnetic texture induced by the implanted ions have been measured by means of hyperfine methods, such as Magnetic Orientation Mössbauer Spectroscopy (Fe), and by Magneto-Optical Kerr Effect and Vibrating Sample Magnetometry. In Co and CoFe an hcp $\rightarrow$ fcc phase transition has been observed under the influence of Xe-ion implantation. For 10^{16} Xe-ions/cm^2, ion beam mixing in the Co/Fe system produces a soft magnetic material with uniaxial anisotropy. The effects have been correlated with changes in the microstructure as determined via X-ray diffraction. The influences of internal and external strain fields, an external magnetic field and pre-magnetization have been studied. A comprehensive understanding of the various effects and underlying physical reasons for the modifications appears to emerge from these investigations.

1. Introduction

Understanding magnetism in thin films has been a challenge for many decades [1]. Ion irradiation of ferromagnetic films, either during the deposition process itself, e.g., via ion-beam assisted deposition (IBAD) or pulsed laser deposition (PLD), or after film deposition is of fundamental scientific interest, but may also be of high technological impact, due to the possibility of nanostructuring magnetic recording devices [2, 3]. In the context of heavy-ion beam mixing experiments on Fe/Si [4] and Fe/Ag [5] bilayers, changes in the magnetism in iron have been investigated. The evolution of magnetic textures in ion-irradiated Ni, permendur or permalloy films has also been observed for ion energies, at which no interface mixing or reactions occur [6–15].

The present article focuses on results of our investigations on the changes in the magnetic and microstructural properties of thin Co and Fe films and Co/Fe bilayers irradiated with noble gas and/or Fe ions [16–20]. Opposite to Ni, Co and permendur films [7–10, 12–16], magnetostriction plays a minor role in Fe. We report on similarities of and differences between irradiation effects relative to ion-irradiated Ni, permalloy and permendur films.

---

* Author for correspondence.
† Permanent address: Institute Instrumentation, Devi Ahilya University, Indore 452017, India.

The main motivations were

(i) summarizing and classifying the phenomena found in the various films as function of the ion mass and fluence and of some external parameters such as substrate, deposition; external magnetic and stress field;
(ii) comparing the magnetic properties studied with hyperfine interaction methods, such as Magnetic Orientation Mössbauer Spectroscopy (MOMS), with those obtained with the Magneto Optical Kerr Effect (MOKE) or Vibrating Sample Magnetometry (VSM); and
(iii) correlating the observed ion-induced magnetic textures with changes in the microstructure related to texture, lattice expansion, strain fields and/or radiation defects.

2. Experiments

Thin polycrystalline films, typically 75-nm thick and 7×10 mm^2 in size, were deposited via electron gun evaporation (Co, Fe, Co/Fe), pulsed-laser deposition (^{nat}Fe, ^{57}Fe, permendur) or evaporation from an effusion cell (^{57}Fe) onto various substrates (crystalline or amorphous Si or SiO_2, MgO). In most experiments, the films were homogeneously irradiated with Xe^+-ions and in some cases also with other noble gas (Ne^+, Kr^+) or Fe^+-ions, at fluences between few 10^{14}/cm^2 and 5×10^{16}/cm^2. In all the experiments, the ion energies were chosen in such a way that the projected ion range was about half the film thickness [21]. Hence the implantation profiles were located within the films and ion beam mixing at the film/substrate interface was avoided [4]. The ion irradiations were carried out by means of the Göttingen ion implanter IONAS (30–900 keV [22]).

The film thickness, implantation profiles of heavy implants (Fe, Kr, Xe) and surface roughness were determined before and after each ion irradiation via Rutherford Backscattering spectroscopy (RBS) with 0.9 MeV α-particles scattered at 165°. These spectra gave access to the effects of sputtering, (generally negligible) interface mixing and ion-induced grain growth (visible as an increased surface roughness [5]). Magnetic hysteresis curves in all the as deposited and irradiated samples were measured by in-plane MOKE with a maximum in-plane polarizing field of $H_{MOKE} = 1.5$ kOe [19]. In addition, VSM analyses were carried out for several samples, using a polarizing field of up to 5.0 kOe.

Several Fe and FeCo films were investigated by Conversion Electron Mössbauer Spectroscopy, either with the γ-beam hitting the sample at normal incidence (CEMS), or at inclined incidence (Magnetic Orientation Mössbauer Spectroscopy = MOMS [17–19]). To enhance the sensitivity 15-nm thin ^{57}Fe layers were deposited at the surface, in the middle or at the Fe/substrate interface of the 75-nm Fe films. The particular geometry of the MOMS analyses allowed us to determine the orientation of the magnetic hyperfine field(s) in the film. XRD analyses in Bragg–Brentano and grazing incidence geometry were carried

Table I. Summary of sample parameters and analysing techniques used

Sample	Ion	Energy [keV]	Fluence Φ [ions/cm^2]	RBS	MOKE	VSM	CEMS, MOMS	XRD
Fe	Ne	35	1E15-2.5E16	x	x	–	x	x
	Fe	90	1E15-5E15	x	x	–	x	x
	Kr	130	1E15-5E16	x	x	–	x	x
	Xe	200	1E14-5E16	x	x	x	x	x
Co	Xe	200	2.5E13-8E15	x	x	x	–	x
Co/Fe/Si	Xe	200	5E14-1E16	x	x	–	–	x

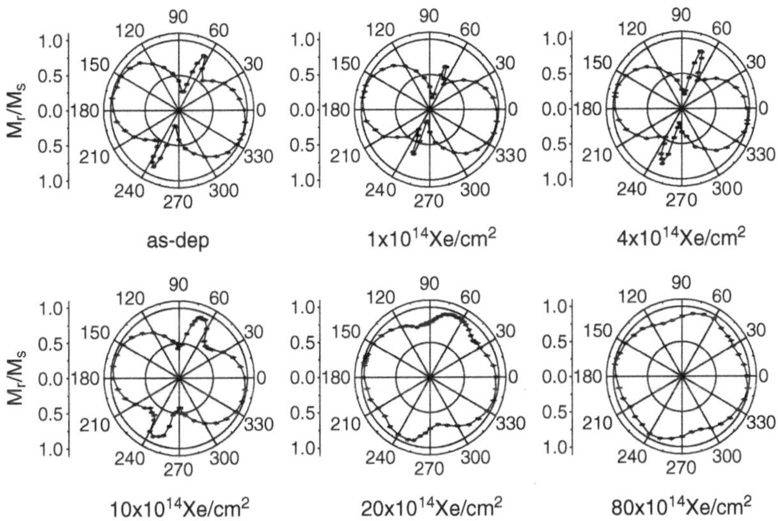

Figure 1. Fluence dependence of the relative remanence M_r/M_s for a 75 nm Co film irradiated with 200-keV Xe-ions. Note the transition from the twofold to the fourfold symmetry occurring at about 2×10^{15} ions/cm^2. The polar plots show the ratio M_r/M_s, where M_r denotes the remanence and M_s the saturation magnetization.

out for all the as-deposited and ion-irradiated films. Further details of the experimental procedures and results are given in several recent publications [6–20]. Table I summarizes the systems discussed here and the analyses performed.

3. Ion-induced hexagonal to face-centered cubic phase transition in cobalt

The high energy density of tens of keV/nm deposited by each impinging heavy ion generates (local) thermal spikes and hence may induce phase transformations during the subsequent rapid cooling process of the spikes [16]. In the case of

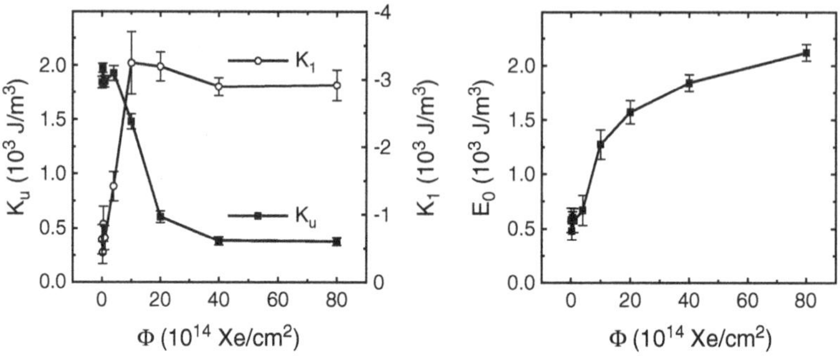

Figure 2. Fluence dependence of the three parameters K_u, K_1 and E_0 describing the energy density (see Equation (1)) during the Xe-ion-bombardment of a 75-nm Co-film.

75-nm thick Co films deposited on SiO_2 substrates via electron-gun evaporation, we studied the hcp → fcc phase transition at room temperature induced during the irradiation with 200-keV Xe^+-ions at fluences of 2.5×10^{13} to 8×10^{15} ions/cm^2. A combination of MOKE, VSM, XRD and RBS was chosen in order to investigate this effect.

The MOKE patterns of the relative remanence M_r/M_s taken for increasing ion fluence are displayed in Figure 1, where M_s denotes the saturation magnetization. They show few changes in the mostly uniaxial distribution of up to about 1×10^{15}/cm^2, but a dramatic switch to a fourfold pattern above this fluence. According to [23], this transition can be readily parametrized by considering that the energy density E_s is composed of a constant term E_0 and an anisotropy term E_a containing a twofold (uniaxial) and a fourfold component,

$$E_s = E_0 + E_a = E_0 + K_u \sin^2(\varphi - \varphi_u) + (K_1/4)\sin^2 2(\varphi - \varphi_1). \quad (1)$$

The polar plot of the quantity $M_r(\varphi)/M_s(\varphi)$ measured via MOKE thus depends on the five parameters K_u, φ_u, K_1, φ_1 and E_0. A fit to the data shown in Figure 1 using Equation (1) results in the parameters K_u, K_1, and E_0 plotted in Figure 2 *versus* the Xe-ion fluence. Apart from a steady increase in the isotropic part E_0, we note a clear change from a uniaxial to a fourfold pattern, which occurs between 1×10^{15} and 2×10^{15} Xe^+-ions/cm^2. The orientation of the MOKE pattern within the film remained rather constant: $\varphi_u = 155(3)°$ and $\varphi_1 = 22(3)°$. If we correlate the uniaxial MOKE pattern with the hcp-phase and the four-fold pattern with the fcc-phase [24], the interpretation of this observation points indeed to an ion-induced phase transition from the original (mostly) hcp-structure to the fcc-phase, which is metastable at room temperature.

In order to test this conjecture, we also performed XRD analyses, which showed a pronounced shift of the hcp (002) peak towards the fcc (111) peak (see Figure 3). Furthermore, we measured VSM curves, which are displayed in

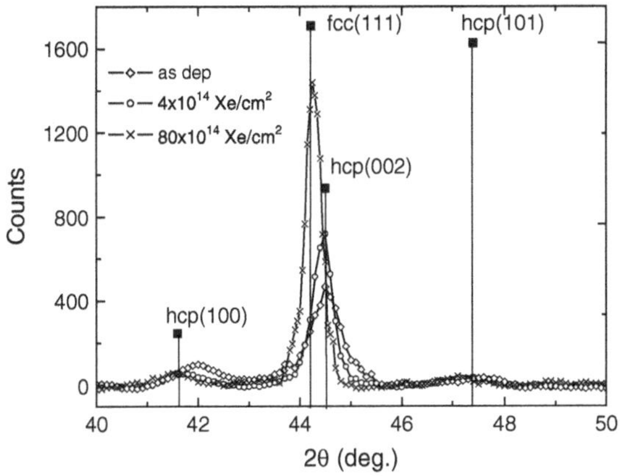

Figure 3. Variation of the XRD spectra with increasing Xe ion fluence. The position of the 44.4° peak shifts from that for hcp-Co (002) to that for fcc-Co (111).

Figure 4. It is interesting to note that the saturation magnetization M_s decreased from the value typical for 100 nm thick hcp-Co films, $M_s = 1{,}375$ emu/cm^3 [25], to values typical of fcc-Co films, $M_s \approx 1{,}230$ emu/cm^3 [26, 27]. Hence, both XRD and VSM support the MOKE results giving evidence for the ion-induced hcp → fcc phase transition occurring at about 1×10^{15} Xe^+-ions/cm^2.

Ion-induced metastable phase formation in alloys has been observed in a number of cases [28–31]. A common mechanism is that the primary ions, besides inducing ballistic transport processes, may deposit their kinetic energy in thermal spikes, i.e., nanometer zones with temperatures above the melting (and Curie) points, which recondensate within picoseconds. For the case of Ag/Fe bilayers irradiated with Xe-ions, Neubauer *et al.* [32, 33] measured the average temperature and diameter of such local spikes to about 2,200 K and 5 nm, corresponding to a spike volume of $V_S \approx 65$ nm^3. In the Co layer the 200-keV Xe ions have a projected range of $R_P = 30$ nm, according to the SRIM simulation [21]. The critical fluence of 2×10^{15} Xe-ions/cm^2, at which the hcp → fcc phase transition has been completed, implies that each volume element of the Co film must have experienced about two collision cascades each involving several spikes in order to be transformed to the metastable fcc phase.

4. Magnetic softening of Xe-ion irradiated Co/Fe bilayers [20]

Magnetic recording devices require soft magnetic materials with perfectly defined uniaxial magnetic anisotropy and high saturation magnetization. Among the known soft magnetic materials, $Fe_{1-x}Co_x$ alloys have the highest saturation magnetization (~2.45 T), but due to their large coercivity they are not suitable for magnetic write heads [34]. Efforts to develop materials with low coercivity have been made, based on FeCoN [34] and laminated FeCoV films [35]. It is

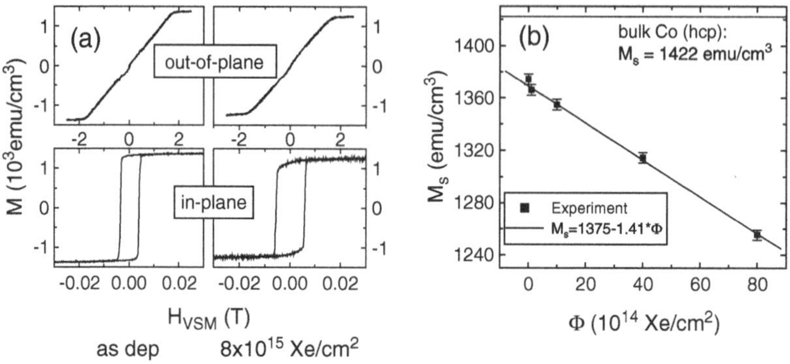

Figure 4. (a) VSM measurements for a 75-nm Co film in the as-deposited state and after irradiation with 200-keV Xe^+-ions at a fluence of $8 \times 10^{15}/cm^2$. (b) Fluence dependence of the saturation magnetization M_s. Also indicated is the value of M_s for bulk Co(hcp) and a linear fit to the data using the expression $M_s(\Phi) = 1{,}375$ emu/cm$^3 - 1.41\ \Phi$ (10^{14} Xe-ions/cm^3).

well known that for Co/Fe bilayers, the microstructure and magnetic properties of the underlayer play an important role during the cooperatively coupled domain-wall movement [36]. Since the magnetic energy density of Co is higher by one order of magnitude than that of Fe, Co/Fe bilayers are generally magnetically hard and isotropic. However, when irradiating them with heavy ions, ion-beam mixing through the Co/Fe interface was found to dramatically change the magnetic properties of the bilayer films.

Thin polycrystalline, magnetically isotropic films of Co(37 nm) / Fe(37 nm), 7×10 mm^2 in size, were deposited by electron-gun evaporation onto Si(100) wafers, at a rate of 0.1 nm/s and a base pressure of about 1×10^{-8} mbar. The films were irradiated at room temperature with 200 keV Xe-ions at fluences of up to $1 \times 10^{16}/cm^2$. As simulated with SRIM2003 [21] and also confirmed by RBS, the Xe-ion profiles covered the Co/Fe films, thus inducing ion beam mixing at the Co/Fe interface, but none at the Fe/Si interface [4]. Figure 5 shows the Co, Fe, Si and Xe concentration profiles in the Co/Fe/Si film, as extracted by means of the program WINDF [37] from the RBS spectra taken after deposition and after irradiation with 1×10^{16} Xe-ions/cm^2. The maximum Xe-concentration in the Co film was about 2.3 at.%. The irradiation induced a 3 nm sputtering effect in the Co top layer and atomic interdiffusion through the Co/Fe interface. We describe the interface diffusion by the mixing rate $k = \Delta\sigma^2(\Phi)/\Phi = [\sigma^2(\Phi) - \sigma_0^2]/\Phi$, where $\sigma^2(\Phi)$ denotes the interface variance of the Co and Fe concentration profiles at the fluence Φ and σ_0^2 the one after layer deposition [28, 29]. If we neglect any ion-induced change in surface roughness, the deduced mixing rate is $k = 22 \pm 3$ nm^4. The profiles indicate some preferential diffusion of Fe into Co, however, due to the similarity in mass number of Co and Fe, CoFe phase formation at the interface, which would be visible as a step function of the Co and Fe profiles, could not be firmly established by RBS. The measured mixing

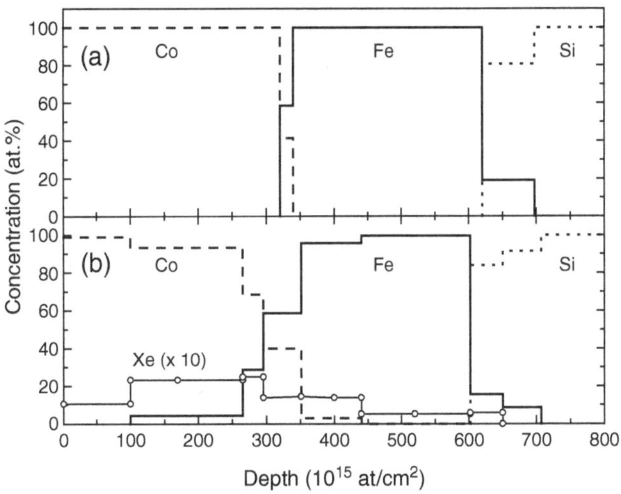

Figure 5. Depth profiles of Co, Fe, Si and implanted Xe in the as-deposited sample and after irradiation with 1×10^{16} Xe-ions/cm^2. Note the sharp Fe/Si interface and the intermixed Co/Fe interface.

rate strongly exceeds the predictions of ballistic mixing ($k_{\text{ball}} \approx 0.4$ nm^4) and local spike mixing ($k_{\text{spike}} \approx 1.5$ nm^4), but is in good agreement with the prediction of spike-enhanced compound formation [4] ($k \approx 25$ nm^4).

Angular scans of the relative remanence M_r/M_s were deduced from the in-plane MOKE hysteresis loops and are shown for increasing ion fluence in Figure 6. The as-deposited Co/Fe bilayer has isotropic magnetization, since the magnetization in the top Co layer reflects the random distribution of the magnetization in the bottom Fe film as found by Park *et al.* [36]. At fluences of up to 5×10^{15} Xe-ions/cm^2, the Co film developed a slight fourfold anisotropy, which became more pronounced with increasing Xe-fluence. The appearance of this fourfold pattern can be attributed to the ion-induced Co-hcp → Co-fcc phase transition, as discussed in Section 3. When the Xe-fluence further increased, the four-fold anisotropy suddenly changed to a uniaxial pattern, which was well pronounced at the highest fluence of 1×10^{16} Xe/cm^2. One possible interpretation of this change from fourfold to uniaxial anisotropy is the formation of CoFe grains due to the mixing effect at the Co/Fe interface. Because of the very high magnetoelastic constant of CoFe ($\lambda = 72.5 \times 10^{-6}$ [38]), the ion-induced internal stress may align the magnetic moments within the intermixed CoFe region and the remaining Co-fcc top layer. Thus, due to the lower saturation field and coercivity of Co and the strong ferromagnetic coupling between Co and CoFe, the parallel magnetic arrangement within the CoFe interlayer influences the orientation of the magnetic moments of the Co top layer and leads to uniaxial magnetic anisotropy.

The anisotropy constants of the films were obtained by fitting the magnetization energy E_s required for saturating the film to Equation (1). Supposing

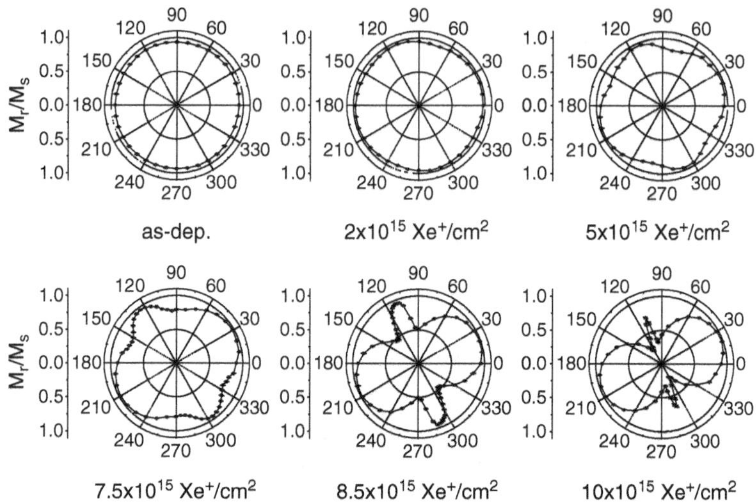

Figure 6. MOKE scans of a Co(37 nm)/Fe(37 nm)/Si film irradiated with 200-keV Xe^+-ions for increasing ion fluence.

that during the MOKE-measurements the external magnetic field changes only the anisotropy energy E_a and isotropy energy E_0, this saturation energy can be determined by numerically integrating the anhysteretic magnetization loops and averaging over both branches of the MOKE magnetization curves as described by Brockmann *et al.* [23]. The fits of the quantities K_u/M_s and K_1/M_s are illustrated in Figure 7. The as-deposited film had almost no magnetic anisotropy ($K_u/M_s = 0.5 \pm 0.2$ Oe, $K_1/M_s = -1.5 \pm 0.5$ Oe). Xe-irradiation at a low fluence of 5×10^{14} ions/cm^2 increased both K_1 and K_u. When increasing the Xe fluence, K_1 increased continuously, while K_u stayed nearly constant up to a fluence of 2×10^{15} ions/cm^2 and then decreased slowly. In this fluence range, the ion-induced hcp → fcc phase transition occurs, which, as we argued previously, is the consequence of local melting in local thermal spikes.

For still higher ion fluences of $\Phi > 8 \times 10^{15}$ Xe-ions/cm^2, the magnitudes of K_1 and K_u began to interchange. For 1×10^{16} ions/cm^2, K_1 had almost vanished ($K_1/M_s = -0.8 \pm 0.5$ Oe), while K_u/M_s reached the high value of 6.7 ± 0.1 Oe, indicating a well-defined uniaxial anisotropy. Such a dramatic change evidently resulted from the Co/Fe intermixing and may indicate the formation of the CoFe intermetallic phase at the Co/Fe interface. During Xe-ion irradiation, the coercive field H_c, measured by MOKE along the easy axis of the films decreased from $H_C = 35$ Oe in the as-deposited film to $H_C = 16$ Oe at 1×10^{16}/cm^2.

5. Ion-irradiated iron film

The studies of ion-induced changes in the magnetic and microstructural properties of Fe films by Müller [19] in his doctoral thesis concentrated mostly

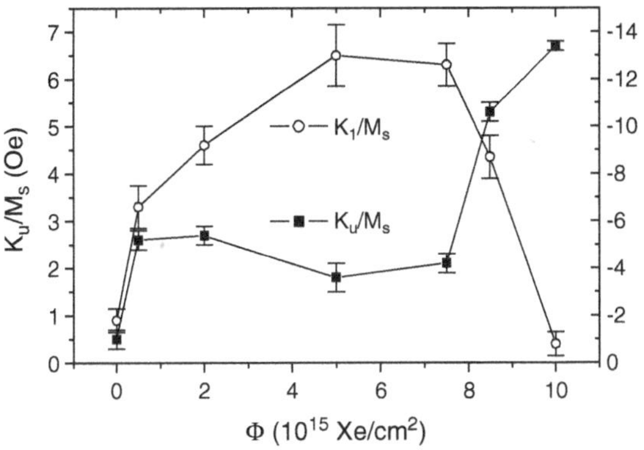

Figure 7. Evolution of the parameters K_u/M_s and K_1/M_s during Xe-irradiation of the Co/Fe film as deduced from the MOKE curves shown in Figure 6 and fitted with Equation (1).

on the influence of the deposition method and type of substrate, the ion mass and fluence and the magnetic history of the samples [12, 15, 18, 19]. Still the situation appears rather ambiguous and complicated for iron. Since the magnetostriction constant and the density of extended defects after ion irradiation of Fe are by one order of magnitude smaller than those in Ni [9, 14, 39], the underlying reasons for the observed magnetic anisotropies are expected to be different in the two materials and hence may allow us to separate the effects of magnetostriction and radiation damage.

For a direct comparison, again 75-nm thin ^{nat}Fe films were deposited via electron evaporation or pulsed laser deposition onto SiO_2, Si(100) or MgO(100) substrates and homogeneously irradiated with 200 keV Xe-ions at 300 K and at fluences of up to $5 \times 10^{16}/cm^2$. During the irradiation a small magnetic field of 104 Oe and/or an external stress (in the form of bending the sample to a curvature of $1/R = 1.0\ m^{-1}$) were applied. The samples were analyzed via RBS, XRD, Mössbauer spectroscopy and MOKE. Similar experiments were also performed with Ne, Fe and Kr ion beams (see Table I), whose energies were chosen to locate the maximum concentration of the implanted ions in the middle of the films.

The main MOKE results obtained for e-gun deposited Fe-films are shown in Figure 8. These films were irradiated with 200 keV Xe-ions in the presence of an external magnetic field of 104 Oe in $\varphi = 0°$ direction. Although the coercivity decreased slightly from $H_C \approx 60$ Oe in the as-deposited samples to $H_C \approx 30$ Oe at a fluence of $1 \times 10^{15}/cm^2$, the magnetic anisotropy started to develop only at a fluence of about $4 \times 10^{15}/cm^2$ and the remanence exhibited a fourfold pattern at the highest fluence. It is interesting to note that for Ni-films a much smaller fluence of some 10^{14} Xe-ions/cm^2 was sufficient to induce an equally strong magnetic texture. One possible reason for this difference is the much smaller density of extended defects in Fe (as compared with Ni) produced by heavy ion implantation and stored in the samples at room temperature [39].

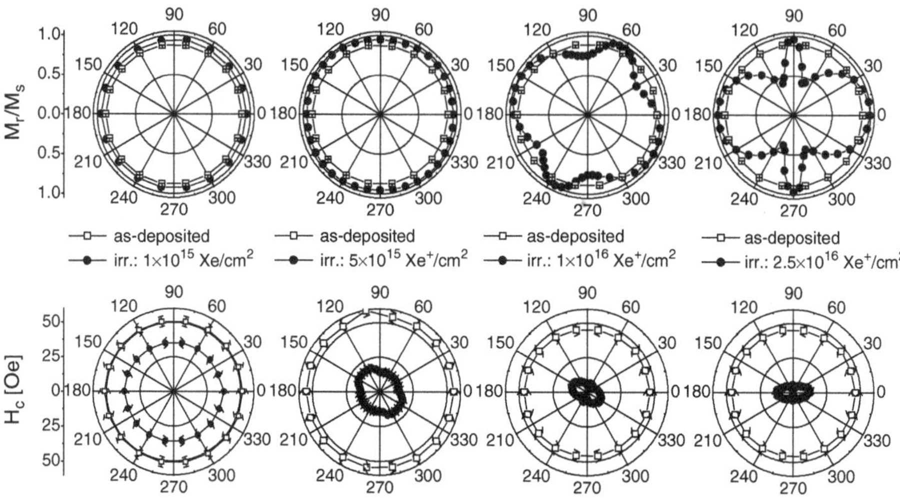

Figure 8. Evolution of the relative remanence M_r/M_s and the coercitivity H_c for 75-nm Fe films irradiated with 200-keV Xe-ions in a magnetic field of 104 Oe at the fluence indicated.

5.1. MOESSBAUER DATA (MOMS)

In order to understand the influence of sample magnetization and stress during irradiation on the magnetic texture as well as the depth-dependent spin distribution of Fe films, we performed a systematic CEMS/MOMS and MOKE study on samples containing thin ^{57}Fe marker layers at different depths [17–19]. Here we only discuss the sample, which had its marker layer in the middle of the film and was irradiated with 1×10^{16} Xe-ions/cm^2. From a CEMS spectrum taken with the γ-beam entering at normal incidence, the hyperfine field(s) was(were) found to be oriented in-plane, but their orientation(s) within the plane could not be determined. To this end, the γ-beam was turned to an angle of $\alpha = 45°$ relative to the film normal, as sketched in Figure 9.

The MOMS spectra shown in Figure 10(a) as function of the rotation angle $\Phi = 0$–$150°$ exhibit a large anisotropy of the intensity ratio $I_2(\varphi)/I_3(\varphi)$ of the 2nd and 3rd component of the Mössbauer sextet (see Figure 10(b)). This intensity ratio can be expressed as [17–19]

$$I_2/I_3 = \sum_i 4\, c_i\left[1 - \sin^2\alpha\, \sin^2(\varphi - \psi_i)\right] \Big/ \left[1 + \sin^2\alpha\, \sin^2(\varphi - \psi_i)\right] \\ +4\, c_{op}\left(1 - \cos^2\alpha\right)/\left(1 + \cos^2\alpha\right). \quad (2)$$

Here it was assumed that the in-plane hyperfine fields i point along the directions ψ_i with the probabilities c_i and that the out-of-plane component has the probability c_{op}, with the normalization $\sum_i c_i + c_{op} = 1$. In our analysis we considered two in-plane field components c_1 and c_2 pointing to the directions ψ_1

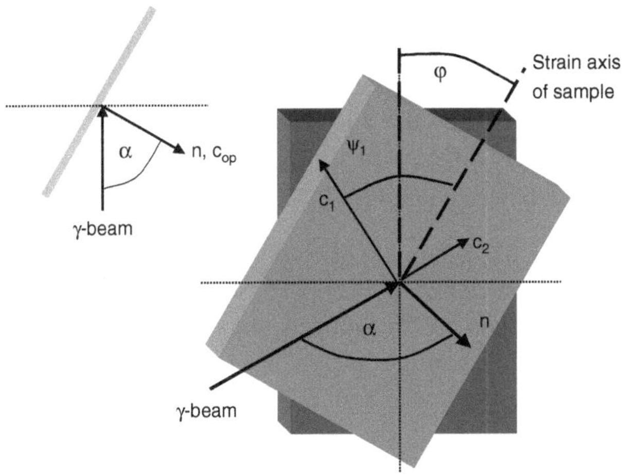

Figure 9. Geometry of the MOMS measurements. The γ-beam enters at the angle $\alpha = 45°$ relative to the film normal n, while the angle φ determines the orientation of the long axis of the sample relative to a direction perpendicular to the γ-beam.

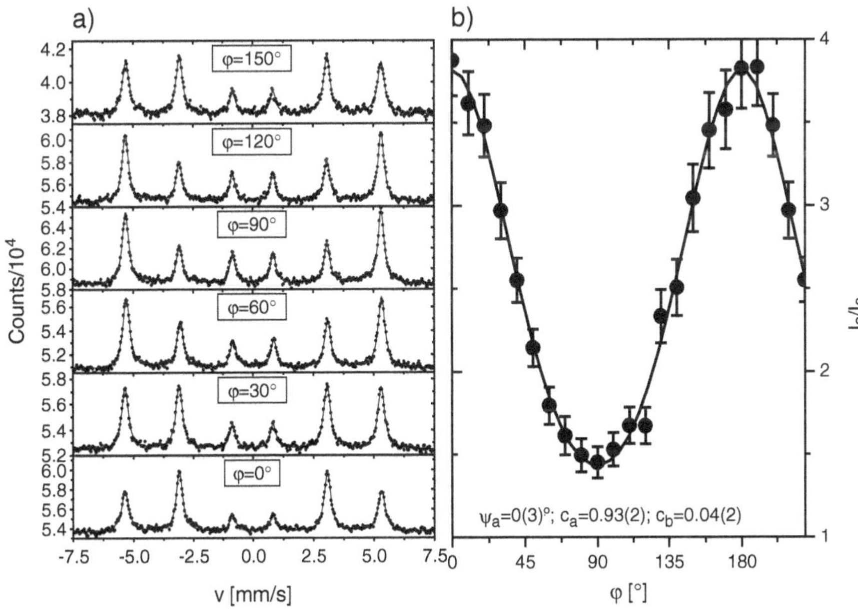

Figure 10. Variation of the MOMS spectrum (a) and the intensity ratio I_2/I_3 (b) with the angle φ.

and $\psi_2 = \psi_1 + 90°$. The fit of the data with Equation (2) resulted in the parameters $c_1 = 0.95(2)$, $\psi_1 = 0(3)°$, $c_2 = 0.05(2)$, and $\psi_2 = 90(3)°$ [17]. Consequently, almost all the ^{57}Fe nuclei had their hyperfine field oriented along the long axis of the sample ($\varphi = 0°$). It is interesting to note that the MOKE analysis gave no magnetic texture in the as-deposited sample and an identical orientation of the magnetization after ion irradiation. In agreement with the very weak magnetostriction in Fe, there was essentially no difference of the anisotropies and hyperfine fields between flat and bent films after releasing the stress.

The advantages of MOMS in comparison with other techniques for the measurement of magnetic anisotropies such as MOKE, VSM and FMR are

(i) MOMS can be used depth-sensitively,
(ii) the incorporation of the hyperfine probes (^{57}Fe) does not affect the film because the probes are introduced already during the deposition process and before the ion bombardment;
(iii) the MOMS measurement itself does not change the magnetization of the film, which may happen during MOKE or VSM analyses, where (rather strong) external magnetic fields are applied.

5.2. INFLUENCE OF MAGNETIC HISTORY

This latter observation allowed us to monitor the ion-induced magnetic texture as function of the magnetic history of the films [12,19] by selecting the order of various film treatment processes, such as deposition, irradiation, bending, RBS, XRD, MOMS and MOKE.

A first example is illustrated in Figure 11. After e-gun deposition, the Fe-film containing a thin ^{57}Fe marker layer in the middle position, was first examined with MOMS and then with MOKE; it showed nearly isotropic magnetization with both techniques (left). The film was then irradiated with 200 keV Xe^{+}-ions to a fluence of 1×10^{16} Xe/cm^2 under the presence of external mechanical stress oriented in $\varphi = 0°$ direction, and again analyzed with MOMS and MOKE. The MOMS intensity ratio I_2/I_3 showed a perfect uniaxial texture, in agreement with the patterns of the coercivity and remanence measured subsequently with MOKE (right). Two effects possibly are the origin of this anisotropy alignment: the external mechanical stress or the remanent magnetization of the film induced in 0° direction by the last MOKE measurement before irradiation.

The second example discussed here refers to a 75-nm Fe-film produced by electron evaporation and containing its ^{57}Fe layer close to the surface. The sample was first analyzed with MOMS and then with MOKE (left hand side). The MOKE scan was isotropic, while MOMS showed an anisotropy symmetric around $\psi_a = 11(5)°$. Then the film was magnetized by an external field of 300 Oe in the direction perpendicular to the hard axis direction previously determined by MOMS (φ=100°) and was then again checked by MOMS. The easy axis had now turned to $\psi_a = 90(4)°$. After Xe-ion irradiation the film was again analysed by MOMS and MOKE, in this order, and both the orientation of the hyperfine field (MOMS, $\psi_a = 88(4)°$) and the uniaxial remanence (MOKE) were preserved as illustrated at the right hand side of Figure 12.

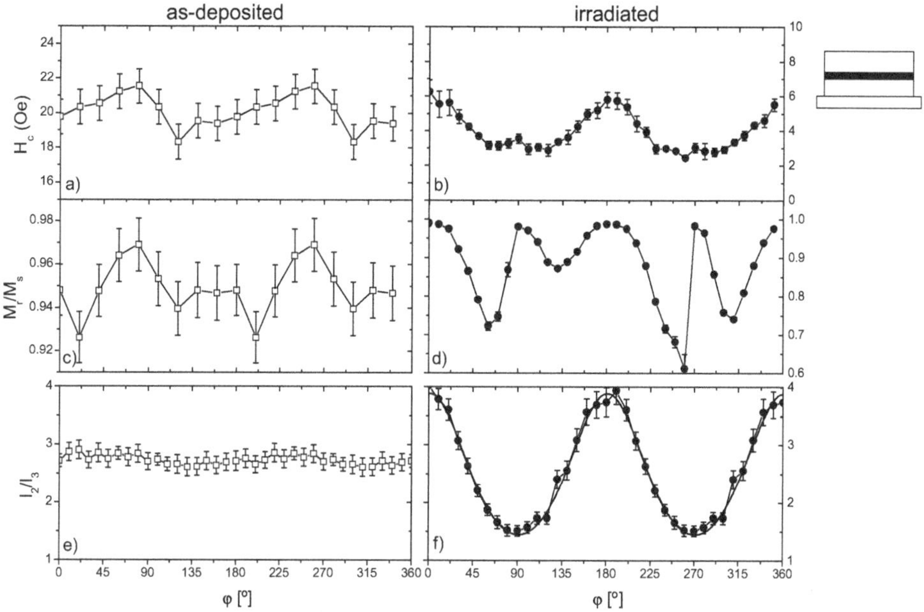

Figure 11. Subsequent MOMS and MOKE analyses of a Fe film, performed after e-gun deposition (left) and Xe-ion irradiation (right). Note the blown-up scale of the MOKE measurement on the left hand side. For details see text.

On the basis of the second example, clearly the origin of the alignment in Figure 11 is the magnetic history of the specimen rather than the mechanical stress. This finding is supported by further experiments carried out on films, not magnetized before ion-irradiation [19].

5.3. VARIATION OF ION MASS AND FLUENCE

As illustrated in Figure 8 for Xe-ions, changes of the coercivity H_C and relative remanence M_r/M_s require a minimum fluence to develop. In general the reduction of H_C sets in at a lower fluence than the anisotropy of the remanence and coercivity. Since the local energy deposition and density of defects increase with the ion mass, a systematic variation of this parameter appeared useful. In addition to the impurity noble gas ions ^{20}Ne, ^{84}Kr and ^{132}Xe, we used ^{56}Fe ions, which of course generate radiation damage, but as self-atoms are not expected to produce clusters and voids.

As usual we fitted the magnetic energy density E_m (normalized to the saturation magnetization M_s) with expression (1), containing an isotropic part K_0/M_s and a uniaxially symmetric part K_u/M_s. No fourfold contribution was considered in the present case. Figure 13 illustrates how the isotropic part

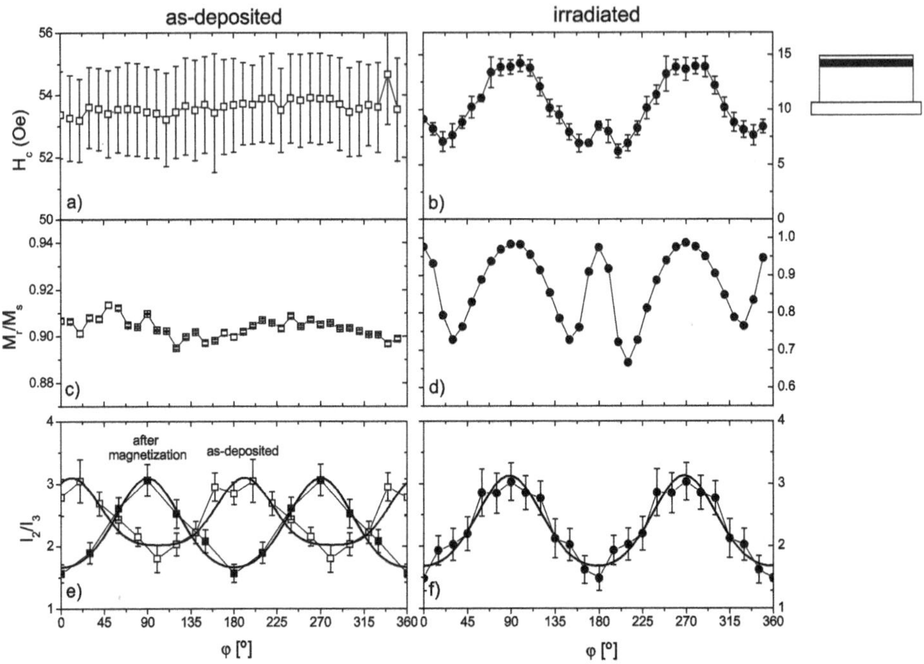

Figure 12. Subsequent MOMS and MOKE analyses of a Fe film, performed after e-gun deposition (left) and Xe-ion irradiation (right). Note the blown-up scale of the MOKE measurement on the left hand side. For details see text.

decreased and the anisotropy increased with increasing ion fluence for all four ion species. On the left-hand side, the parameter

$$\kappa \equiv [K_0/M_s]_{\mathrm{irr}}/.[K_0/M_s]_{\mathrm{as-dep}} \tag{3}$$

was introduced, which measures the relative change in the isotropic component. Clearly κ decreased much faster for the heaviest ion species (Xe) than for the lightest one (Ne), and with intermediate rates for the medium-mass ions Fe and Kr. A similar trend was observed for the evolution of the anisotropic part K_u/M_s, which increased most steeply for Xe, least for Ne ions and again at intermediate rates for Fe and Kr.

Let us now compare the evolution of the magnetic properties in relation to the structural properties in these films as obtained by X-ray diffraction in Bragg–Brentano (XRD) or grazing incidence (GIXRD) geometry. From these data, the stress σ as well as the lattice constant a were derived [19] and are plotted in Figure 14(a,b). Since the heavy ion beam enters at normal incidence and therefore generates a non-isotropic lattice expansion, the lattice parameter a_{op} parallel to the beam (out-of-plane) is slightly larger than the mean lattice parameter a_0 of the stress-free lattice. For all four ions, we observed that a_0 first decreased by about 0.015 Å for small fluences, but above 1×10^{15} ion/cm^2 a_0

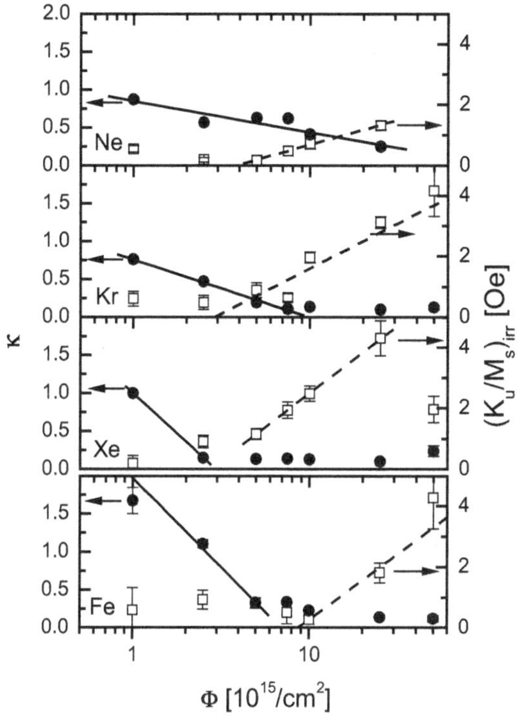

Figure 13. Mass and fluence dependence of the isotropic part κ and uniaxial part K_u/M_s of the magnetization measured as function of the ion mass and fluence.

started to increase again and reached the value typical for the as-deposited films, a_0 = 2,880 Å (see Figure 14b). Concerning the evolution of stress, ion irradiation first led to a relaxation of the tensile stress in all the as-deposited films ($\sigma \approx +3{,}8$ Gpa) and then to a rather constant compressive stress saturating at $\sigma \approx -1$ Gpa at fluences above 5×10^{15} ions/cm^2 (see Figure 14a). This evolution is clearly visible by the increase of a_{op}. It is interesting to note that at least for Fe, Kr and Xe ions the compressive stress reached the saturation value for increasing ion fluence, which did not depend on the ion mass. A different behaviour was found for the Ne irradiation, concerning the compressive stress and lattice constant: both parameters increased steadily with the fluence Φ. For details see [19].

The microstructural data thus allowed us to distinguish between two ranges:

(i) At low ion fluences, stresses of the films generated during deposition are relaxed leading to a decrease of the lattice constant a_0

(ii) For fluences above some 10^{15} ions/cm^2, a slight and saturating compressive stress of -1 Gpa accumulates, which is due to the implanted ions and which is accompanied by an increased out-of-plane lattice constant a_{op}.

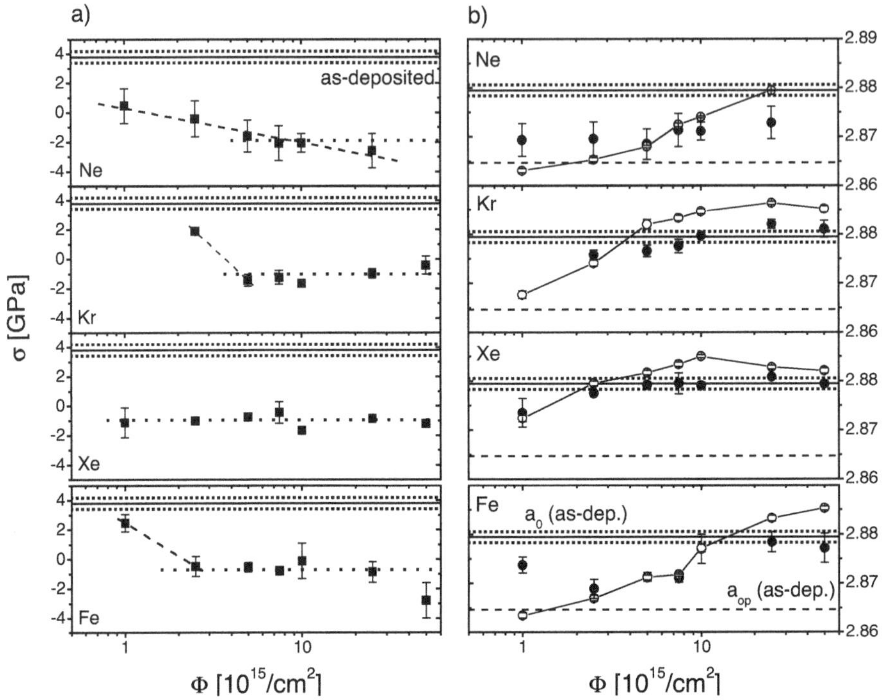

Figure 14. Evolution of the stress σ (a) and lattice constant a (b) with the fluence Φ for Ne, Kr, Xe and Fe-ion irradiations. The values for the as-deposited samples are indicated. The symbols in (b) refer to the mean lattice constant a_m (circles) and the out-of-plane lattice constant a_{op} (dots). For details see text.

Both ion-beam induced processes, relaxation and build-up of stress, compensate each other at about 3×10^{15} ions/cm^2, i.e., at the fluence where the magnetic texture starts to develop.

6. Conclusions and outlook

Our systematic studies of ion-induced effects in polycrystalline films of ferromagnetic 3d-elements as well as permalloy and permendur have given insight into a number of rather dramatic changes in the magnetic properties and their possible relations to modifications of the microstructure. Clearly, it is desirable to develop concepts, which relate the radiation damages and pinning centers microscopically to these magnetic texturing effects. The most crucial point not understood so far is that, despite perpendicular ion bombardment of the films, the magnetic texture clearly shows a well defined orientation and the mechanism of this symmetry breaking is still not understood.

We are grateful to J. Faupel and Y. Luo (Göttingen) for their help with the film deposition and several VSM measurements as well as to D. Purschke for providing the IONAS beams of α-particles (RBS) and implanted heavy ions. We acknowledge discussions with M. Farle, U. Gradmann, M. Münzenberg and K. Samwer, which have contributed in understanding some details of this work, which was funded by Deutsche Forschungsgemeinschaft (DFG).

References

1. Gradmann U., *Magnetism in Ultrathin Transition Metal Films, Handbook of Magnetic Materials*, Elsevier, Amsterdam, 1993 pp. 1–96; Heinrich B. and Bland J. A. C. (eds), *Ultrathin Magnetic Structures*, Vols. I and II, Springer, Berlin, 1994; Huber A. and Schäfer R., *Magnetic Domains – The Analysis of Magnetic Microstructures*, Springer, Berlin, 1998.
2. Lewis W. A., Farle M., Clermens B. M. and White R. C., *J. Appl. Phys.* **75** (1994), 5644.
3. Chappert C., Bernas H., Ferré J., Kottler V., Jamet J.-P., Chen Y., Cambril E., Devolder T., Rousseaux F., Mathet V. and Launois H., *Science* **280** (1998), 1919; Aign T. *et al.*, *Phys. Rev. Lett.* **81** (1998), 5656.
4. Milosavljevic M., Bibic N., Dhar S., Lieb K. P., Schaaf P., Huang Y.-L. and Seibt M., *J. Appl. Phys.* **90** (2001), 4474; Dhar S., Schaaf P., Bibic N., Hooker E., Milosavljevic M. and Lieb K. P., *Appl. Phys.* **A76** (2003), 773.
5. Crespo-Sosa A., Schaaf P., Bolse W., Lieb K.P., Gimbel M., Geyer U. and Tosello C., *Phys. Rev. B* **53** (1996), 14795.
6. Neubauer M., Reinecke N., Uhrmacher M., Lieb K. P., Münzenberg M. and Felsch W., *Nucl. Instr. Meth. B* **139** (1998), 332.
7. Zhang K., Lieb K. P., Schaaf P., Uhrmacher M., Felsch W. and Münzenberg M., *Nucl. Instr. Meth. B* **161–163** (2000), 1016.
8. Zhang K., Lieb K. P., Müller G. A., Münzenberg M. and Uhrmacher M., *Eur. Phys. J.* **42** (2004), 193.
9. Zhang K., Doctoral Thesis, Universität Göttingen (2001); *Nucl. Instr. Meth.*, in press.
10. Gupta R., Müller G. A., Schaaf P., Zhang K. and Lieb K. P., *Nucl. Instr. Meth. B* **216** (2004), 350.
11. Gupta R., Lieb K. P., Luo Y., Müller G. A., Schaaf P. and Zhang K., submitted.
12. Müller G. A., Kulinska A., Zhang K., Gupta R., Schaaf P., Uhrmacher M. and Lieb K. P., *Hyperfine Interact.* **151/152** (2003), 223.
13. Kulinska A., Lieb K. P., Müller G. A., Uhrmacher M. and Zhang K., *J. Magn. Magn. Mat.* **272-276** (2004), 1149.
14. Lieb K. P., Zhang K., Müller G. A., Schaaf P., Uhrmacher M., Felsch W. and Münzenberg M., *Acta Phys. Polon.* **100 A** (2001), 751.
15. Lieb K. P., Gupta R., Müller G. A., Schaaf P., Uhrmacher M. and Zhang K. In: Jokic S., Milosevic I. and Savic I. (eds.), *Proceedings of the 5th General Conference of the Balkan Physical Union, Vrnjacka Banja*, Serbian Physical Society, 2004, pp. 113–134.
16. Zhang K., Gupta R., Lieb K. P., Luo Y., Müller G. A., Schaaf P. and Uhrmacher M., *Europhys. Lett.* **64** (2003), 668.
17. Schaaf P., Müller G. A. and Carpene E. In: Miglierini M., Mashlan M. and Schaaf P. (ed.), *Mössbauer Spectroscopy in Materials Science II*, NATO Science Series, Vol. 94, Kluwer Academic, Dordrecht, 2003, p. 127.
18. Müller G. A., Gupta R., Lieb K. P. and Schaaf P., *Appl. Phys. Lett.* **82** (2003), 73.
19. Müller G. A., Doctoral Thesis, Universität Göttingen (2003); http://webdoc.sub.gwdg.de/diss/2004/mueller_georg/index.html; Müller G. A., Lieb K. P., Carpene E., Faupel J., Gupta R., Zhang K. and Schaaf P., *Eur. Phys. J.*, in press.

20. Zhang K., Gupta R., Müller G. A., Schaaf P. and Lieb K. P., *Appl. Phys. Lett.* **84** (2004), 3915.
21. Ziegler J. F., Biersack J. P. and Littmark U., *The Stopping and Range of Ions in Solids*, Vol 1, Pergamon, New York, 1999; *Stopping Range of Ions in Matter,* SRIM 2003 for Windows: www.SRIM.org/SRIM/SRIM2003.html.
22. Uhrmacher M., Pampus K., Bergmeister F. J., Purschke D. and Lieb K. P., *Nucl. Instr. Meth. B* **9** (1985), 234.
23. Brockmann M., Miethaner S., Onderka R., Köhler M., Himmelhuber F., Regens-burger H., Bensch F., Schweinböck T. and Bayreuther G., *J. Appl. Phys.* **81** (1997), 5047.
24. Suzuki T. *et al.*, *Appl. Phys. Lett.* **64** (1994), 2736.
25. Bubendorff J. L., Beaurepair E., Mény C., Panissod P. and Bucher J. P., *Phys. Rev. B* **56** (1997), R7120.
26. Gu B. X. and Wang H., *J. Magn. Magn. Mater.* **187** (1997), 47.
27. Wolf J. A., Krebs J. J. and Prinz G. A., *Appl. Phys. Lett.* **65** (1994), 1057.
28. Johnson W. L., Cheng Y. T., van Rossum M. and Nicolet M. A., *Nucl. Instr. Meth. B* **7/8** (1985), 657.
29. Bolse W., *Mat. Sci. Eng. R* **12** (1994), 52; *Mat. Sci. Eng. A* **253** (1998), 194.
30. Kulińska A., Wodniecki P., Uhrmacher M. and Lieb K. P., *Hyperfine Interact.* **120/121** (1999), 337.
31. Chang G. S., Lee Y. P., Rhee J. Y., Jeong K. and Wang C. N., *Phys. Rev. Lett.* **87** (2001), 067208; *J. Korean Phys. Soc.* **37** (2000), 438.
32. Neubauer M., Schaaf P., Uhrmacher M. and Lieb K. P., *Phys. Rev. B* **53** (1996), 10237.
33. Neubauer M., Lieb K. P., Uhrmacher M. and Wodniecki P., *Europhys. Lett.* **43** (1998), 177.
34. Wang S. X., Sun N. X., Yamaguchi M. and Yabukami S., *Nature* **407** (2000), 150; Sun, N. X. and Wang S. X., *J. Appl. Phys.* **93** (2003), 6468.
35. Pan G. and Du H., *J. Appl. Phys.* **93** (2003), 5498.
36. Park M. H., Hong Y. K., Gee S. H., Mottern M. L., Jang T. W. and Burkett S., *J. Appl. Phys.* **91** (2002), 7218.
37. Barradas N. P., Jeynes C. and Webb R. P., *Appl. Phys. Lett.* **71** (1997), 291; Barradas N. P., Knights A. P., Jeynes C., Mironov O. A., Grasby T. J. and Parker H. C., *Phys. Rev. B* **59** (1999), 5097.
38. Chikazumi S., *The Physics of Ferromagnetism*, Oxford University Press, 1997.
39. Kirk M. A., Robertson I. M., Jenkins M. L., English C. A., Black T. J. and Vetrano J. S., *J. Nucl. Mater.* **149** (1987), 21.

Hyperfine Interactions (2005) 160:57–65
DOI 10.1007/s10751-005-9149-5

Exchange Interactions and Curie Temperatures in Dilute Magnetic Semiconductors

K. SATO[1,2,*], P. H. DEDERICHS[1] and H. KATAYAMA-YOSHIDA[2]
[1]*Institut fuer Festkoerperforschung, Forschungszentrum Juelich, D-52425 Juelich, Germany; e-mail: p.h.dederichs@fz-juelich.de*
[2]*The Institute of Scientific and Industrial Research, Osaka University, 8-1 Mihogaoka, Ibaraki, Osaka 567-0047, Japan; e-mail: ksato@cmp.sanken.osaka-u.ac.jp, hiroshi@sanken.osaka-u.ac.jp*

Abstract. We discuss the origin of ferromagnetism in dilute magnetic semiconductors based on ab initio calculations for Mn-doped GaN, GaP, GaAs and GaSb. We use the Korringa–Kohn–Rostoker method in connection with the coherent potential approximation to describe the substitutional and moment disorder. Curie temperatures (T_C) are calculated from first-principles by using a mapping on a Heisenberg model in a mean field approximation. It is found that if impurity bands are formed in the gap, as it is the case for (Ga, Mn)N, double exchange dominates leading to a characteristic $\sqrt{c}$ dependence of T_C as a function of the Mn concentration c. On the other hand, if the d-states are localized, as in (Ga, Mn)Sb, Zener's p–d exchange prevails resulting in a linear c-dependence of T_C. In order to have more precise estimations of T_C, effective exchange coupling constants J_{ij}'s are calculated by using the formula of Liechtenstein *et al.* It is found that the range of the exchange interaction in (Ga, Mn)N is very short. On the other hand, in (Ga, Mn)As the interaction is weaker but long ranged. Monte Carlo simulations show that the T_C values of (Ga, Mn)N are very low since percolation is difficult to achieve for small concentrations and the mean field approximation strongly overestimates T_C. Even in (Ga, Mn)As the percolation effect is still important.

Key Words: Curie temperature, dilute magnetic semiconductor, double exchange, Monte Carlo simulation, p–d exchange.

1. Introduction

It has been attempted to make use of not only the charge but also the spin degree of freedom in modern semiconductor electronics for information processing. This new developing field is called spintronics [1]. The discovery of the ferromagnetism in (In, Mn)As [2] and (Ga, Mn)As [3] promoted dilute magnetic semiconductors (DMS) to be fundamental materials for spintronics because of a compatibility with semiconductors used in present electronics. However, their Curie temperatures (T_C) are not high enough for real applications. From

* Author for correspondence.

application point of view, it is indispensable to realize room-temperature ferromagnetism in DMS systems. In this paper, T_C's of several DMS are evaluated from first-principles and materials design for high-T_C ferromagnetic DMS is proposed.

The origin of the ferromagnetism in DMS has been investigated not only by first-principles approach [4–9] but also by a model Hamiltonian [10, 11]. Despite of basic differences, both methods gave similar predictions for the ferromagnetism. Nevertheless, no consensus has been reached about the origin of the ferromagnetism. In fact, Dietl *et al.* proposed Zener's p–d exchange interaction to describe the magnetism [10]. This model explains many physical properties of (Ga, Mn)As successfully, as is shown by MacDonald *et al.* [11]. On the other hand, Akai pointed out by first-principles calculations that Zener's double exchange mechanism is responsible for the ferromagnetism in (In, Mn)As [4]. Similar arguments were also given by Sato *et al.* [5]. Thus, despite the fact that the ferromagnetism of DMS is one of the most important topics in spintronics, this is still a controversial issue. In this paper, we investigate exchange interactions in Mn-doped III-V DMS by analyzing concentration dependences of T_C's. We discuss the relation between the electronic structure and the dominant exchange mechanism and show that the two mechanisms lead to very different concentration dependences of T_C.

Recently, (Ga, Mn)N has been mentioned as the most promising high-T_C DMS [5, 8–11], and many groups have tried to fabricate ferromagnetic (Ga, Mn)N. However the experimental results are very confusing [12] and the magnetism in (Ga, Mn)N is still an open question which we reconsider in this paper. *Ab-initio* calculations show that the magnetism in these systems is dominated by double exchange mechanism and the ferromagnetism is stabilized by the broadening of the impurity band [5, 8, 9]. In the mean field approximation (MFA), high T_C values have been predicted in these systems. We will show that a general obstacle for ferromagnetism exists in these dilute systems. Due to the large band gap the wave function of the impurity state in the gap is well localized, leading to a strong but short ranged exchange interaction. Therefore, for low concentrations, the percolation of a ferromagnetic cluster cannot be achieved, so a ferromagnetic alignment of the impurity moments cannot occur.

2. Calculation

The electronic structure of DMS is calculated in the local density approximation (LDA) by using the Korringa–Kohn–Rostoker (KKR) method. The substitutional and moment disorder is described in the coherent potential approximation (CPA) [13]. In the CPA, configuration averaged properties of alloys are calculated within a single-site approximation [13]. For example, the ferromagnetic state III–V DMS is described as $(\mathrm{III}_{1-c}, \mathrm{TM}_c^{\mathrm{up}})\mathrm{V}$, where TM refers to transition metal impurities, and the spin-glass III–V DMS is simulated as $(\mathrm{III}_{1-c}, \mathrm{TM}_{c/2}^{\mathrm{up}},$

$TM_{c/2}^{down}$)V, where up and down indicate the direction of local magnetic moment of TM impurities [13]. For the present KKR-CPA calculations, we use the package MACHIKANEYAMA2000 coded by Akai [14].

Stability of the ferromagnetism is discussed from total energy difference $\Delta E = TE$(spin-glass state)$-TE$(ferromagnetic state). As shown in [8], by mapping the total energy results ΔE on a Heisenberg model, realistic estimations of T_C are achieved in the mean field approximation, i.e., $k_B T_C = (2/3)\Delta E/c$, where c is the concentration of magnetic impurities. This equation allows us to estimate T_C from first-principles. In the present calculations, local lattice distortions are neglected and the experimental lattice constants of host compounds are used. The crystal potential is approximated by using a muffin–tin potential. Relativistic effects are considered in the scalar relativistic approximation.

To calculate Curie temperatures beyond the mean field approximation (Monte Carlo simulation), the exchange interaction J_{ij} between two impurities at sites i and j in the ferromagnetic CPA medium is explicitly calculated by using the frozen potential approximation [15] and apply a formula by Liechtenstein *et al.* [16]. According to this formula, the total energy change due to infinitesimal rotations of the two magnetic moments at sites i and j is calculated using the magnetic force theorem, and the total energy change is mapped on the classical Heisenberg model $H = -\Sigma_{i \neq j} J_{ij} \mathbf{e}_i \mathbf{e}_j$, there $\mathbf{e}_i$ is a unit vector parallel to the magnetic moment at site i, thus resulting in the effective exchange coupling constants J_{ij}.

3. Exchange mechanism

First, we show calculated Curie temperatures of (Ga, Mn)N, (Ga, Mn)P, (Ga, Mn)As and (Ga, Mn)Sb in mean field approximation to discuss exchange mechanisms in these ferromagnetic DMS. Results are shown in Figure 1. We can see a rich chemical trend in concentration dependences. In (Ga, Mn)N, T_C's are approximately proportional to the square root of the Mn concentration c for low concentrations and become maximum approximately at 5%. Due to this $\sqrt{c}$ dependence, T_C reaches room-temperature already at low concentrations. In (Ga, Mn)P, T_C goes up to 300 K and saturates. T_C of (Ga, Mn)As shows very similar dependence to (Ga, Mn)P, but T_C increase more moderately for low concentrations and still goes up to room-temperature for high concentrations. We can recognize a $\sqrt{c}$ behavior in T_C's of (Ga, Mn)P and (Ga, Mn)As for low concentrations. In contrast to these $\sqrt{c}$ dependencies, T_C's vary almost linearly with the Mn concentration in (Ga, Mn)Sb. As a whole, the concentration dependence shows a dramatic transition from a $\sqrt{c}$ dependence to a linear behavior.

We consider the two extreme cases of (Ga, Mn)N and (Ga, Mn)Sb as model cases to understand the different origin of ferromagnetism and the dominating

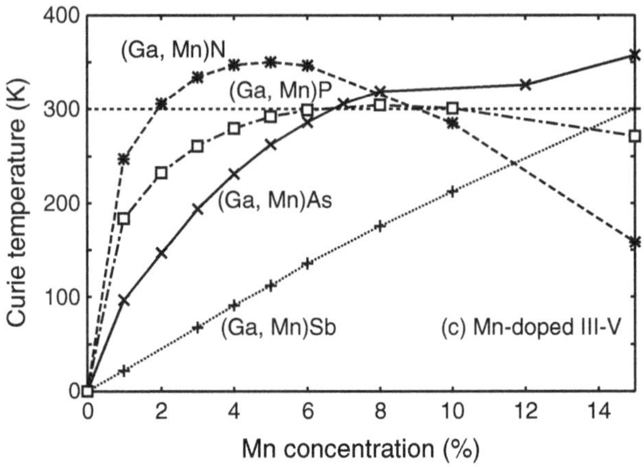

Figure 1. Curie temperatures of Mn-doped GaN, GaP, GaAs and GaSb calculated from first-principles in the mean field approximation.

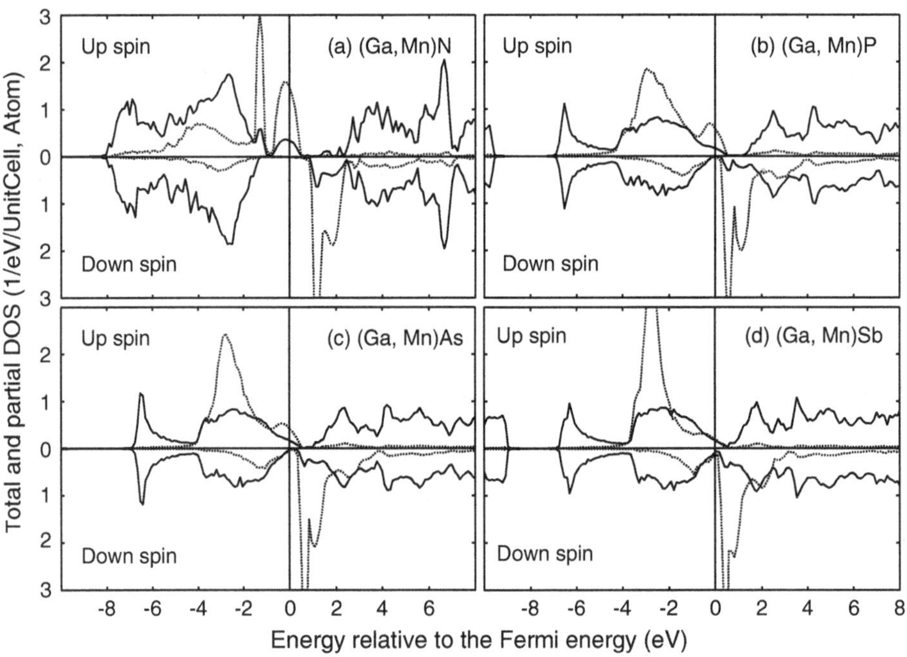

Figure 2. Total density of states per unit cell (*solid line*) and partial density of d-states per Mn atom at Mn site (*dotted line*) in (a) (Ga, Mn)N, (b) (Ga, Mn)P, (c) (Ga, Mn)As and (d) (Ga, Mn)Sb in the ferromagnetic state. Mn concentration is 5%.

exchange mechanisms in DMS. In Figure 2, we show calculated density of states (DOS) in Mn-doped III–V DMS. As shown in the figure, the Mn d-level changes drastically with respect to the valence p-band from a position in the gap (for (Ga, Mn)N) to a position below the p-band (for (Ga, Mn)Sb). (Ga, Mn)As and (Ga, Mn)P are intermediate cases. We can recognize small peaks around the Fermi level (E_F), but the impurity states are almost merged into host valence bands showing broad resonances.

As shown in Figure 2(a), the electronic structure of (Ga, Mn)N is characterized by an impurity band in the gap, and E_F is in the upper part of this band, leaving one state per Mn empty. With increasing concentration the impurity bands broaden, and the broadening of the partially filled band stabilizes the ferromagnetism (double exchange interaction) [4, 5, 8, 9]. The energy gain is proportional to the effective band width, being proportional to $\sqrt{c}$. In case of the spin-glass state, half of Mn neighbors have moments parallel to the Mn moment at site 0. Therefore the energy gain due to the double exchange is always smaller than the one of the FM state. The other half of the Mn neighbors, being anti-parallel aligned, gain energy by the super-exchange interaction [4, 8], which is expected to scale with the concentration $c/2$ of anti-aligned Mn-pairs. Therefore for small concentrations always the double exchange interaction wins due to the $\sqrt{c}$ behavior and stabilizes the ferromagnetism. This explains the behavior of T_C for (Ga, Mn)N shown in Figure 1.

For (Ga, Mn)Sb, the majority d-states can be regarded as localized and holes exist in the majority valence band of GaSb. This behavior is well described by Zener's p–d exchange model [10, 11]. In the FM state the hybridization between the Mn d levels and the Sb p-states pushes the lower d-levels down and the higher p-levels up [17], and holes appear in the majority host band so that the GaSb host becomes polarized with a moment anti-parallel to the local Mn moment. This polarization favors the ferromagnetic coupling of the Mn moments, by an energy proportional to the host polarization, scaling linearly with c. In the spin-glass state the average host polarization vanishes, so that this state is unfavorable. In conclusion, ferromagnetism is stabilized by p–d hybridization and T_C increases linearly with c.

4. Percolation effect

It is well known that the Curie temperature in the mean field approximation T_C^{MFA} is calculated as $k_B T_C^{MFA} = (2c/3)\,\Sigma_{i \neq 0} J_{0i}$, where k_B is Boltzmann constant. As shown in this equation, evaluation of T_C^{MFA} does not require any information on the interaction range, because only the sum of the coupling constants appears in the equation. This simplification leads to significant error in calculated T_C of a dilute system. For example, in the nearest neighbor Heisenberg model, we have perfect ferromagnetic network when the concentration of magnetic sites is 100%. However, due to the dilution the magnetic network is weakened and below a percolation threshold, the network cannot spread all over the system leading to a paramagnetic state. This effect is not counted in the mean field equation, because the dilution effect is included only as a concentration factor c. Due to this percolation problem, the exact T_C values could be much lower than the mean field values, if the exchange interaction is short ranged.

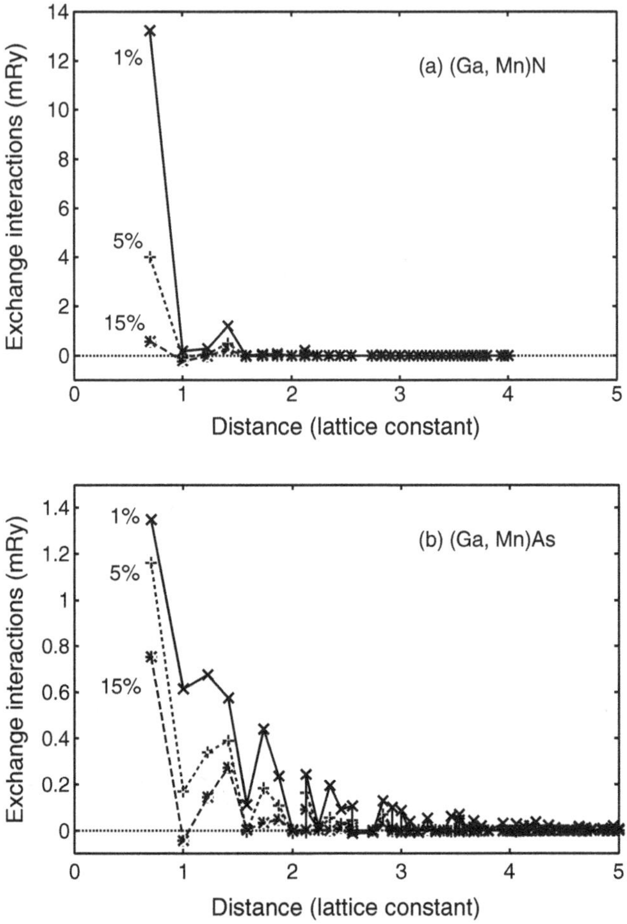

Figure 3. Calculated exchange interaction in (a) (Ga, Mn)N and (b) (Ga, Mn)As as a function of distance.

In order to discuss the exchange interaction range in DMS, we calculate effective exchange coupling constants as a function of distance. Figure 3 shows the calculated exchange interactions in (Ga, Mn)N and (Ga, Mn)As. As shown in the figure, in (Ga, Mn)N the interaction is strong, but the interaction range is short, so that the exchange coupling between nearest neighbors dominates. Therefore, in this case the very large mean field value of T_C is mostly determined by J_{01}. Concerning to the mechanism of the ferromagnetism, it has already been pointed out that the double exchange mechanism dominates in (Ga, Mn)N where pronounced impurity bands appear in the gap [5, 8, 9]. It is intuitively understood that the exchange interaction in (Ga, Mn)N becomes short ranged due to the exponential decay of the impurity wave function in the gap. In contrast to (Ga, Mn)N, the exchange interaction has long tails in (Ga, Mn)As in particular for low concentrations. The qualitative difference between (Ga, Mn)N and (Ga, Mn)As is

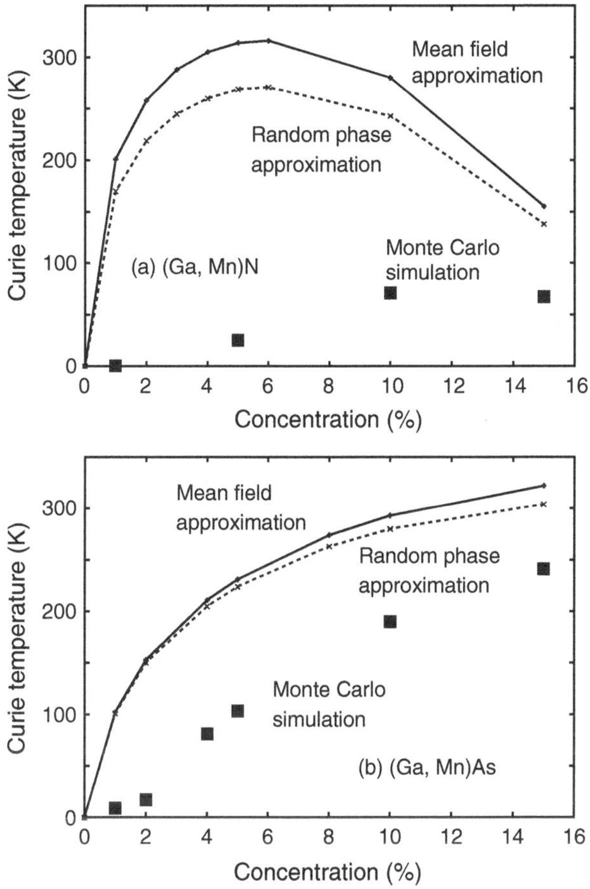

Figure 4. Curie temperatures of (a) (Ga, Mn)N and (b) (Ga, Mn)As calculated by the mean field approximation (*solid lines*), the random phase approximation (*dotted lines*) and the Monte Carlo simulations (*filled squares*).

apparent from the figure. In (Ga, Mn)As the p–d exchange interaction becomes important as discussed in [9].

In order to calculate Curie temperatures with taking the percolation effect into account, we perform Monte Carlo simulations (MCS) for the effective classical Heisenberg model. The thermal average of magnetization M and its powers are calculated by means of the Metropolis algorithm [18]. Due to the finite size of super cells used in the simulation, it is difficult to determine T_C from a temperature dependence of $M(T)$. To avoid this difficulty, we use the cumulant crossing method proposed by Binder [18]. Our cell sizes are 6*6*6, 10*10*10 and 14*14*14 and for each temperature we perform 240,000 Monte Carlo steps using every 20th step for averaging. Thirty configurations of Mn atoms are considered for averaging and J_{ij}-interactions up to 15 shells are included.

The calculated T_C values are shown in Figure 4. As shown in the figure, very small T_C values are predicted for low concentrations in (Ga, Mn)N, and MFA and random phase approximation (RPA) values are too large. Thus we find that the magnetism is strongly suppressed due to the missing percolation of the strong nearest neighbor interactions. On the other hand, due to the longer ranged interaction in (Ga, Mn)As, the reductions from the MFA are not very large, but still significant. The T_C value of 103 K for 5% of Mn is in good agreement with the experimental values of 118 K reported by Edmonds *et al.* [19].

5. Summary

In this paper, we have presented first-principles calculations of Curie temperatures for several DMS systems. Concentration dependences of T_C's in Mn-doped GaN, GaP, GaAs and GaSb are investigated in detail to clarify underlying mechanisms of the ferromagnetism in these compounds. It is found that double exchange mechanism dominates if impurity bands are formed in the gap (for example in (Ga, Mn)N). In this case, T_C increases proportional to $\sqrt{c}$, where c is Mn concentration. On the other hand, if the d-states of Mn are nearly localized and below the host valence band, the p–d exchange mechanism dominates (for example in (Ga, Mn)Sb), leading to linear c dependence of T_C.

We have also shown by using the Monte Carlo simulations that (Ga, Mn)N shows no high-T_C ferromagnetism for low Mn concentrations. The strong ferromagnetic interaction of Mn nearest neighbors is not effective below the nearest neighbor percolation limit. Therefore the experimentally observed very high-T_C values do not refer to a homogeneous ferromagnetic phase, but have to be attributed to small ferromagnetic MnN clusters. Our results are of relevance for all DMS systems with impurity bands in the gap. To obtain higher Curie temperatures one needs longer ranged interactions and/or higher concentrations. The latter requirement naturally points to II–VI semiconductors, having a large solubility for transition metal atoms. Recent experimental T_C value of 300 K for (Zn, Cr)Te with 20% of Cr [20] is in line with these arguments.

Acknowledgements

This research was partially supported by JST-ACT, JST-CREST, NEDO-nanotech, a Grant-in-Aid for Scientific Research on Priority Areas, 21st Century COE and special coordination funds for promoting Science and Technology from the Ministry of Education, Culture, Sports, Science and Technology. This work was also partially supported by the RT Network Computational Magneto-electronics (Contract RTN1-1999-00145) of the European Commission.

References

1. Wolf S. A., Awschalom D. D., Buhrman R. A., Daughton J. M., von Molnar S., Roukes M. L., Chtchelkanova A. Y. and Treger D. M., *Science* **294** (2001), 1488.

2. Munekata H., Ohno H., von Molnar S., Segmuller A., Chang L. L. and Esaki L., *Phys. Rev. Lett.* **63** (1989), 1849.
3. Ohno H., Shen A., Matsukura F., Oiwa A., Endo A. and Iye Y., *Appl. Phys. Lett.* **69** (1996), 363.
4. Akai H., *Phys. Rev. Lett.* **81** (1998), 3002.
5. Sato K. and Katayama-Yoshida H., *Semicond. Sci. Technol.* **17** (2002), 367.
6. Kudrnovsky J. *et al.*, *Phys. Rev., B* **69** (2004), 115208.
7. Sandratskii L. M. and Bruno P., *Phys. Rev.* **B66** (2002), 134435.
8. Sato K., Dederichs P. H. and Katayama-Yoshida H., *Europhysics Lett.* **61** (2003), 403.
9. Sato K., Dederichs P. H., Katayama-Yoshida H. and Kudrnovsky J., *Physica B* **340–342** (2003), 863.
10. Dietl T., Ohno H., Matsukura F., Cibert J. and Ferrand D., *Science* **287** (2000), 1019.
11. Jungwirth T., König J., Sinova J., Kucera J. and MacDonald A. H., *Phys. Rev.* **B66** (2002), 012402.
12. Matsukura F., Ohno H. and Dietl T., *Handbook of Magnetic Materials* **14** (2002), 1.
13. Akai H. and Dederichs P. H., *Phys. Rev.* **B47** (1993), 8739.
14. Akai H., KKR-CPA program package, http://sham.phys.sci.osaka-u.ac.jp/~kkr/.
15. Oswald A., Zeller R., Braspenning P. J. and Dederichs P. H., *J. Phys. F. Met. Phys.* **15** (1985), 193.
16. Liechtenstein A. I., Katsnelson M. I., Antropov V. P. and Gubanov V. A., *J. Magn. Magn. Mater.* **67** (1987), 65.
17. Kanamori J., *Trans. Magn. Soc. Japan* **1** (2001), 1.
18. Binder K. and Heermann D. W., *Monte Carlo Simulation in Statistical Physics*, Springer, Berlin, Heidelberg, New York, 2002.
19. Edmonds K. W., Wang K. Y., Campion R. P., Neumann A. C., Farley N. R. S., Gallagher B. L. and Foxon C. T., *Appl. Phys. Lett.* **81** (2002), 4991.
20. Saito H., Zayets V., Yamagata S. and Ando K., *Phys. Rev. Lett.* **90** (2003), 207202.

Hyperfine Interactions (2005) 160:67–79
DOI 10.1007/s10751-005-9150-z

Sr_2FeMoO_6: A Prototype to Understand a New Class of Magnetic Materials

SUGATA RAY[1] and D. D. SARMA[1,2,*]
[1]*Solid State and Structural Chemistry Unit, Indian Institute of Science, Bangalore 560 012, India*
[2]*Jawaharlal Nehru Center for Advanced Scientific Research, Bangalore, India;*
e-mail: sarma@sscu.iisc.ernet.in

Abstract. We propose a new mechanism to explain the magnetic structure of a recently discovered magnetoresistive double perovskite oxide system, Sr_2FeMoO_6, with the help of detailed experimental and theoretical results. This model, based on a strong antiferromagnetic coupling between the local moment and the charge carriers arising from local hopping interactions, can give rise to ferromagnetic metallic as well as ferromagnetic insulating ground states. The relevance of this mechanism in understanding the magnetism in dilute magnetic semiconductors such as $Ga_{1-x}Mn_xAs$, is also discussed.

Key Words: dilute magnetic semiconductor, double perovskite, magnetoresistance.

1. Introduction

Following the discovery [1] of the spectacular decrease in electrical resistivity on application of a magnetic field, known as Colossal Magnetoresistance (CMR), there has been tremendous increase in research activities on doped perovskite manganites because of the fundamental issues involved and also because of their potential technological implications in magnetic storage devices and spintronics. However, technological applications of the CMR property of manganites are strongly limited by the requirements of low temperature and high applied magnetic field for achieving appreciable CMR effect in these compounds. Recently, a double perovskite oxide, Sr_2FeMoO_6, has been reported to exhibit an appreciable negative CMR even at room temperature and low magnetic fields [2]; thereby increasing the excitement once again in the community. Interestingly, the type of dominant *MR* response observed in this compound is not really of an intrinsic nature and rather depends on many external parameters. This particular *MR* response arises from a spin dependent tunneling mechanism across insulating barriers and often termed as tunneling magnetoresistance (TMR). An important criterion for realizing such a response is a complete spin polarization of the conduction electrons in the system, which strongly depends on the

* Author for correspondence.

magnetic state of the compound. Therefore, the observation of significant *MR* response of this compound at a relatively higher temperature can easily be connected to the fact that Sr_2FeMoO_6 has a surprisingly high magnetic Curie temperature (~415 K) [3]. This high magnetic transition temperature is the most unusual observation concerning this compound, because of the fact that the magnetic Fe ions are far separated by other normally nonmagnetic ions, Mo and O in the crystal structure of this compound. In such a situation, one would expect only a weakly antiferromagnetic system, as indeed found in the case of the analogous compound, Sr_2FeWO_6, with a Néel temperature, $T_N \sim 37$ K [4, 5]. It is to be noted that there are several other examples of both ferromagnetic and antiferromagnetic compounds within the $A_2BB'O_6$, double perovskite oxide series; for example, Sr_2CrMoO_6 and Sr_2FeReO_6 are ferromagnetic, while Sr_2FeWO_6, Sr_2MnMoO_6 and Sr_2CoMoO_6 are antiferromagnetic [6–9]. Thus, an explanation of the magnetic structure of Sr_2FeMoO_6 must also be consistent with such diverse properties observed within the same double perovskite family.

2. Magnetoresistance and magnetization

In Figure 1, we show the percentage magnetoresistance (%*MR*) of Sr_2FeMoO_6 where *MR* is defined as

$$MR(T,H) = 100 * [\rho(T,H) - \rho(T,0)]/\rho(T,0)$$

where, $\rho(T, H)$ is the resistivity of the sample at a temperature, T and in presence of an applied magnetic field strength of H.

Figure 1(upper panel) shows the results obtained at $T = 4.2$ K and Figure 1 (lower panel) at $T = 300$ K. The magnetization curves in the insets establish the room temperature ferromagnetism in this system. Consistent to its magnetic responses, this compound exhibits sharp and pronounced *MR* responses in the low-field regime, though the magnitude of the *MR* is considerably higher at the lower temperature, suggesting that TMR is the dominant mechanism in the total *MR* response though beyond 1 Tesla, the *MR* indeed exhibits a slower change without showing any sign of saturation up to the highest magnetic field (7 Tesla), possibly indicating the presence other of contributions in the total *MR* response. The idea of stronger TMR contribution is also supported by an absence of the sharp low-field *MR* response in single crystalline bulk [10] and epitaxial [11–13] samples of Sr_2FeMoO_6.

3. Crystal and electronic structure of Sr_2FeMoO_6

We show the crystal structure of Sr_2FeMoO_6 in Figure 2. The unit cell dimensions are $a = b = 5.57$ Å, and $c = 7.90$ Å with a space group of I4/mmm [2, 14, 15]. As can be seen from this figure, Fe and Mo sites alternate along all

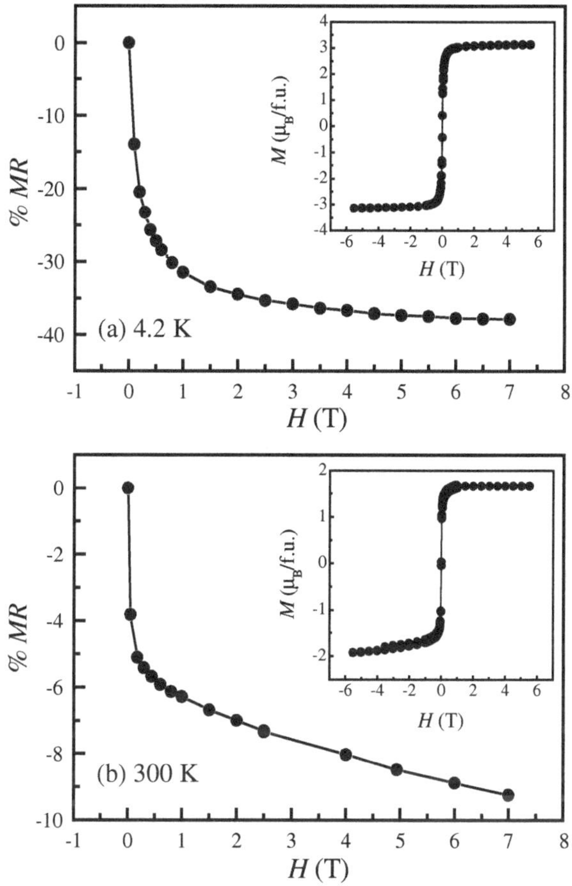

Figure 1. MR vs. *H* data for polycrystalline Sr_2FeMoO_6 at two different temperatures, 4.2 K (*upper panel*) and 300 K (*lower panel*). The corresponding *M*(*H*) *curves* are also shown in the corresponding *insets*. (Adopted from Ref. [29].)

the three (*x*, *y*, and *z*) directions appearing at the corners of small cubes, separated by intervening oxygen ions at the edge–center positions. Extensive band structure calculations have been carried out on this system and the resultant total density of states (DOS) along with the partial Fe *d*, Mo *d* and O *p* DOS are shown in Figure 3. The spin integrated DOS and partial DOS are shown in Figure 3(a), while the corresponding spin-up and spin-down components are shown in Figure 3(b) and (c), respectively. The complete spin-polarization or in other words the half-metallic property of this system can be essentially seen from Figure 3(b,c), where a substantial gap in the spin-up DOS across the Fermi energy, E_F is observed; in contrast, the spin-down channel shows finite and continuous DOS across the E_F in agreement with the metallic state of this system.

This particular situation, termed half-metallic, where one spin channel (the up-spin channel) behaves like an insulator with a finite gap at E_F, while the other

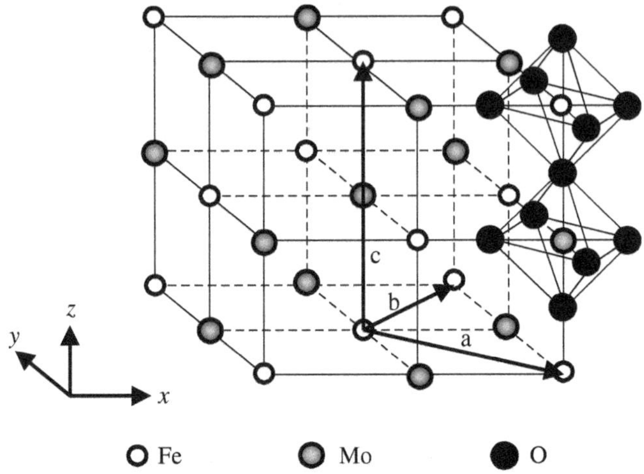

Figure 2. Crystal structure of Sr_2FeMoO_6. The unit cell axis are shown by *thick arrows*. Sr ions, appearing at the body center of each *small cube* and most of the oxygen atoms are not shown to maintain clarity. (Adopted from Ref. [29].)

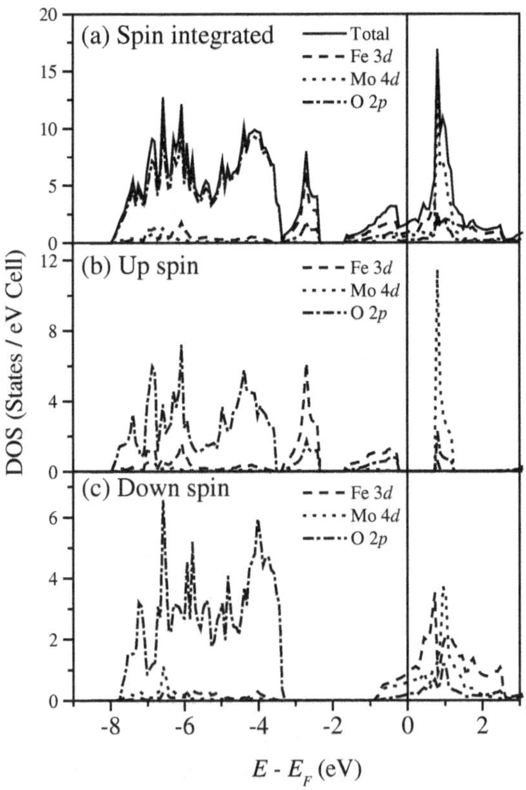

Figure 3. Total density of states (DOS) of Sr_2FeMoO_6 along with partial Fe *d*, Mo *d* and O *p* density of states. (a) Spin integrated densities. (b), (c) The corresponding DOS for the up- and down-spin channels, respectively. (Adopted from Ref. [29].)

spin channel (the down-spin one) has finite DOS at E_F leads to a 100% spin-polarization of the charge carriers; this is an essential condition for achieving substantial TMR response in a ferromagnetic compound.

4. Description of valence and conduction bands of Sr_2FeMoO_6

In order to understand the electronic and magnetic structures of this compound, it is essential to carefully analyze the respective valence and conduction bands. We have carried out detailed high resolution photoemission and Bremsstrahlung isochromat spectroscopy (BIS) studies on this compound, which provide a complete description of the electronic structure of Sr_2FeMoO_6.

Synthesis and details of the high resolution photoemission experiment will be found elsewhere [16]. We show a set of high resolution valence band spectra of Sr_2FeMoO_6, obtained with different incident photon energies, in Figure 4. Consistent with the overall metallic transport property of the system, all the experimental spectra exhibit finite DOS at the E_F (the zero line in the energy scale). Also, spectra collected with different incident photon energies appear very similar with certain subtle differences in the intensities of different valence band features. There are four main features in the experimental valence band spectrum of Sr_2FeMoO_6 appearing at 0.4, 1.8, 5.6 and 8.3 eV binding energies, as shown by the vertical dotted lines. The relative variations in the intensities of these features as a function of photon energy provide a pathway to identify their origin by using the idea of resonance photoemission spectroscopy.

The most effective way to determine the origin of these spectral features is to find out the dependencies of their normalized spectral intensities on the photon energies (Figure 5a,b). We have carried out experiments with several photon energies spanning the range of Fe $3p$ absorption threshold near 55 eV and the range of Mo Cooper minimum at about $h\nu \sim$ 100–110 eV [17]. Localized Fe related features are expected to exhibit pronounced variation in the intensity with photon energy sweeping through the Fe $3p$ threshold; likewise, localized Mo $4d$ related features are expected to show rapid changes in the intensity leading to a minimum, as a function of the photon energy near the Cooper minimum. We found that the two features appearing at 1.8 and 8.3 eV exhibit significant non-monotonic $h\nu$ dependence in resonance with Fe $3p$ absorption threshold (Figure 5a), clearly establishing substantial Fe $3d$ contributions in these states.

More interestingly, the nature of the dependencies is different for the two features. The 8.3 eV feature exhibits about 30% resonant enhancement, characteristic of correlation driven satellite or incoherent features, while, in contrast, the 1.8 eV feature shows approximately a 20% dip in the intensity near the on-resonant condition, as is characteristic of more band-like or coherent features [18].

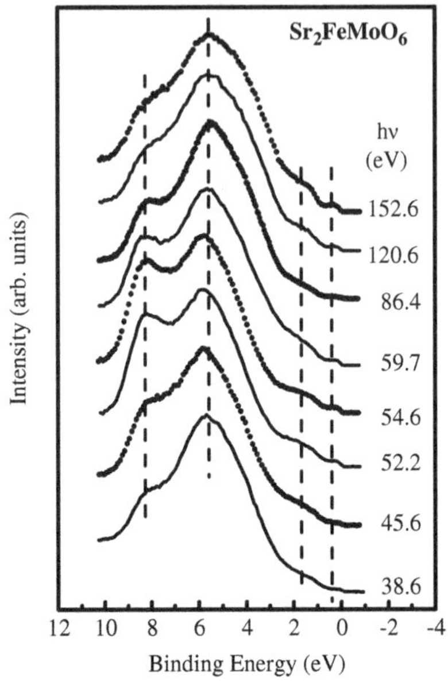

Figure 4. Valence band photoemission spectra from polycrystalline Sr_2FeMoO_6, collected with different incident photon energies. (Adopted from Ref. [16].)

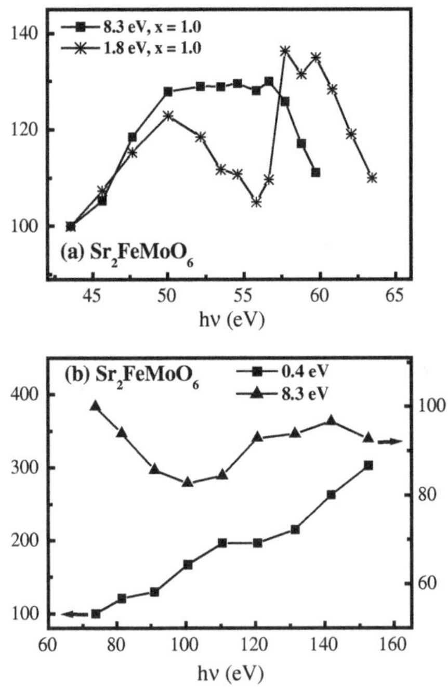

Figure 5. Percentage variation in the relative intensities of various spectral features in the valence band spectra of Sr_2FeMoO_6 as a function of photon energy (a) between 45 and 65 eV (b) between 70 and 152 eV. All intensities are normalized with the intensity of the most intense peak at 5.8 eV and then for easier comparison, the normalized intensity of each feature recorded with the lowest *hν* is always scaled to 100%. (Adopted from Ref. [16].)

This particular Fe feature corresponds to the fully filled up-spin e_g state as found also from the band structure results [2].

In Figure 5(b), we plot the normalized intensities of the features at the E_F and 8.3 eV for $h\nu$ between 75 and 152 eV, which clearly shows a significant minimum in the relative intensity of the 8.3 eV feature (triangles) at about $h\nu \sim$ 100 eV, exactly where the Cooper minimum for the localized Mo 4*d* states are expected. The significant contribution of Mo 4*d* states into the 8 eV spectral region is thereby established, while the delocalized Mo contribution near the E_F do not exhibit any Cooper minimum [17]. Instead, the relative intensity of the near Fermi edge feature, compared to the O 2*p* states at around 5.8 eV binding energy, shows a rapid, monotonic increase with increasing photon energy, $h\nu$, due to the changes in the relative Mo 4*d*: O 2*p* cross-sections [19] (see square data in Figure 5b), establishing a dominant contribution from delocalized Mo 4*d* down-spin t_{2g} states in this energy region.

Next we discuss the results of Bremsstrahalung Isochromat Spectroscopy (BIS) experiments, which provide detailed information about the unoccupied electron states in the system. We show the experimental BI spectrum of Sr_2FeMoO_6 with filled circles in Figure 6. The Fermi energy defining the zero of the energy scale was determined from the BI spectrum of a clean Cu metal in electrical contact with the sample. It is evident that there is a finite spectral intensity at E_F; this is consistent with the metallic behavior of this compound. The BI spectrum consists of three main features with peaks positioned at 2.5, 7.9 and 14.9 eV. In order to understand the nature and origin of the features, we calculated density of states by Linearized Muffin–Tin Orbital (LMTO) method within Generalized Gradient Approximation (GGA) along with the partial densities of states. We found that the band-structure results systematically underestimated the energies of different experimental features by about 15%. Such effects have been reported earlier in the literature [20, 21] and were attributed to the intrinsic limitations of such linearized calculations.

In order to make the comparison between the experiment and the theory meaningful, we expand the energy scale of the calculated data by 15%. The total DOS is shown with the thin solid line at the bottom of Figure 6. The partial DOS of Fe, Mo, Sr and O are also shown in the same figure with broken lines. The total DOS was broadened with an energy dependent Lorentzian to account for life-time effects and with a Gaussian to account for instrumental resolution present in the experimental data; the broadened DOS (thick solid line) is shown just below the experimental spectrum in Figure 6. We find a reasonable agreement between the experiment and theory for the features at 7.9 and 14.9 eV, both in terms of peak positions and spectral shapes. The 7.9 eV region is dominated by Sr states (primarily *s* and *d* states) admixed with O *p* and Mo 4*d* hybridized states, while the higher energy feature at about 15.5 eV is dominated by O states, as shown by the partial DOS in Figure 6. It is interesting to note that there is a very strong disagreement between the experiment and theory in the

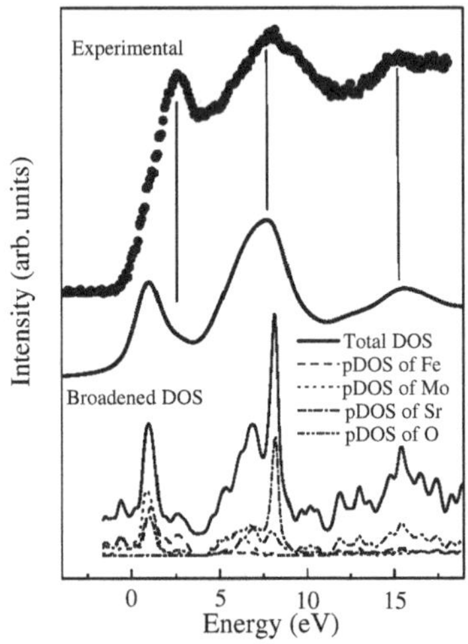

Figure 6. BI spectrum of Sr_2FeMoO_6 (*solid circles*). The LMTO DOS and pDOS are shown at the bottom of the figure. The broadened DOS is also presented for the sake of comparison with the experiment.

0–4 eV region. The partial DOS shown in Figure 6 establishes that this energy region is contributed primarily by Fe 3*d* states and Mo 4*d* states. It is well known that Coulomb interaction is strong within such transition metal *d* states, particularly in the first row transition metal 3*d* states; however, effective single particle theories, such as the band structure calculations presented here, cannot take into account the Coulomb interaction effects appropriately. This is the main reason for the disagreement between the experiment and theoretical results in Figure 6; the strong disagreement suggests that the Coulomb interaction effects are indeed dominant [16] in determining the low-energy electronic structure of this compound.

A qualitative comparison of the experimental spectrum with the calculated partial DOS (pDOS), provides a better understanding of the electronic structure of the unoccupied states. The near E_F feature is derived from the admixture of all Fe, Mo and O states. This feature is strongly dominated by Mo down-spin t_{2g} and e_g states, especially considering much larger photoemission cross-sections of Mo states compared to Fe and O at this energy range; this effect is not incorporated in the calculated DOS, as shown in Figure 6. Fe down spin and O *p* states also contribute significantly at this energy range. The second major peak at 7.9 eV has significant contributions from Sr *s* and *d* states, along with contributions coming from Mo e_g and O *p* states. The unoccupied states beyond 10 eV above E_F are essentially featureless, continuum states, contributed dominantly by O related states, as found from the calculated pDOS (see Figure 6). The above result suggests that the excitation spectrum of Sr_2FeMoO_6 is rather well described by the band calculation except for a narrow energy window close to E_F, which is a signature of strong electron correlation effects, present in the system.

5. Basic concepts on the magnetic structure of Sr_2FeMoO_6

In accordance with the basic crystal structure described for Sr_2FeMoO_6 and within a fully ionic description, this compound nominally consists of $Fe^{3+}3d^5$ $S = 5/2$ and $Mo^{5+}4d^1$ $S = 1/2$ ions alternating along the cubic axes, where the Fe and Mo sublattices are ferromagnetically coupled within each sublattice with the two sublattices coupled antiferromagnetically to give rise to an overall $S = 2$ state. But, such an extremely ionic description is inconsistent with the highly metallic properties of the compound, suggesting the presence of delocalized electrons, rather than electrons with localized moments; moreover, the occurrence of such a high ferromagnetic T_C in this compound remains to be a puzzle with such descriptions. In close analogy to the case of manganites, a double exchange type mechanism has been often postulated [22–25] as the origin of the ferromagnetic coupling between the Fe sites. There are, however, some important distinctions between the physics of the manganites and that of Sr_2FeMoO_6. Unlike manganites, the localized (Fe up-spin) and the delocalized spins (Mo down-spin) in Sr_2FeMoO_6 orient antiparallel to each other; as a consequence, the Hund's coupling strength, I, between the parallel oriented localized and delocalized electrons, which provides the energy scale of the on-site spin coupling in the double exchange mechanism for the manganites, becomes irrelevant for Sr_2FeMoO_6. Here, it has to be noted that the ferromagnetic ordering of the Fe sublattice is inevitable as long as there is a well-defined spin-coupling between the localized moment and the delocalized at every local moment site, irrespective of the sign of the spin-coupling. Therefore, we believe that in the magnetic structure of Sr_2FeMoO_6, the most important point to understand is the nature and origin of the spin-coupling of the mobile electrons and the localized ones. In some reports, it has been implicitly assumed [26] that an antiferromagnetic coupling between the Fe and the Mo sites via superexchange is responsible for the observed magnetic structure. This assumption again does not appear to be correct, since such a superexchange mechanism, coupling the Fe site to the delocalized and highly degenerate Mo d states, is expected to be weak at best, and therefore, not compatible with the unusually high ordering temperature ~420 K. Therefore, it is evident that we need a very different model to explain the magnetic structure of Sr_2FeMoO_6.

6. A new mechanism

Molybdenum systems are not generally strongly correlated system and are mostly known to be nonmagnetic. Therefore, a well-defined spin ordering between the delocalized electrons and the localized Fe electrons is unexpected, as such an ordering requires a large spin splitting of the delocalized band, derived from the Mo d and oxygen p states. A novel mechanism has been recently proposed to explain this new type of magnetic interaction between the localized electrons and the conduction electrons, leading to a strong polarization of the mobile charge carriers [27]. Here, we try to describe the mechanism responsible for the large exchange splitting of the delocalized band with the help of a schematic diagram (Figure 7). In the upper panel, we show the spin-polarized electronic structure in Sr_2FeMoO_6 in absence of hopping interaction between the local exchange–split up-spin and down-spin Fe states and the unpolarized valence states. Our valence band study shows that the localized up-spin Fe states lie much below the Fermi level (~2 eV binding energy), while the delocalized states at the E_F are mainly comprised of hybridized down-spin Mo *d* and O *p* states with some admixture of Fe *d* down-spin states. The analysis of the conduction band exhibits a larger contribution of Mo down-spin t_{2g} and e_g states in the near Fermi level feature while the Fe down-spin states also appear nearby. Accordingly in the schematic figure, the localized Fe up-spin states are shown as a black, thick vertical line at higher binding energy with the vertical dotted line denoting the E_F. The up- and down-spin delocalized bands are shown around the E_F, ensuring an overall metallic property of the system.

As this band has the highest contribution from Mo states, it is expected to be normally nonmagnetic and therefore, exchange splitting of these states can be neglected. The localized down-spin band is shown again as a vertical line above this delocalized band. Now, in the presence of hopping interactions coupling the Fe states to the delocalized states derived from the Mo 4*d*–O 2*p* states, there is a finite coupling between the same symmetry and spin states at the Fe and the delocalized electrons, leading to shifts in these bare energy levels. This effect is schematically shown in the lower panel of Figure 7. It is then easily seen that the opposite shifts of the up- and down-spin conduction states, therefore, induce a spin-polarization of the mobile electrons due to purely hopping interactions between the localized electrons and the conduction states. The low energy configurations of the up-spin localized and the down-spin delocalized states, therefore, ensures an antiferromagnetic coupling between them. It is also evident that the extent of the spin-polarization of the conduction electrons is primarily governed by the effective hopping strength and the charge-transfer energy between the localized and the delocalized states [27], identical conclusions have also been arrived at in subsequent papers [28] based on perturbative arguments. It is clear from the schematic diagram in Figure 7 that a significant extent of shift arising from a sizable hopping strength can even lead to half-metallicity, as is indeed observed in the case of Sr_2FeMoO_6. Detailed many-body calculations [27] have shown that the spin-polarization of the conduction band in this double perovskite compound is as large as 1–1.5 eV and the strength of the anti-ferromagnetic coupling between the conduction band and the localized electrons at the Fe site is of the order of 18 meV which is larger than that in the doped manganites, explaining the high T_C in this compound.

Interestingly, this mechanism appears to be of quite a general nature and can be applied to many exciting magnetic systems. Dilute magnetic semiconductors,

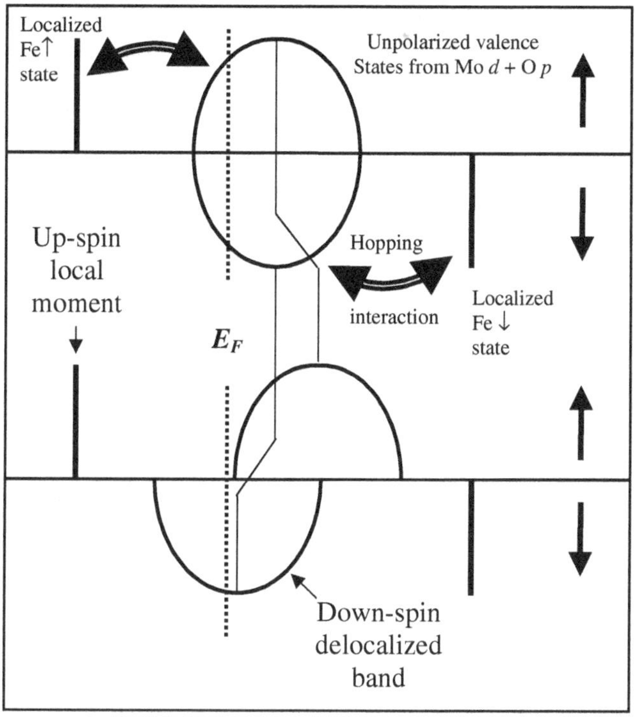

Figure 7. Schematic presentation of various energy bands in absence of hopping interactions (*upper part*) and in presence of hopping interactions (*lower part*).

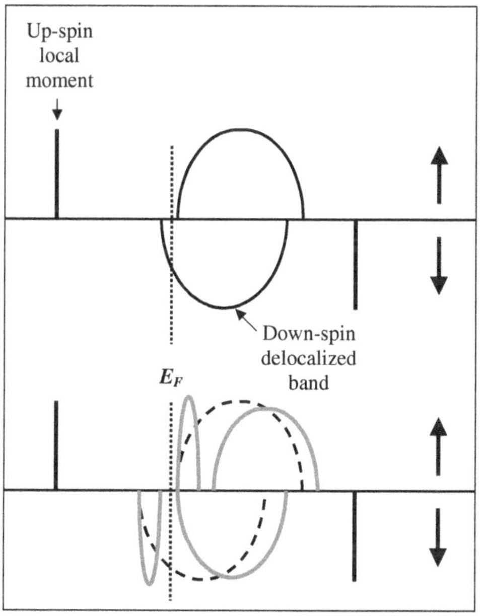

Figure 8. Schematic presentation of the possible scenario for ferromagnetic insulators in presence of different effects.

such as $Ga_{1-x}Mn_xAs$, are most likely examples of this new class of magnetic systems. It should be noted here that metallic property is not essential to realize such a magnetic interaction leading to the ferromagnetic state. For example, in systems with a low density of charge-carrier, as illustrated in Figure 8 (upper panel), it is clearly conceivable that the states near E_F are Anderson localized due to disorder effects. This will give rise to an insulating ferromagnetic ground state; this is possibly the case for small extent of Mn doping in $Ga_{1-x}Mn_xAs$. Structural distortion and/or electron correlation effects can also give rise to an insulating ferromagnetic ground state within this mechanism, by opening a finite gap at the E_F, as illustrated in the lower panel of Figure 8. The figure schematically shows the original semi-elliptical spin-polarized bands (dashed lines) split into two parts each, opening the gap at E_F. In all these systems, the conduction band is polarized antiferromagnetically with respect to the localized moment at the transition metal site due to the hopping interactions between the two and a large exchange splitting of the localized state, thereby driving a ferromagnetic arrangement of the localized moments. But, the extent of interaction can vary from system to system, giving rise half-metallic, semiconducting or even insulating ground states.

References

1. von Helmholt R., Wecker J., Holzapfel B., Schultz L. and Samwer K., *Phys. Rev. Lett.* **71** (1993), 2331; for a review: Phys. Today Special issue on Magnetoelectronics (1995).
2. Kobayashi K.-I., Kimura T., Sawada H., Terakura K. and Tokura Y., *Nature (London)* **395** (1998), 677.
3. Patterson F. K., Moeller W. C. and Ward R., *Inorg. Chem.* **2** (1963), 196; Galasso F. S., *J. Chem. Phys.* **44** (1966), 1672.
4. Nakagawa T., Yoshikawa K. and Nomura S., *J. Phys. Soc. Jpn.* **27** (1969), 880.
5. Kawanaka H., Ease I., Toyama S. and Nishihara Y., *Physica B* **281–282** (2000), 518.
6. Arulraj A., Ramesha K., Gopalakrishnan J. and Rao C. N. R., *J. Solid State Chem.* **155** (2000), 233.
7. Kobayashi K.-I., Kimura T., Tomioka Y., Sawada H., Terakura K. and Tokura Y., *Phys. Rev. B* **59** (1999), 11159.
8. Hamada N. and Moritomo Y., *The 3rd Japan–Korea Joint Workshop on First-Principles Electronic Structure Calculations*, 2000 p. 56.
9. Moritomo Y., Xu S., Machida A., Akimoto T., Nishibori E., Takata M. and Sakata M., *Phys. Rev. B* **61** (2000), R7827.
10. Tomioka Y. *et al.*, *Phys. Rev. B* **61** (2000), 422.
11. Westerburg W., Reisinger D. and Jakob G., *cond-mat/0001398* **27** (Jan 2000).
12. Yin H. Q. *et al.*, *Appl. Phys. Lett.* **75** (1999), 2812.
13. Yin H. Q. *et al.*, *J. Appl. Phys.* **87** (2000), 6761.
14. Sarma D. D. *et al.*, *Solid State Comm.* **114** (2000), 465.
15. Itoh M., Ohta I. and Inaguma Y., *Mater. Sci. Eng. B* **41** (1996), 55.
16. Ray S. *et al.*, *Phys. Rev. B* **67** (2003), 085109.
17. Abbati I., Braicovich L., Carbone C., Nogami J., Yeh J. J., Lindau I. and del Pennino U., *Phys. Rev. B* **32** (1985), 5459.

18. Fujimori A., Mamiya K., Mizokawa T., Miyadai T., Sekiguchi T., Takahashi H., Mori N. and Suga S., *Phys. Rev. B* **54** (1996), 16329; Walker K.-H., Kisker E., Carbone C. and Clauberg R., *ibid.* **35** (1987), 1616.
19. Yeh J. J. and Lindau I., *At. Data Nucl. Data Tables* **32** (1985), 1.
20. Sarma D. D., Shanthi N., Barman S. R., Hamada N., Sawada H. and Terakura K., *Phys. Rev. Lett.* **75** (1995), 1126.
21. Sarma D. D., Shanthi N. and Mahadevan Priya, *Phys. Rev. B* **54** (1997), 1622.
22. Kim T. H., Uehara M., Cheong S. W. and Lee S., *Appl. Phys. Lett.* **74** (1999), 1737.
23. Marti′nez B., Navarro J., Balcells L. and Fontcuberta J., *J. Phys., Condens. Matter* **12** (2000), 10515.
24. Garcia Landa B. *et al.*, *Solid State Comm.* **110** (1999), 435.
25. Chattopadhyay A. and Millis A. J., *cond-mat/0006208* (13 June 2000).
26. Ogale A. S., Ramesh R. and Venkatesan T., *Appl. Phys. Lett.* **75** (1999), 537.
27. Sarma D. D., Mahadevan P., Saha-Dasgupta T., Ray S. and Kumar A., *Phys. Rev. Lett.* **85** (2000), 2549.
28. Kanamori J. and Terakura K., *J. Phys. Soc. Jpn.* **70** (2001), 1433; Fang Z., Terakura K. and Kanamori J., *Phys. Rev. B* **63** (2001), 180407(R).
29. Sarma D. D., *Curr. Opin. Solid State Mater. Sci.* **5** (2001), 261.

Hyperfine Interactions (2005) 160:81–94
DOI 10.1007/s10751-005-9151-y

Semiconductor Nanoparticles

M. BANGAL[1], S. ASHTAPUTRE[1], S. MARATHE[1], A. ETHIRAJ[1], N. HEBALKAR[1], S. W. GOSAVI[1], J. URBAN[2] and S. K. KULKARNI[1,*]
[1]*Department of Physics, University of Pune, Pune 411007, India;*
e-mail: skk@physics.unipune.ernet.in
[2]*Fritz Haber Institute der Max Planck Gesellschaft, Berlin D-97074, Germany*

Abstract. Semiconductor nanoparticles exhibit size dependent properties, when their size is comparable to the size of Bohr diameter for exciton. This can be exploited to increase fluorescence efficiency or increase the internal magnetic field strength in doped semiconductors. Nanoparticles are usually unstable and can aggregate. It is therefore necessary to protect them. Surface passivation using capping molecules or by making core–shell particles are some useful ways. Here synthesis and results on doped and un-doped nanoparticles of ZnS, CdS and ZnO will be discussed. We shall present results on core–shell particles using some of these nanoparticles and also discuss briefly the effect of Mn doping on hyperfine interactions in case of CdS nanoparticles.

Key Words: core–shell, doping, ESR, semiconductor nanoparticles, TEM.

1. Introduction

Since last two decades there has been an explosive growth of research in the field of nanomaterials. Several review articles and books have been published on different aspects of nanomaterials. See for example [1–8]. Nanomaterials may be in the range of 1 to 100 nm size and one can reduce the dimension in one, two or all three directions to obtain thin films, wires or dots, respectively. Nanomaterials can be of various shapes too and properties may change according to size and/or shape. These materials may be metals, semiconductors, metal oxides, organic materials or biomaterials. Thus there is a tremendous scope to design new materials with unusual properties.

Amongst the various types of nanomaterials, semiconductor nanoparticles have been widely investigated. This is quite understandable. Semiconductors have been useful in making devices. The drive towards miniaturization of electronic components and integration to accommodate huge number of them in small volume has been there for decades. This has enabled to have very compact digital watches, calculators, computers, laptops etc. In fact Moors' law

* Author for correspondence.

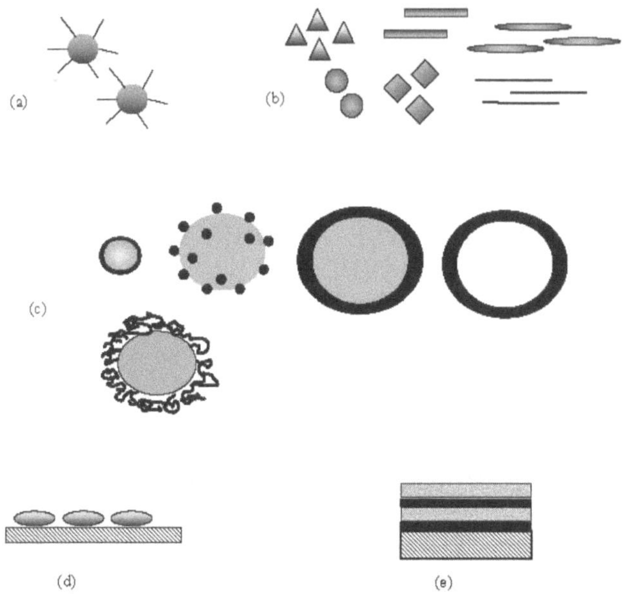

Figure 1. Some possible nanostructures: (a) surface passivated nanoparticles (b) different shapes of nanoparticles (c) core–shell particles (d) self assembly, and (e) multilayers.

predicted in 1960 has now reached its limits in such a way that any further reduction in size changes the materials properties. Dimensions of some of the devices are now in the nanometer range. Electronic structure of nanomaterials may be different compared to corresponding bulk material. This has lead to interesting devices like single electron transistors, tunnel junctions, magnetic spin valves etc. which do not have bulk counterparts. Besides this, at nano scale, semiconductor materials like silicon that are not optoelectronic materials due to indirect band gap have showed strong luminescence in visible range [9]. Moreover they exhibit emission which is size dependent luminescence. Some groups have showed [10–13], for II–VI and III–V semiconductors, the change in band gap with particle size.

Materials with zero dimensions i.e. the ones whose all three dimensions are reduced to nanometer regime are known as quantum dots or simply nano-particles. There is a variety of nanoparticles. A few possibilities are illustrated schematically in Figure 1. We have synthesized a large number of oxides, sulphides, core–shell particles, self-organized islands of Ge/Si, fullerenes, Diluted Magnetic Semiconductor (DMS) nanoparticles like TiO_2 doped with cobalt and metallic multilayers like Fe/Si, Cu/Ni Co/Pt etc. Here we will discuss a variety of semiconductor nanoparticles viz. ZnS, CdS and ZnO, their synthesis and some optical properties. We shall then discuss how nanoparticles can be encapsulated inside some robust shell of silica or how to anchor such particles on silica for exploiting large surface area of silica particles. We also discuss how doping with manganese can alter the splitting of energy levels in CdS nanoparticles causing changes in the internal magnetic field of the nanoparticles.

2. Experimental

Synthesis of semiconductor nanoparticles was carried out using a chemical route known as chemical capping. In this procedure, some inert organic long chain molecules are used to passivate the surface of the particles so that the particles do not agglomerate or ripen to form larger particles. This ensures that the nanoparticles synthesized using some particular set of parameters yields stable nanoparticles with uniform size distribution. Figure 2 shows the flow chart of synthesis procedure used to synthesize ZnS, CdS and ZnO nanoparticles having <8 nm particle size. Details of synthesis are given below.

2.1. SYNTHESIS OF NANOPARTICLES

2.1.1. *ZnS nanoparticles*

Zinc sulphide nanoparticles were synthesized using zinc acetate [Zn $(CH_3CO_2)_2$ $2H_2O$], *N*, *N*-dimethylformamide [$HCON(CH_3)_2$], sodium sulphide [Na_2S], thioglycerol [$HSCH_2CH(OH)CH_2OH$], sodium hydroxide [NaOH] and ethanol [C_2H_5OH]. Synthesis was carried out as follows. Zinc acetate (2.2 mM in 50 ml ethanol) was mixed with thioglycerol (3.5 mM in 50 ml ethanol) in DMF medium and both were stirred together for 30 min. Sodium sulphide (0.55 mM in 8 ml ethanol) was slowly mixed over a period of 4 h. The reaction was carried out at room temperature under nitrogen atmosphere. The solution was constantly stirred using a magnetic stirrer. Sodium hydroxide solution was then added till pH of the solution was 8. The whole solution was then refluxed for 14 h. Nanoparticles of ZnS were then obtained by size selective precipitation method [14] described as follows. The colloidal solution of nanoparticles in ethanol was concentrated using roto-evaporator. To this concentrated solution acetone, a non-solvent, was added slowly till flocculation was obtained which was then allowed to settle down. The supernatant was removed and flocculate was centrifuged to get precipitate enriched with largest crystallites in the sample. Precipitate was washed repeatedly using methanol. Dispersion of precipitate in ethanol and size selective precipitation was repeated till no further sharpening in sample's absorption spectrum had been seen.

2.1.2. *CdS nanoparticles*

Non-aqueous solutions of cadmium acetate [Cd $(CH_3CO_2)_2 \cdot 2H_2O$], thioglycerol and sodium sulfide prepared in ethanol were used to synthesize CdS nanoparti-

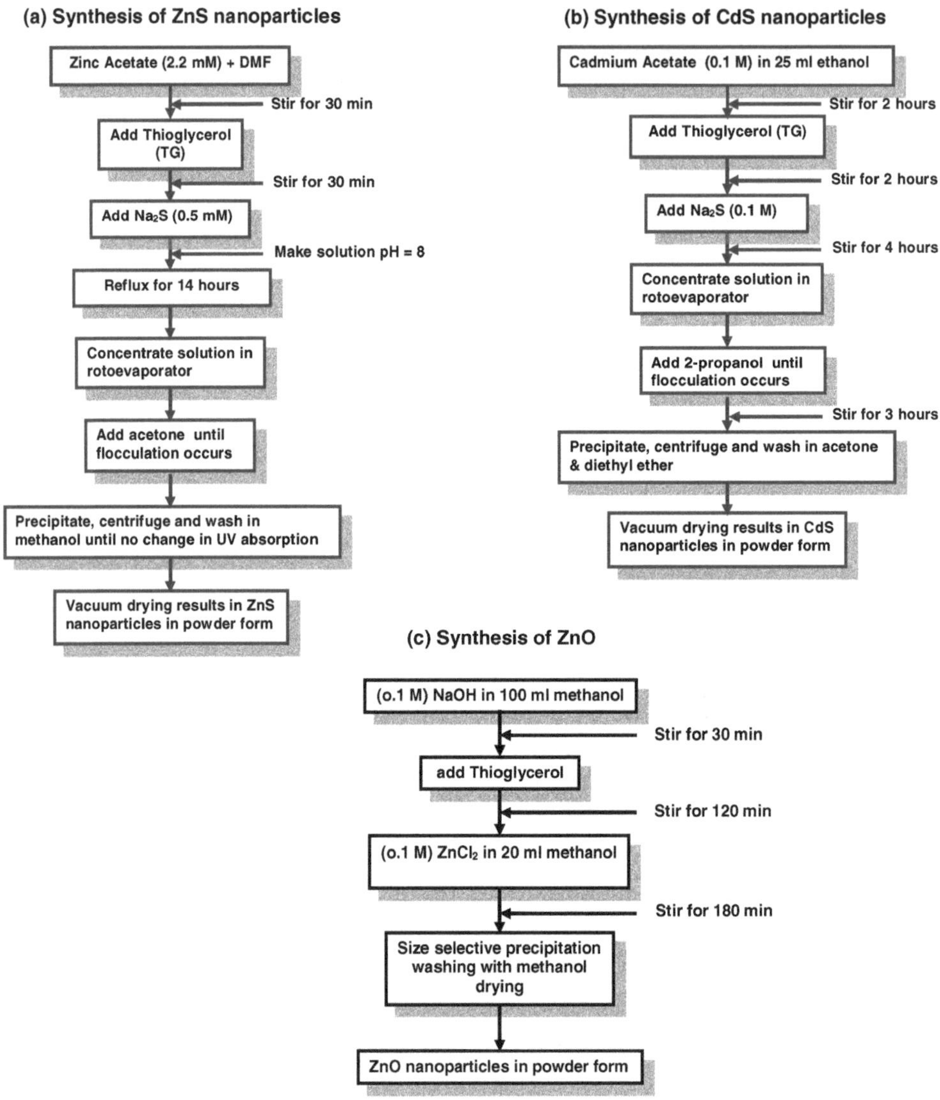

Figure 2. Procedure for synthesis of different nanoparticles (a) ZnS, (b) CdS and (c) ZnO nanoparticles. Variation of thioglycerol gives particles of different sizes.

cles. Thioglycerol was used as a capping agent. Cadmium acetate solution (0.1 M in 25 ml ethanol) was taken in reaction vessel and stirred continuously using magnetic stirrer. After two hours, thioglycerol (0.95 M in 25 ml ethanol) was added dropwise to cadmium acetate solution and stirred for two more hours. Finally sodium sulfide solution (0.1 M in 25 ml ethanol) was added and the solution was stirred continuously for 4 h. The synthesis was carried out at room temperature under nitrogen atmosphere to prevent the oxidation of nanoparticles. Size-selective precipitation method discussed above was used to obtain the monodispersed particles. Precipitate was separated by centrifugation and washed with acetone and diethyl ether. It was then vacuum dried to get CdS nanoparticles in powder form.

2.1.3. *ZnO nanoparticles*

Zinc oxide nanoparticles were synthesized using zinc chloride [$ZnCl_2 \cdot 2H_2O$], methanol [CH_3OH], sodium hydroxide and thioglycerol. The synthesis was carried out in non-aqueous medium. The solutions were prepared in methanol. Thioglycerol was used as capping agent. Sodium hydroxide (0.1 M in 100 ml methanol) was stirred for 30 min. Thioglycerol (0.1 M in 20 ml methanol) was added drop wise to above solution and stirring was continued further for 60 min. Zinc chloride (0.1 M in 20 ml methanol) solution was then slowly added to this solution. The solution was then stirred for 180 min. The precipitate obtained by size selective procedure was washed with methanol and dried to get ZnO nanoparticles in powder form.

2.1.4. *Doping of nanoparticles*

For manganese doping both in ZnS and in CdS nanoparticles, manganese salt ($MnCl_2 \cdot H_2O$) was added to the respective metal salts in the solution form and resulting solution was stirred for 2 h before adding capping agent to the solution. After that the same synthesis procedure described above was used to synthesize doped nanoparticles. Doping of other elements can be achieved by using corresponding salts.

2.2. CORE–SHELL PARTICLES

2.2.1. *Synthesis of SiO_2 particles*

Figure 3 shows the flow chart explaining various steps used in synthesizing silica particles, encapsulation of nanoparticles or coating of nanoparticles in general.

Following the procedure developed by Stöber *et al.* [15], we synthesized SiO_2 particles. Here we use tetraethylorthosilicate, water and ammonia. A solution of ethanol, water and ammonium hydroxide in the molar ratio 49:31:4 was used. The solution was stirred at room temperature for 30 min and TEOS (1M) was added to it. Solution was further stirred for 3 h. Centrifugation and drying in ambient resulted in silica powder. By varying the reaction time, spherical silica particles of different sizes (50 nm–0.1 μm) could be obtained. In most of the cases particles with very narrow size distribution (1–3%) could be obtained.

2.2.2. *Encapsulation of nanoparticles inside SiO_2*

The synthesis procedure used to obtain silica particles can be further extended to encapsulate the nanoparticles inside silica. Here we have encapsulated man-

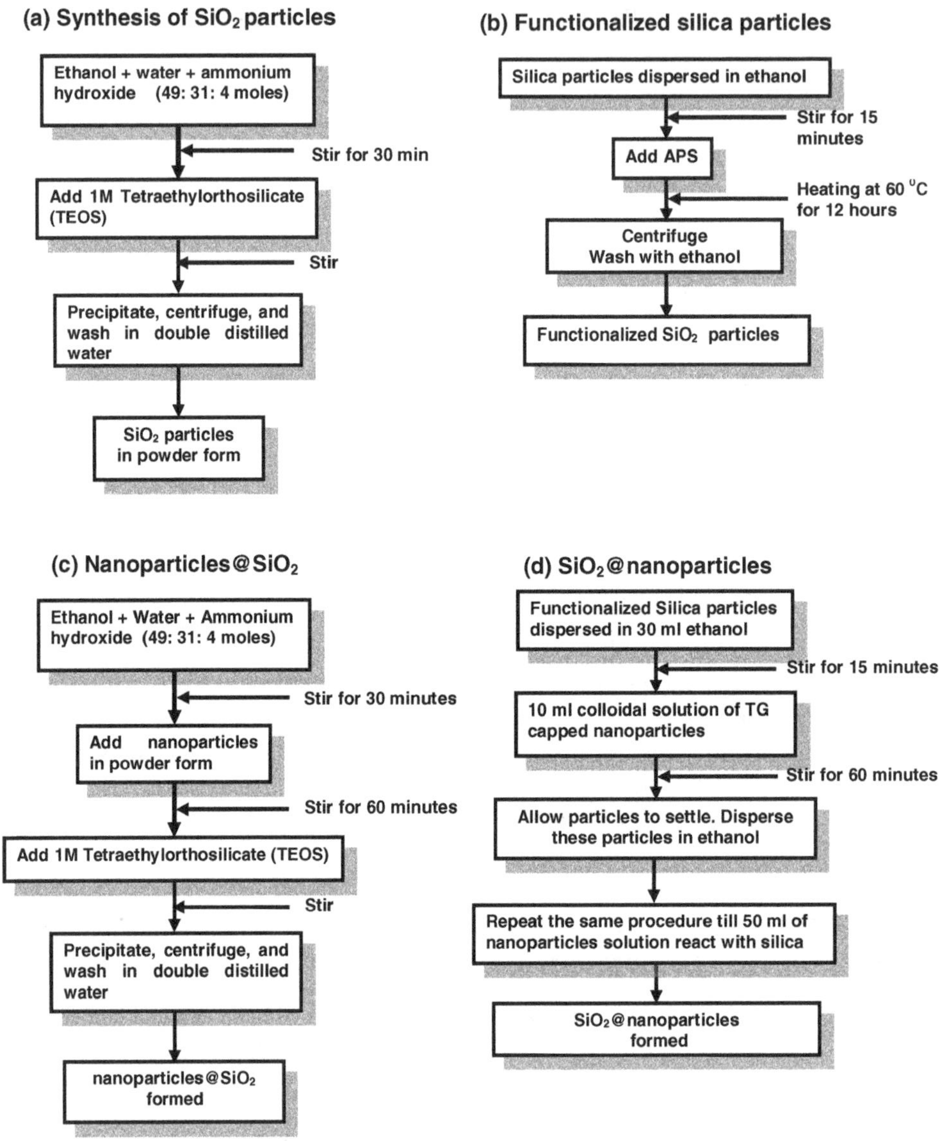

Figure 3. Procedure to synthesize (a) spherical SiO_2 particles, (b) functionalized SiO_2 (c) nanoparticles@SiO_2 core–shell particles and (d) SiO_2@nanoparticles core–shell particles. Variation of stirring period after TEOS addition gives particles of different sizes.

ganese doped zinc sulphide nanoparticles inside spherical silica particles. The synthesis procedure can be briefly outlined as follows. A solution of ethanol, methanol and water is made. It is continuously stirred for 30 min. Thioglycerol capped nanoparticles in the form of powder are added to the solution and further stirred for 60 min. By adding TEOS to this solution silica spheres start forming with nanoparticles trapped inside the core. By removing the solution at different intervals core–shell particles of different sizes can be obtained. It may be pointed out that in this case many nanoparticles are trapped inside each silica sphere.

2.2.3. *Functionalization of SiO_2 particles*

Bifunctional molecule 3-aminopropyltrimethoxysilane (APS) has three [$-OCH_3$] groups and one [$NH_2-(CH_2)_3$-] group bonding to a silicon atom. Therefore it can link with two different types of molecules or particles. This has been used by various groups to bind nanoparticles to larger particles of silica or polymers and has been reviewed in [16, 17]. Here we use similar approach and attach TG capped ZnS nanoparticles and CdS nanoparticles to SiO_2 particles. One can control the amount of particles added to the shells using the procedure used here.

Silica particles are dispersed in ethanol and stirred for 15 min. To this solution APS (1×10^{-3} M) was added. The solution was then heated to 60°C and stirred continuously at this temperature for 12 h. Solution was allowed to cool to room temperature, and was centrifuged to obtain a precipitate. Precipitate was finally washed in ethanol.

2.2.4. *Coating SiO_2 particles with nanoparticles*

APS functionalized silica particles were mixed in ethanol along with TG capped nanoparticles, either of CdS or ZnS. Using different concentration of nanoparticles one can get thin or thick coating of nanoparticles on silica. See Figure 3 for flow chart.

2.3. CHARACTERIZATION

Characterization of above mentioned nanoparticles and core–shell particles has been carried out using different techniques like optical absorption, fourier transform infra red spectroscopy, photoluminescence, X-ray diffraction, electron spin resonance and transmission electron microscopy. The description of samples (liquid or powder) and instruments is briefly given below.

2.3.1. *Optical absorption*

The samples were analyzed using Shimadzu 300 double beam spectrometer by dispersing them in ethanol in case of ZnS, CdS and methanol in case of ZnO. Pure ethanol and methanol were used as reference for ZnS, CdS and ZnO, respectively. Spectra were recorded from 600 to 200 nm.

2.3.2. *Fourier transform infra red spectroscopy*

Samples for FTIR analysis were prepared in the form of pellets. Powder samples of nanoparticles or core–shell were mixed with KBr powder and pellets were formed. Shimadzu 8400 spectrometer was used to record the spectra in the 300–4000 cm^{-1} frequency region.

2.3.3. *Photoluminescence spectroscopy*

Spectra were acquired using Perkin-Elmer LS-50 model. The dry powder samples were held in a powder sample holder. Spectra were recorded from 200 to 350 nm in excitation mode and 350 to 700 nm in emission mode using appropriate filters.

2.3.4. *X-ray diffraction*

X-ray Diffraction of powder samples was carried out on Philips PW1840 powder diffractometer. Copper radiation (CuK_α 1.54 Å) was used in the analysis with nickel filter.

2.3.5. *Electron spin resonance*

The powder samples were analyzed using Brucker EMX-X band ESR spectrometer. The microwave source frequency was 9.5 GHz and 100 kHz modulation was used.

2.3.6. *Transmission electron microscopy*

Philips CM 200 FEG microscope equipped with a field emission gun was used for TEM analysis of the samples. EDAX analysis also could be carried out using this microscope. For the analysis, powder samples were dispersed in ethanol and a drop of solution was placed on ~5 nm thick carbon film on copper grid. After air drying, the grid was inserted in the microscope and using 200 KeV energy images were acquired.

3. Results and discussion

Nanoparticles of II–VI semiconductors have been investigated for a long time for their optical properties [18–20]. As the particle size reduces below the Bohr size of exciton, the energy gap increases and can be observed as shift in the absorption edge in UV–VIS absorption spectra. Band gaps for zinc sulphide, zinc oxide and cadmium sulphide bulk materials are 3.68 eV, 3.35 eV and 2.4 eV respectively [21]. The absorption edges are expected at 337 nm for ZnS, 370 nm for ZnO and 517 nm for CdS. As showed in Figure 4, the absorption spectra for ZnS, ZnO and CdS nanoparticles are well shifted from their bulk values. In case of ZnO the binding energy of exciton is 60 meV. This is quite large compared to

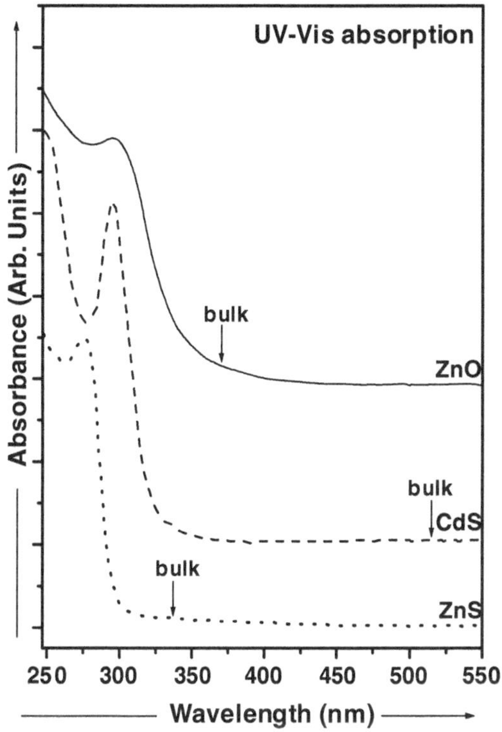

Figure 4. UV–Vis absorption spectra for ZnS (1.1 nm), CdS (3.6 nm) and ZnO (3.1 nm) nanoparticles.

thermal energy available at room temperature (~25 meV) to destroy the exciton. Therefore in ZnO, exciton is detectable even at room temperature. However ZnS (29 meV) and CdS (27 meV) excitons with very low binding energies are destroyed and are not detectable for bulk semiconductors of CdS or ZnS. However one can detect well defined excitonic peaks even for ZnS and CdS at room temperature, when CdS and ZnS are in the form of nanoparticles. Average particle sizes can be determined using the widths of X-ray diffraction peaks. In the present case the particle sizes are 1.1 nm for ZnS, 3.1 nm for ZnO and 3.6 nm for CdS particles. The presence of excitonic peak is expected theoretically too [22, 23]. Depending upon the preparation method there are variations in the quality of spectra obtained.

Semiconductor nanoparticles are quite useful as biological labels, barcoades, display screens etc [24–26]. Advantage with semiconductor nanoparticles is that they are photostable i.e., unlike many organic dyes they do not show degradation of luminescence intensity after some time. Besides they can be excited with any of the wavelengths in UV range rather than some fixed wavelength. It has also been showed [2, 7, 28] that efficiency for luminescence is large in case of nanoparticles compared to their bulk materials. Thus it is possible to excite photoluminescence in different nanoparticles using a single wavelength. In

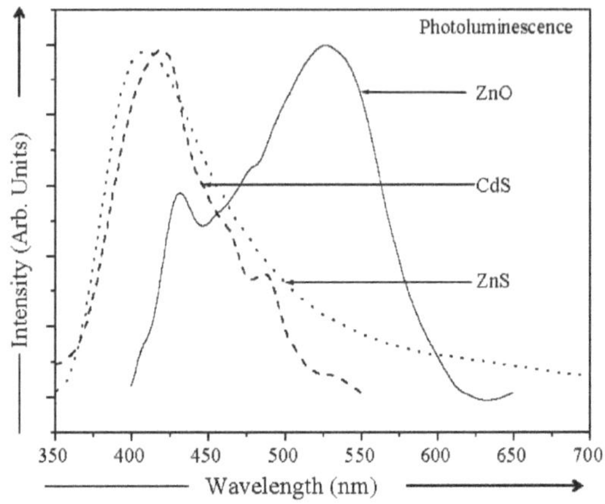

Figure 5. Photoluminescence of different nanoparticles.

Figure 5, photoluminescence for ZnS, ZnO and CdS nanoparticles has been showed. Nanoparticles in general have higher quantum efficiency of luminescence as compared to their bulk counterparts. Mechanism of luminescence of ZnS, CdS and ZnO nanoparticles has been discussed widely in the literature, see e.g. [27, 28].

Interestingly it is possible to coat the chemically capped nanoparticles on SiO_2 particles. SiO_2 particles of 50–500 nm size are well suited to adsorb such 1–5 nm size particles. Figure 6 (a) shows TEM images of SiO_2 particles while Figure 6 (b) and (c) show SiO_2 particles coated with CdS and ZnS respectively. Such coated particles often known as core–shell particles are written as core (material)@shell (material). Thus SiO_2 core and CdS coating is written as SiO_2@CdS. If CdS would get encapsulated inside SiO_2 then this would be written as CdS@SiO_2. The details are reported elsewhere [29, 30]. However it can be mentioned here that there is no significant change in the properties of nanoparticles due to such coating of nanoparticles. FTIR analysis (not shown here) indeed showed the presence of TG molecules as well as APS [29, 30]. This type of core – shell particles would be useful for some applications where large number of well dispersed nanoparticles are required.

In some applications, specially the biological applications, it is necessary to use water soluble materials. SiO_2 particles are water soluble. Therefore fluorescent nanoparticles can be encapsulated inside SiO_2 particles forming core–shell particles. Advantage with this type of coating is that, nanoparticles can be excited with UV radiation as SiO_2 are found to be nearly transparent to UV radiation. We have found [31] that by encapsulating nanoparticles inside silica particles the photoluminescence increases. Increase in luminescence is due to protection provided by silica shell.

Transition metal doping, specially in II–VI semiconductors, is further interesting for the applications in spintronics [32]. Recently there have been reports about observation of giant magnetic field in nanoparticles of CdS doped with Mn

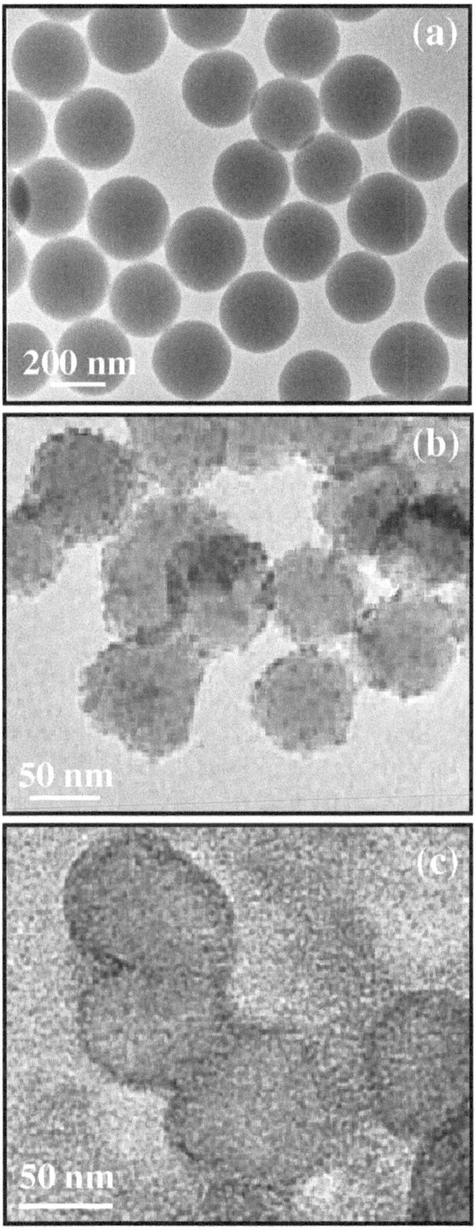

Figure 6. Transmission electron micrographs of (a) silica particles, (b) core–shell particles of SiO_2@CdS and (c) core–shell particles of SiO_2@ZnS.

in absence of zero external magnetic field [33]. This leads to observation of large splitting of exciton sub levels which has been interpreted as due to increase in the electron–hole interaction in nanoparticles. Also it is possible to observe ferro-magnetism in nanoaprticles as discussed for (GaMn)As system [34]. Pileni group [35] has showed that there are concentration dependent changes in ESR spectra

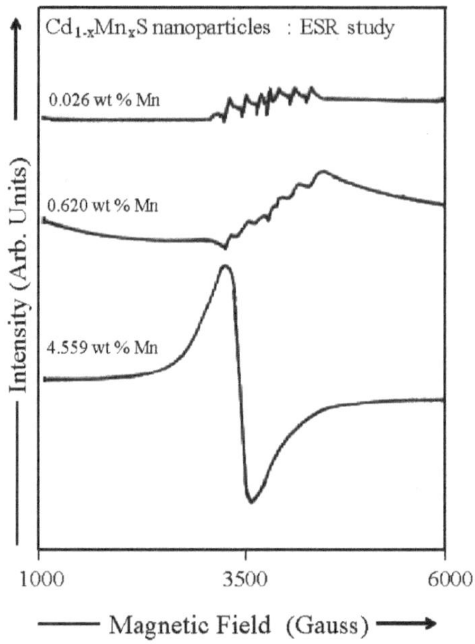

Figure 7. Electron spin resonance study of Mn doped CdS nanoparticles.

for CdS doped with Mn ions, synthesized using inverse miceller route. It is thus interesting to dope II–VI semiconductor nanoparticles to obtain ferromagnetism or even large paramagnetic moment in them. We have therefore initiated the investigations of Mn doped nanoparticles. Figure 7 shows ESR spectra for Mn doped CdS nanoparticles (3.8 ± 0.2 nm). It can be seen that six line pattern characteristic of isolated Mn ion doping, converts to a single line characteristic of strong Mn–Mn ion interaction similar to that discussed earlier [36, 37] for Mn doped in ZnS. Note that in Figure 7 the spectra are for CdS nanoparticles having same size but varying concentration of Mn ions. These results are in qualitative agreement with some earlier work on differently prepared nanoparticles of CdS doped with Mn [33, 35]. In order to assess the utility of doped II–VI semiconductor nanoparticles as spintronics materials, more experiments would however be necessary.

4. Conclusions

Stable semiconductor nanoparticles and core–shell particles can be synthesized using chemical routes. Quality of nanoparticles and core–shell particles is quite good and such particles would be useful for biological labeling, barcodes or fluorescent materials for display screens. Doped nanoparticles may find applications in spintronics too. However more experiments would be necessary in future to evaluate this aspect.

Acknowledgement

SKK thanks UGC, India for a continuous support. This work has been supported by DST, India and Volkswagenstiftung, Germany.

References

1. Jacak L., Hawrylak P. and Wojs A., *Quantum Dots*, Springer, Berlin Heidelberg New York, 1997.
2. Gaponenko S. V., *Optical Properties of Semiconductor Nanocrystals*, Cambridge, 1997.
3. Davis J. H., *Physics of Low Dimensional Structures*, Cambridge, 1998.
4. Banyai L. and Koch S. W., Semiconductor Quantum Dots, *World Scientific Series on Atomic, Molecular and Optical Physics*, Vol. 2, 1993.
5. Kelly M. J., *Low Dimensional Semiconductors*, Clarendon, 1995.
6. Jaros M., *Physics and Applications of Semiconductor Microstructures*, Clarendon, 1989.
7. Moriarty P., *Rep. Prog. Phys.* **64** (2001), 297.
8. Gleiter H., *Prog. Mater. Sci.* **33** (1989), 223.
9. Canham L. T., *Appl. Phys. Lett.* **57** (1990), 1046.
10. Brus L. E., *J. Chem. Phys.* **80** (1984), 4403.
11. Lee J., Sunder V. C., Heine J. R., Bawendi M. G. and Jensen K. F., *Adv. Mater.* **12** (2000), 1102.
12. Henglein A., *Top. Curr. Chem.* **143** (1988), 113.
13. Wang W. and Herron N., *J. Phys. Chem.* **95** (1991), 525.
14. Murray C. B., Norris D. J. and Bawendi M. G., *J. Am. Chem. Soc.* **115** (1993), 8706.
15. Stöber W. and Fink A., *J. Coll. Int. Sci.* **26** (1968), 62.
16. Caruso F., *Adv. Mater.* **13** (2001), 11.
17. Liz-Marźan L. M., Correa-Durate M. A., Pastorza-Santos I., Mulvaney P., Ung T., Giersig M. and Kotov N. A., Handbook of Surfaces and Interfaces of Materials, Nalwa H. S. (ed.), *Nanostructured Material, Micelles and Colloids*, Vol. 3, 2001, p. 189.
18. Brus L. E. and Trautman J. K., *Phil. Trans. R. Soc. Lond. A* **353** (1995), 313.
19. Chestnoy N., Hull R. and Brus L. E., *J. Chem. Phys.* **85** (1986), 2237.
20. Bawendi M. G., Steigerwald M. L. and Brus L. E., *Ann. Rev. Phys. Chem.* **41** (1990), 477.
21. Sze S. M., *Physics of Semiconductor Devices*, Wiley, Delhi, 1981.
22. Lippens P. E. and Lanoo M., *Phys. Rev., B* **39** (1989), 10935.
23. Ramakrishna M. V. and Friesner R. A., *Phys. Rev. Lett.* **67** (1991), 629.
24. Bruchez M., Moronne M., Gin P., Weiss S. and Alivisatos A. P., *Science* **281** (1998), 2013.
25. Gaponik N., Radtchenko I. L., Sukhorukov G. B., Weller H. and Rogasch A. L., *Adv. Mater.* **14** (2002), 879.
26. Battersby B. J., Bryant D., Meutermans W., Matthews D., Smythe M. L. and Trau M., *J. Ame. Chem. Soc.* **122** (2000), 2138.
27. Masumoto Y., In: Shionoya S. and Yen W. M. (eds.), *Phosphor Handbook*, CRC, Boston, p. 71.
28. Kulkarni S. K., In: Nalwa H. S. (ed.), *Encyclopedia of Nanoscience and Nanotechnology*, Vol. X, Ame. Sci. Pub., USA.
29. Hebalkar N., Kharrazi S., Ethiraj A., Urban J., Fink R. and Kulkarni S. K., *J. Coll. Int. Sci.* **278** (2004), 107.
30. Ethiraj A. S., Hebalkar N, Sainkar S. R., Urban J. and Kulkarni S. K., *Surface Engineering* **20** (2004), 367.
31. Ethiraj A. S., Hebalkar N., Kulkarni S. K., Pasricha R., Urban J., Dem C., Schmitt M., Kiefer W., Weinhardt L., Joshi S., Fink R., Heske C., Kumpf C. and Umbach E., *J. Chem. Phys.* **118** (2003), 8945.

32. Sato K. and Yoshida H. K., *Semicond. Sci. Tech.* **17** (2002), 367.
33. Hofmann D. M., Meyer B. K., Ekimov A. I., Merculov I. A., Efros A. L., Rosen M., Couino G., Gacoin T. and Boilot J. P., *Sol. State Comm.* **114** (2000), 547.
34. Sapra S., Sarma D. D., Sanvito S. and Hill N. A., *Nano Lett.* **2** (2002), 605.
35. Levy L., Feltin N., Ingert D. and Pileni M. P., *J. Phys. Chem. B* **101** (1997), 9153.
36. Borse P. H., Srinivas D., Date S. K., Vogel W. and Kulkarni S. K., *Phys. Rev., B* **60** (1999), 8659.
37. Kennedy T. A., Glasser E. R., Klein P. B. and Bhargava R. N., *Phys. Rev. B* **52** (1995), R14356.

Hyperfine Interactions (2005) 160:95–106
DOI 10.1007/s10751-005-9152-x

Nanostructuring by Energetic Ion Beams

D. K. AVASTHI
Nuclear Science Centre, P.O. Box 10502, New Delhi 110067, India; e-mail: dka@nsc.ernet.in

Abstract. Development of nanostructured materials has become of wide interest due to their exotic properties and interesting physics aspects. Energetic ions play a crucial role in the development of nano materials. Ions of different energy regimes have different roles in growth of nano particles. Low energy ions (typically up to a (kiloelectronvolt) keV) in plasma, have been in use for growth of nano particle thin films. Low energy ions (typically a few hundred (kiloelectronvolt) keV) from ion implanters are used for growth of nano particles in a matrix. High energy heavy ions (swift heavy ions) have been in use in recent years for growth of nanostructures and also for modifying nanostructures. Highly charged slow moving ions and focused ion beams too, have potential for creating nanostructures. Out of these several possible roles of energetic ions, there have been developments at NSC Delhi in growth of nanostructures by RF plasma, low energy ions and swift heavy.

1. Introduction

Nanotechnology has become very popular and attracting field of research due to the possibilities of a wide range of promising futuristic applications. Physicists find it interesting as systems of confined low dimensions, representing a picture of quantum mechanically confined potentials, where the confining dimension is represented by the size of nanostructures. The ability of altering the properties (physical, electronic, optical, etc.) by merely changing the size of the nano particles makes it extremely suitable for a variety of applications. Most common and economical approach of generating nanostructures is the chemical rout, where the precipitation in a solution is arrested by capping the particles when they are in the region of suitable nano dimensions. If such a process is not arrested, one gets a colloidal solution and then precipitation in form of insoluble matter getting settled in the solution. One can produce powder and thin films of nano particles embedded in certain matrices by chemical routes. In most of the applications, one requires thin films in a suitable matrix, which is sometimes not achievable by a chemical route. There are other routes of creating nano structures like vapor phase condensation, thermal co-deposition, arc discharge, RF plasma sputtering, RF plasma co-sputtering, ion beam co-sputtering, ion implantation, other ion beam based methods, pulsed laser deposition etc. Two basic approaches in synthesis of nanostructures are the bottom up and top to down. In the bottom up

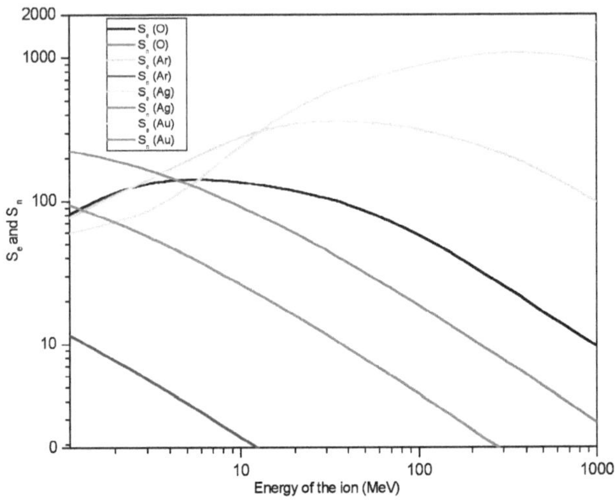

Figure 1. The electronic and nuclear energy losses for O, Ar and Au ions as calculated by TRIM. The values of S_e and S_n are in eV/A. The curves show that (1) the S_e at higher energies (about 100 MeV or higher) is always higher (two orders of magnitude or more) than S_n and (2) the value of S_e can be varied from about a few tens of eV/A to about 2 keV/A.

approach the atoms are brought together to form particles of nano dimension whereas in the top–down approach, the large size grains are broken to form nano size particles. In the ion beam based methods, mostly the bottom up approach is used.

In the present article, the methods related to energetic ions only, are discussed for synthesis of nanostructures. It is therefore essential to have some basic idea of ion–matter interaction.

During the passage through the material, energetic ions loose energy, which is quantitatively governed by the ion mass and its energy. Energy loss in the material can be either by elastic collisions or by in-elastic collisions. A typical curve for the energy loss in elastic collisions (nuclear stopping) and that in in-elastic collisions (electronic stopping) is given in Figure 1 as estimated by TRIM code [1]. Important features to be noted in these curves is that (1) the electronic energy loss dominates at high energies (>100 kiloelectronvolt (keV)/nucleon) and nuclear energy loss dominates at the low energies (<10 kiloelectronvolt (keV)/nucleon) and (2) ions of mass have higher energy losses (nuclear as well as electronic). It will be shown in the present article that the energetic ions have different types of role in nanostructure synthesis depending on the ion energies. This can be divided in three energy regimes e.g., (1) very low energies up to a few kiloelectronvolt (keV) suitable for sputtering purpose (2) energies from tens of kiloelectronvolt (keV) to a few megaelectronvolt (MeV)

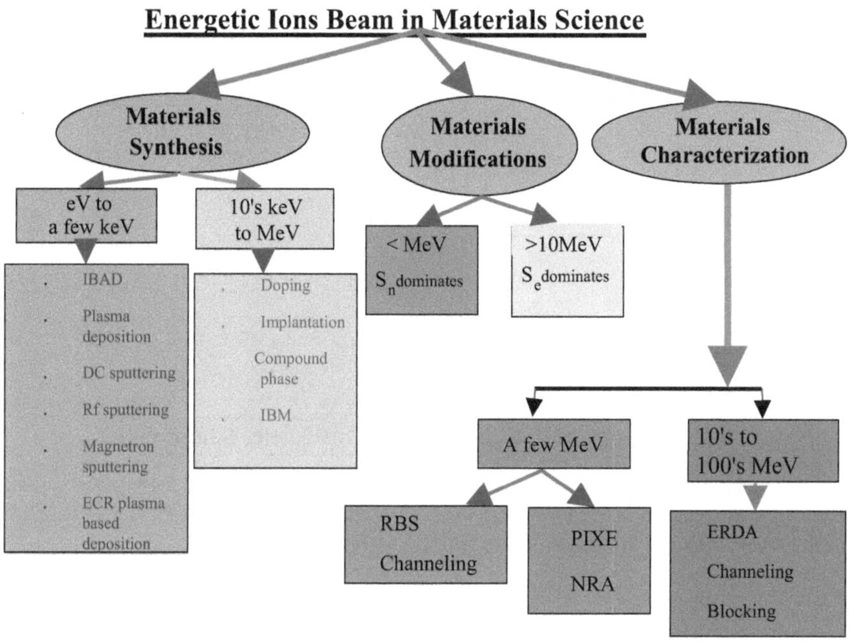

Figure 2. Schematic view to show the role of energetic ions in materials science in synthesis, modification and characterization of materials. A few electronvolts (eV) to a few kiloelectronvolts (keV) energy ions from ion sources or from plasmas play a crucial role in making thin films. IBAD stands for Ion Beam Assisted Deposition, which is a good tool to deposit films free from voids as compared to other conventional methods. ECR stands for Electron Cyclotron Plasma. The ECR source are popular due to their capability of providing high charge state of ions and high ion current. Low energy of 10 keV to about a MeV ions revolutionalized the field of semiconductor technology for doping semiconductors. IBM represents Ion Beam Mixing, a technique to produce compound phase by ion irradiation. These energies from ion implanters also find application in compound phase formation. Modification of materials have been shown both by S_n and S_e. Work on latter is mostly from mid eighties onwards. Rutherford backscattering (*RBS*), channeling, proton induced X-ray emission (*PIXE*), nuclear reaction activation analysis (*NRA*), elastic recoil detection analysis (*ERDA*), blocking etc. have been widely used characterization tools of ion beams.

and (3) energies (>1 megaelectronvolt (MeV)/nucleon), where electronic energy loss is dominant.

Energetic ions in general are of use to materials science as given in Figure 2. in different energy regions. They play a crucial roles in the synthesis, modification and characterization of materials. In the field of nano materials, the energetic ions are of use for synthesis and modification. The important issues in nano materials are the (1) control of the size of particles and (2) and the size distribution. This being a review will give a brief account of various aspects of the ion beam related to synthesis and modification of nano materials and the

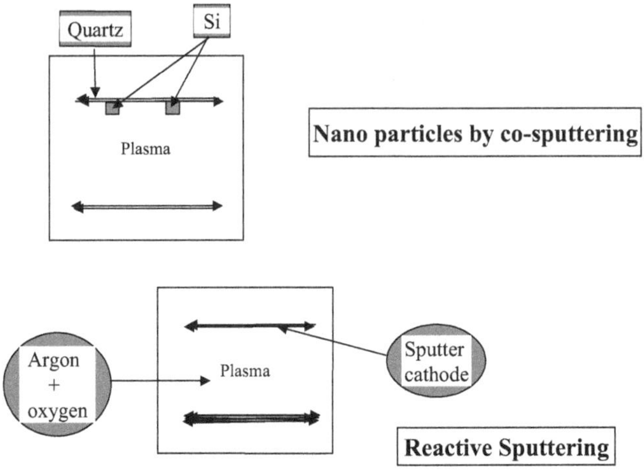

Figure 3. Schematic of (a) RF reactive sputtering and (b) RF co-sputtering. The Argon and oxygen gas ratios are controlled by mass flow controllers. The electrode at the bottom side is used to keep the substrate at which the film is to be deposited.

issues of control of size and size distribution will be discussed with examples of the work at NSC [3–7] and elsewhere [8–13].

2. RF sputtering and co-sputtering

Sputtering is a phenomenon of erosion of solids under ion impingement. Erosion of solids can be in the form of neutral atoms, ions, clusters etc. In contrast to thermal evaporation, where the vapors/atoms arrive at the substrate (for growth of film) with a typical energy of a fraction of an electronvolt (eV), the sputtered atoms arrive at the substrate with an energy of a few electronvolt (eV). This helps in the atomic motions/surface diffusion for better growth of the film and in the nucleation and growth of the nano particles in a matrix. A RF sputtering/co-sputtering set up (schematic as shown in Figure 3) has a sputtering cathode housed in a vacuum chamber, grounded electrode, RF power amplifier, matching network, mass flow controller, and a vacuum system. The substrate on which the nano particles in a matrix are grown in the form of a film is kept on the grounded electrode. The process of sputtering or co-sputtering can also be performed by an ion source or an atom source.

2.1. GROWTH OF NANO PARTICLES OF NITRIDE AND OXIDES OF COPPER BY REACTIVE SPUTTERING

An indigenous RF sputtering unit has been set up [2] at NSC. It has been used for reactive sputtering for the growth of copper nitride [3] and copper oxides [4].

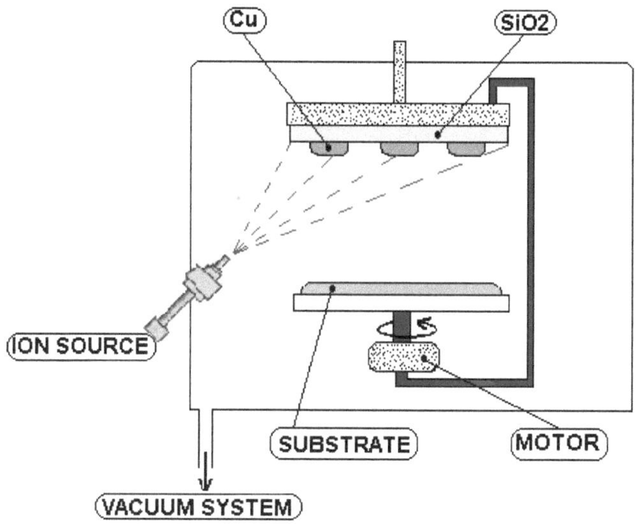

Figure 4. A sketch of atom beam sputtering. An ion source (in present case an atom source) is mounted in the chamber. Gas and electrical connections are provided by appropriate feedthroughs. The atoms from source strike the metal and silica, which sputter and get deposited on the substrate. The substrate is rotated by a motor for uniform deposition.

The operating pressure for copper nitride was 50 mtorr and the gas flow of N_2 was 15 sccm. RF power used was 100 W. Grain size of the copper nitride grown was 10 to 50 nm depending on the substrate temperature. More details of the deposition and characterization by XRD, ERDA and UV–Vis absorption spectroscopy are given in the work by Ghosh *et al.* [3].

The operating pressure for two oxides was 50 mtorr and oxygen gas flow was 7.5 sccm. RF power used was 200 W. The films were characterized for the size by atomic force microscopy and the X-ray diffraction. The later gives the average size distribution determined by the Scherer formula using the width of XRD peak. Average grain size of 10 to 45 nm was obtained depending on the substrate temperature [4].

In the above two examples, oxygen and nitrogen gases do the sputtering as well as provide the reactive environment of plasma for the formation of oxides and nitrides respectively.

2.2. GROWTH OF NANO PARTICLES IN A MATRIX BY RF CO-SPUTTERING

The experimental set up in this method remains the same except the sputtering cathode, which in this case is a disc of a dielectric matrix (for example silica) with a few pieces of the material (whose nano particles are to be formed) glued to the disc. These are spatially distributed uniformly for growth of uniform particle density. It is advisable to have substrate rotation during the deposition. Nanoparticles of ZnO in an SiO2 matrix by RF co-sputtering have been prepared at NSC [5].

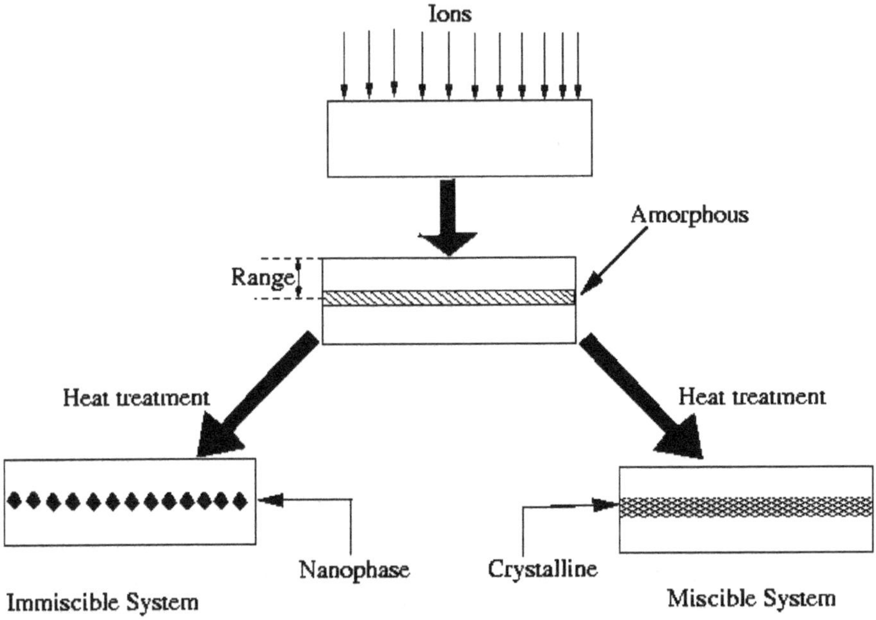

Figure 5. Schematic of creating nano particles buried in a matrix by ion implantation and subsequent annealing. Implanted species on subsequent annealing result in a crystalline phase (in case of a missible system) or in nano particles (in case of im-miscible system).

2.3. GROWTH OF Cu METAL CLUSTERS IN SILICA BY ATOM BEAM CO-SPUTTERING

A fast atom source is used to sputter the metal and silica simultaneously to obtain the embedded Ag and Cu particles in a silica matrix. The atom beam sputtering set up has recently been assembled at NSC. An atom source is preferred to an ion source to get rid of the problem of space charge near the insulating sputtering cathode. The characterizations of these metal particles embedded in silica matrix are in progress by UV–vis absorption spectroscopy, which reveals the surface plasmon resonance peak, a characteristic of metal particles in an insulating matrix. This aspect is briefed in Section 4.4 (Figure 4).

3. Nano particles by low energy ion irradiation

3.1. EMBEDDED Si NANO PARTICLES IN SILICA BY ION IMPLANTATION

Low energy ions typically up to a few tens of kiloelectronvolt (keV) to hundreds of kiloelectron (keV) are employed to implant the ions whose nano particles are to be formed in the substrate (in which the particles are to be embedded). The condition for such a process is that the implanted ion should not make a phase or compound with the substrate material. Energy of the ions decides the mean depth at which particles are embedded. A schematic of the process shown in Figure 5, gives in brief the route to form nano particles by ion implantation and subsequent annealing. In a recent work by Tanuja *et al.* [7], Si nano crystal in silica matrix are grown by implantation of 100 keV Si in silica and subsequent annealing. These samples of Si nano crystals in silica were characterized by photo luminescence. Normally size distribution achieved by this method is rather wide. The efforts are going on to achieve a narrower size distribution by SHI irradiation. It is expected that each SHI will produce a defected zone of certain diameter and therefore Si nano particle size will be narrowed down to the diameter of the track during annealing of SHI irradiated, Si implanted samples.

3.2. FORMATION OF C DOTS

Carbon dots are shown to be formed by irradiation of Si based polymers (Methyltriethoxysiulane) by 3 MeV Au ions. These were characterized by high resolution transmission electron microscopy. It was shown that the dot size (typically 5 nm) is decreasing with the increase of the fluence [8]. The irradiated films exhibit photo luminescence at 2.1 eV and is maximum at the fluence of 10^{14} ions/cm^2. It was also shown that the size of the particles were dependent on the fluence.

4. Swift heavy ions (SHI) in nanostructuring

4.1. GENERATION OF ALIGNED C CLUSTERS IN NANO SIZE CYLINDERS

It was shown by Pivin *et al.* [8] and Seki *et al.* [9] that the irradiation of Si based polymers by swift heavy ions creates C clusters all along the ion path in the entire sub-micron thick film. The cylinders were however filled only up to 60% or so with the clusters [9]. If this filling can be increased to 100% it will be like a quantum wire. Irradiation by 100 MeV Au ions, using the NSC Delhi Pelletron facility, was performed on the polycarbosilane (PCS) and Allylhydridopolycarbosilane (HP CS). TEM and EFTEM revealed the alignment of carbon clusters all along the ion path. Photoluminescence observed in these samples was more or less the same as in the case of low energy ion irradiation (3 MeV Au ions).

4.2. CREATION OF UNIFORM SIZE DISTRIBUTION OF NANO PARTICLES

It has been shown by Valentine *et al.* [10] that a more uniform size distribution can be achieved by the use of 30 MeV Si ions. Normally Cu and copper oxide nano particles in glass matrix are obtained by annealing of the glass containing Cu. If the system is irradiated before annealing a better size distribution is achieved. This was correlated to the electronic energy loss estimated by TRIM. As the ions penetrates the material they loose energy and therefore the electronic energy loss changes at different depths. It was shown by cross sectional TEM that such a phenomenon takes place up to a certain depth which indicates the role of electronic energy loss, in achieving better size distribution as compared to the conventional tools. It suggests that there is a threshold of electronic energy loss above which it helps in achieving a better size distribution.

4.3. CONDUCTING C RODS IN DIAMOND LIKE CARBON

The irradiation of diamond like carbon (DLC) film leads to the increase in conductivity. It is therefore expected that all along each ion track created by individual ion impact will have higher conductivity than the surrounding. It is therefore possible to generate conducting C rods or nano wires in the DLC matrix by high energy heavy ion irradiation of DLC film as demonstrated recently [11]. The DLC films were irradiated by 350 MeV and 1 GeV Au ions and the resistance measurement at the ions' impacts is carried out by scanning force microscopy. It gave a significant increase in conductivity, revealed by the large current flowing from the scanning force microscopy tip to the substrate via the ion path in the film. There is depletion of H along the ion path and structural modifications making the zone along the ion path more like graphite, which is cause of the high conductivity along the ion path. Possibility of creating conducting carbon tracks in fullerene matrix is planned at NSC by swift heavy ion irradiation of thin fullerene films.

4.4. EFFECT OF SHI IRRADIATION ON METAL PARTICLES IN INSULATING MATRIX

Metal nano particles in an insulating transparent matrix like silica have the property of absorbing visible light in a narrow band owing to a surface plasmon resonance. The collective motion of free electrons in the particle induced by the oscillating electric field vector of incident electromagnetic wave is called as surface plasmons. The resonance frequency of the spherical metal nano particles depends on the size of the particles and the dielectric constants of the particle and the surrounding medium. In cases of metal nano particles in an insulating transparent matrix, the UV–Vis spectroscopy becomes an excellent tool to characterize the nano particles by surface plasmon resonance. There are possibilities of interacting nanoparticles for miniature waveguides [12]. The effect of SHI on metal nano particles is of interest (1) to understand the ion and low dimension matter (2) to look into the details of modification to probe the possibilities of engineering the nano structures.

4.4.1. *Growth of Au particle size under SHI irradiation*

Au nano particles were grown in a Teflon matrix by tandem deposition [13]. Size of Au particles (in as deposited film) was not spherical as investigated by TEM.

The samples were prepared specially on TEM grids [13] for the ease of characterization. The samples on TEM grids were irradiated by 100 MeV Au ion beam at different fluences. It was observed [14] that the shape of particles became spherical and the size starts growing with fluence. The formation of perfect spherical particles indicates the possibility of the existence of a transiently molten state, as in the molten state the metal will like to have spherical shape like a liquid drop. In general the average size (about 9 nm before irradiation) of the particle became about 14 nm after irradiation. It shows that the particles coalesce together to form a bigger particle under SHI irradiation.

4.4.2. *Effect of SHI irradiation on Ag particles in insulating matrix*

Ag nano particles were also made in the similar way as stated above on TEM grids and were irradiated at different fluences. The TEM study of the irradiated samples indicated different type of behavior as observed in Au particle case. The average size of particles is reduced slightly. It appeared that there is loss of Ag content which is attributed to the electronic sputtering of Ag particles. Smaller the size of particle, greater is the confinement of electrons [15–17] (produced in the wake of passage of incident high energy ion) within the particle, which causes higher temperature rise resulting in transient molten state during which the metal atoms escape.

Randomly distributed Ag particles in silica matrix are shown [18] to be aligned along the beam direction as a result of 30 MeV Si ion irradiation. This was investigated by plan-view TEM. This is an excellent approach to align the existing particles in a matrix.

In an another experiment Ag particles were grown in silica by magnetron co-sputtering and then the response of irradiation of 100 MeV Ag was studied [19]. It was observed that the Ag atoms were released from the clusters as a result of irradiation indicating dissolution of clusters in silica. This conclusion was based on the decreasing integral intensity of the plasmon resonance peak with the fluence. The decrease in particle size in the work by Biswas *et al.* is also consistent with the present observation. It is apparent that the matrix in which the particles are existing also plays a role in deciding the effect of swift heavy ions on the nano particles. In this work, the aspect of aligning of the clusters along the beam direction was not studied, which requires cross sectional TEM.

4.5. EFFECT OF SHI ON COPPER OXIDE NANO PARTICLES

Copper oxide nano particles were prepared by the activated reactive evaporation method [20]. The effect of large electronic excitation (24 keV/nm) generated by 120 MeV Ag ions was examined by X-ray diffraction, Raman spectroscopy and photoluminescence [21]. The XRD revealed that the pristine films were having both the phases CuO and Cu_2O and the irradiation induced a phase change from CuO to Cu_2O. It is most likely due to release of oxygen. The crystalline quality of the films appeared to have improved with irradiation. It was observed that the PL intensity increased in the irradiated sample, which was attributed to the removal of surface states. This observation is similar to the results [22, 23] of 1 MeV proton irradiation of CdS nanoparticles, where the irradiation leads to increase of the PL intensity which was attributed to the removal of surface states.

4.6. EFFECT OF SHI ON SEMICONDUCTOR NANO PARTICLES IN POLYMER MATRIX

Nano particles of ZnO in polyvinyl alcohol matrix were prepared and irradiated by 100 MeV Cl ions. It was observed [24] that the particle size increased from about 1 nm for a pristine sample to 4 nm at the fluence of 5×10^{12} ions/cm^2. The size went up sharply at the fluence of 10^{13} ions/cm^2. The characterizations were performed by UV–Vis absorption spectroscopy and electron microscopy.

4.7. SHI INDUCED FORMATION OF Si NANO PARTICLES

Thin films of SiOx were irradiated with swift heavy ions [25], as a result of which the formation of Si nano crystals was reported. This was investigated by PL and TEM. The irradiation causes evolution of oxygen from the film. The possible reaction due to SHI was formation of Si nano crystal with SiO_2 accompanied by release of oxygen. This could be explained on the basis of the occurrence of transient temperature spike during which a molten state is created all along the ion path, resulting in release of oxygen and the formation of Si nanocrystals. Creation of such latent tracks in the insulating matrix is already known.

5. Future promising aspects of ion beams in nanostructuring

There is a variety of possibilities, in which energetic ions can play a crucial role in the engineering of nanomaterials. Most of these possibilities are touched upon as stated in above sections. A key issue is presently to generate the nano dots in a regular pattern. A possibility in this direction is to use focused ion beams of nano dimensions (current state of art is beams of 10 to 20 nm size) for creation of defects in a regular pattern and then growing quantum dots on these defected zones. Since the defected regions are excellent nucleation sites, this is likely to produce nanostructure as desired. The nano beam will also play a crucial role in characterization of nano particles, where the composition and structure can be probed at a nano scale. Another interesting way of achieving nanostructures at a surface is by irradiation of highly charged ions, which is not discussed in this article.

6. Conclusion

The present article gives a very brief overview of the interesting aspects of ions in different energy regimes for nano structuring, with examples of the work carried out at NSC. The ion energies from a few tens of electronvolt (eV) (in atom sputtering and RF sputtering) to a few hundred megaelectronvolt (MeV) are presented in context of synthesis and modification of nano-structures.

Acknowledgements

The author is thankful to the Department of Science and Technology for creating the facilities at NSC and to Indo French Centre for Promotion of Advanced Research (IFCPAR) for the financial assistance to the project on nano phase generation by energetic ion beams. The author is also thankful to the efforts made by D. Kabiraj, Abhilash and V.V. Sivakumar in creating the necessary facilities of atom beam sputtering and RF sputtering.

References

1. Ziegler J. F., Biersack J. P. and Littmark U., *The Stopping and Ranges of Ions in Solids*, Vol. I, Pergamon, New York, 1985.
2. Sivakumar V. V., Venkataraman S., Ajithkumar B. P., Avasthi D. K., Sarangi D., Bhattacharya R. and Gupta A., *Proceedings of National Symposium on Vacuum Science and Technology*, 1997, p. 19.
3. Ghosh S., Singh F., Choudhery D., Avasthi D. K., Genaesan V., Shah P. and Gupta A., *Surf. Coat. Technol.* **142–144** (2001), 1034.
4. Ghosh S., Avasthi D. K., Shah P., Ganesan V., Gupta A., Sarangi D., Bhattacharya R. and Assmann W., *Vacuum* **57** (2000), 377.
5. Sivakumar V. V. and Singh F., *DAE Solid State Symposium*, Dec. 26–30, Gwalior.
6. Kabiraj D., *Internal Report* (Private Communication).
7. Mohanty T., Misra N. C., Pradhan A. and Kanjilal D., *13th International Conference on Surface Modifications by Ion Beams SMMIB 2003, Texas, USA*; Mohanty T., Satyam P. V., Misra N. C. and Kanjilal D., *Rad. Meas.* **36** (2003), 137.
8. Pivin J. C., Pippel E., Woltersdorf J., Avasthi D. K. and Srivastava S. K., *Z. Met.kd.* **92** (2001), 712.
9. Seki S., Maeda K., Tagawa S., Kudoh H., Sugimoto M., Morita Y. and Shibata H., *Adv. Mater.* **13** (2001), 1876.
10. Valentine E., Bernas H., Riscolleau C. and Creuzet F., *Phys. Rev. Lett.* **86** (2001), 99.
11. Krauser J., Zollondz J. H., Weidinger A. and Trautmann C., *J. Appl. Phys.* **94** (2003), 1959.
12. Maier S. A., Kik P. G., Atwater H. A., Meltzer S., Harel E., Koel B. E. and Requuicha A. A. G., *Nat. Matters* **2** (2003), 229.
13. Biswas A., Marton Z., Kanzov J., Kruse J., Zaporojtchenko V., Faupel F. and Strunskus T., *Nano Lett.* **3** (2003), 69.
14. Biswas A., Avasthi D. K., Fink D., Kanazov J., Schurmann U., Ding S. J., Aktas O. C., Saed U., Zaporojtchenko V., Faupel F., Gupta R. and Kumar N., *Nucl. Instrum. Methods B* **217** (2004), 39.

15. Avasthi D. K., Ghosh S., Srivastava S. K. and Assmann W., *Nucl. Instrum. Methods B* **219–220** (2004), 206.
16. Ghosh S., Tripathi A., Som T., Srivastava S. K., Ganesan V., Gupta A. and Avasthi D. K, *Radiat. Eff. Defects Solids* **154** (2001), 151.
17. Gupta A. and Avasthi D. K., *Phys. Rev. B* **64** (2001), 155407.
18. Penninkhof J. J., Polman A., Sweatlock L. A., Maier S. A., Atwater H. A., Vredenberg A. M. and Kooi B. J., *Appl. Phys. Lett.* **20** (2003).
19. Pivin J. C., Roger G., Garcia M. A., Singh F. and Avasthi D. K., *Nucl. Instrum. Methods B* **215** (2004), 373.
20. Balamurugan B. and Mehta B. R., *Thin Solid Films* **396** (2001), 90.
21. Balamurugan B., Mehta B. R., Avasthi D. K., Singh F., Arora A. K., Rajalakshmi M., Raghavan G., Tyagi A. K. and Shivaprasad S. M., *J. Appl. Phys.* **92** (2002), 3304.
22. Asai K., Yamaki T., Ishigure K. and Shibata H., *Thin Solid Films* **227** (1996), 169.
23. Asai K., Yamaki T., Seki S., Ishigure K. and Shibata H., *Thin Solid Films* **284** (1996), 169.
24. Mohanta D., Nath S. S., Bordoloi A., Doloi S. K., Misra N. C. and Choudhury A., Annual Report, NSC, 2001–2002, p. 116.
25. Prajakta S., Bhave T. M., Kanjilal D. and Bhoraskar S. V., *J. Appl. Phys.* **93** (2003), 3486.

Hyperfine Interactions (2005) 160:107–121
DOI 10.1007/s10751-005-9153-9

Xenon-ion Induced Magnetic and Structural Modifications of Ferromagnetic Alloys

RATNESH GUPTA*,†, K. P. LIEB, G. A. MÜLLER, P. SCHAAF and K. ZHANG
II. Physikalisches Institut, Universität Göttingen, Friedrich-Hund-Platz 1, 37077 Göttingen, Germany; e-mail: gratnesh_ioi@yahoo.com

Abstract. Thin polycrystalline films of permalloy ($Ni_{79}Fe_{21}$) and permendur ($Co_{50}Fe_{50}$) have been irradiated with Xe-ions to fluences of 10^{14}–10^{16} ions/cm^2. Ion-induced structural and magnetic modifications have been measured by grazing angle X-ray diffraction, Rutherford backscattering and magneto-optical Kerr effect. In the case of permendur, the Xe-ion implantation first reduced the coercivity, because of stress relaxation, while higher ion fluences increased the coercivity due to pinning centers generated in the film. The ion irradiation aligned the in-plane easy axis of the magnetization along the direction of the external magnetic field during implantation. Phase shifts obtained from magnetic force microscopy confirmed these modifications. The effects of Xe-ion irradiation in permalloy films are much weaker and underline the importance of magnetostriction in the variation of the coercivity and anisotropy.

PACS: 61.80 Jh, 76.80+y, 75.50.Bb.

1. Introduction

Research on low-dimensionally magnetic materials are of high fundamental and applied interest, for instance in connection with the development of nano-structured magnetic recording and sensor devices [1, 2]. Patterning materials such as surfaces, interfaces, thin films and multilayers with nm and sub-nm precision has also given access to novel phenomena. Magnetic multilayers and alloys containing 3d-transition metals are being especially intensively investigated. Since the magnetic properties of a material are strongly correlated with its microstructure, crystallographic modifications can be used as a tool for tuning the magnetic properties. The reduced symmetry at a surface may cause an increase of the magnetic anisotropy energy, as compared to that of bulk materials.

Ion-beam irradiation is a highly efficient tool to modify magnetic properties of thin films and multilayered structures. Chappert *et al.* [3, 4] investigated

* Author for correspondence.
† Present address: School of Instrumentation, Devi Ahilya University, Khandwa Road, Indore 452017, India.

ion-beam mixing of Co/Pt bi- and multilayers by He^{+} ions and found the coercivity of the films progressively to decrease for increasing ion fluence. Hysteresis loop measurements as a function of the sample temperature showed that this is primarily due to a strong irradiation-induced decrease of the Curie temperature. At suitable ion fluences, the intermixing triggered a spin-reorientation transition from easy axis out-of-plane to easy axis in-plane. The system Fe/Tb is another example, in which the out-of-plane magnetic anisotropy was reduced due to ion irradiation, leading to changes of the interface roughness in the multilayer film [5]. Generation of in-plane stress during ion-beam-assisted deposition of iron and nickel films was found to induce an in-plane magnetic anisotropy due to inverse magnetostriction [6]. On the other hand, the production of magnetic textures in ion-irradiated Ni and Fe films on various substrates was observed for ion energies, at which no interface mixing or reactions occur [7–10]. In thin Co films, a Xe-induced phase transformation from hexagonal to face-centered-cubic structures was found [11, 12].

The present article reports on the modifications of typically 75-nm thick permalloy and permendur films by Xe-ion irradiation and intends to compare the effects in these ferromagnetic alloys, having very different magnetostriction constants, with results obtained in iron and nickel films. The corresponding magnetostriction constants in polycrystalline films vary from $\lambda_{avg} = 1.2 \times 10^{-6}$ (permalloy) over $\lambda_{avg} = -4.4 \times 10^{-6}$ (iron), $\lambda_{avg} = -33 \times 10^{-6}$ (nickel) to $\lambda_{avg} \approx 65 \times 10^{-6}$ (permendur). Note that among all the 3d-ferromagnets, permendur has the largest magnetic moment. Preliminary results on permendur layers have been recently communicated in Ref. [13].

2. Experimental details

In order to study the influence of the preparation conditions, we considered several preparation techniques to deposit the ferromagnetic alloy films. Electron-beam evaporation was used for the preparation of the permendur ($Fe_{50}Co_{50}$) films. The alloy was produced by arc melting the proper amounts of both metals. Permalloy ($Ni_{79}Fe_{21}$) films were deposited by either DC-magnetron sputtering in an Ar-atmosphere or with the help of pulsed laser deposition (PLD). For the latter, a KrF excimer laser ($\lambda = 248$ nm) with an energy density of 6–7 J/cm^2 at a repetition rate of 9 Hz was used. The 7×10 mm^2 rectangular films of 30–90 nm thickness were deposited at a pressure of 1×10^{-8} mbar in the chamber.

The irradiations with 200-keV Xe^{+}-ions were performed with the Göttingen ion implanter IONAS [14]. The ion energy was adjusted in order to deposit the ions near the center of the typically 75 nm thick films. Consequently, any ion-beam mixing with the Si substrate was avoided, even at large fluences, where sputtering reduced the film thickness. The ion flux was maintained at 0.7 μA/cm^2 in order to avoid beam heating, and an XY-beam sweeping system provided for homogeneous implantation over the full sample area. The irradiations were performed in the presence of an external magnetic field of 104 Oe, which was oriented in-plane along the short axis of the samples.

Rutherford backscattering spectrometry (RBS) was used to monitor the thickness of the films as well as sputtering and (negligible) interface mixing. The RBS spectra were taken with the 900-keV He^{2+} beam of IONAS and two surface barrier detectors positioned at ±165° to the beam and having an energy resolution of 12 keV full width at half maximum (FWHM). Most samples were analyzed by X-ray diffraction at glancing angle incidence (3°). Magnetic hysteresis loops were measured at room temperature by the magneto-optical Kerr effect (MOKE) with the linearly polarized beam of a He–Ne laser (λ = 632 nm) in the longitudinal configuration, which probes the in-plane magnetization. The samples were mounted in an automated MOKE goniometer with 1° precision and a maximum field of 1.5 kOe [15]. Angular scans of the magnetization were recorded, where the MOKE field was oriented the angles φ relative to the long axis of the specimen, which was considered as reference line $\varphi = 0°$. From the hysteresis loops the coercivity H_C and the magnetization ratio M/M_S, relative to the saturation value M_S, were deduced. The remanence ratio $\Delta(M_r/M_S)$ between the easy and hard axis serves as a measure of the anisotropy. The irradiated areas and surroundings in some of the samples were characterized simultaneously by atomic force microscopy (AFM) and magnetic force microscopy (MFM). The instrument was operated at room temperature in the "tapping/lift" scanning mode, to separate short-range topographic effects from the long-range magnetic signal. Thin film pyramidal MFM tips [type magnetic etched silicon probe] coated with a magnetic CoCr alloy magnetized perpendicular to the sample surface have been used as a scanning probe. The tip used was a high sensitivity one with high coercivity response [16, 17].

3. Results

3.1. AS-DEPOSITED FILMS

Figure 1 shows MOKE hysteresis curves obtained for 75 and 100 nm thick specimens of permalloy prepared by DC-magnetron sputtering. Plotted is the quantity M/M_S versus the MOKE field. Clearly, the coercivity of the 100-nm thick specimen is larger than that of the 75-nm film. When annealing the 75 nm sample for 1 h at 100°C, the coercivity increased, as did the linewidth of the fcc-(111) peak in the XRD spectrum (not shown). Both findings indicate that the coercivity increased due to grain growth in the specimens. The MOKE pattern of a 75-nm thick sample prepared by PLD exhibited no significant difference in comparison with that of the sample prepared by DC sputtering.

MOKE magnetization loops of an as-deposited permalloy film as a function of azimuthal angle φ are shown in Figure 2. The shape of the loop varied from

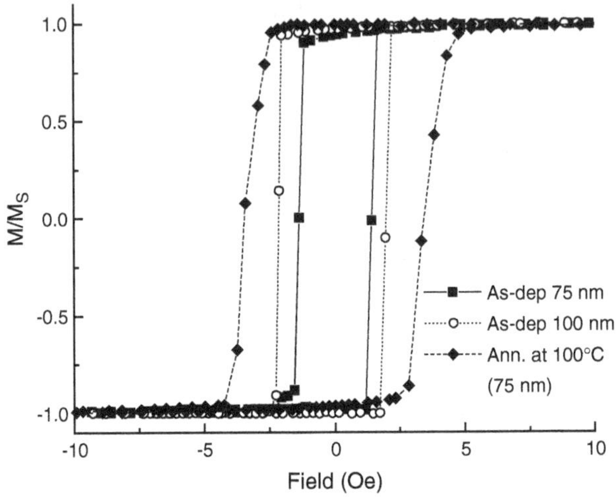

Figure 1. Longitudinal hysteresis loops of permalloy films of the thicknesses 75 nm, 100 nm and after annealing the 75-nm film for 1 h at 100°C.

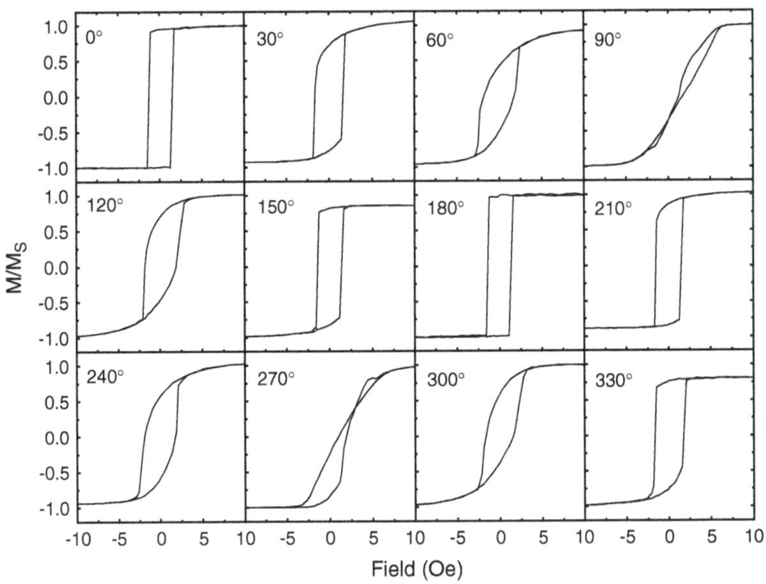

Figure 2. Hysteresis loops of the as-deposited permalloy specimen as a function of the azimuthal angle φ.

perfect square ($\varphi = 0°$) to spindle ($\varphi \approx 60°$) and then to a straight line ($\varphi \approx 90°$). Along the long axis of the specimen ($\varphi \approx 0°$), the perfect square shape and high values of the coercivity H_C and remanence M_r favoured this direction to be an easy axis. The magnetization curve measured along the short axis ($\varphi \approx 90°$) almost showed no hysteresis and values of M_r and H_C were close to zero. These

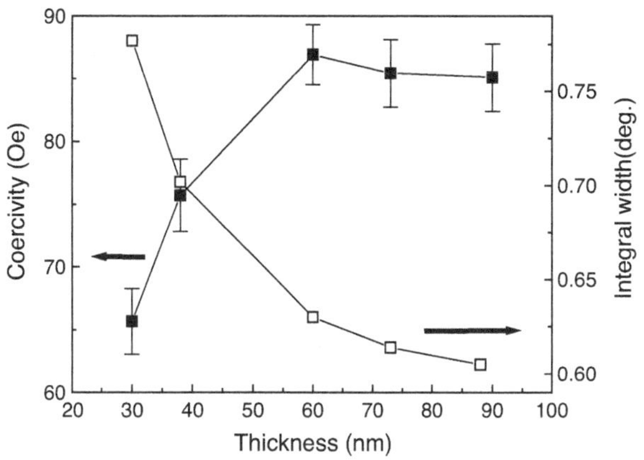

Figure 3. Thickness dependence of the coercivity (MOKE) and total line width of the bcc-(110) peak (XRD) of FeCo/Si(100) films measured at room temperature.

observations clearly proved that an intrinsic magnetic anisotropy was already present in the as-deposited specimens. It may be due to shape anisotropy, which arises from anisotropic grain growth of the film [18]. For permalloy samples, it is common practice to anneal the samples to remove any anisotropy. If no such annealing is performed, an "intrinsic anisotropy" remains [19]. We thus annealed the specimen for 1 h at about 100°C and a pressure of 1×10^{-3} mbar, but found no significant change in the magnetization. In the case of the PLD specimens, the MOKE scans gave fairly isotropic magnetization loops.

We now turn to the analysis of the e-beam-deposited permendur films. Figure 3 shows the dependence of its coercivity on the film thickness, along with the integral width of the bcc-(110) peak in the corresponding XRD spectra. Up to about 60 nm, the coercivity increased with the film thickness and then reached saturation. The MOKE angular scans of the as-deposited films revealed a small in-plane anisotropy in the 30-nm thick specimen, but isotropy in all the thicker films. The corresponding variation of the line width of the (110) peak suggested again grain growth as the reason for the increase in coercivity. Each grain boundary may act as local barrier, which enhances the coercivity of the relatively thicker films. According to Freeland *et al.* [20], the average grain size is an important parameter to determine the coercive field of thin films. In this reasoning, no account was taken for any possible inhomogeneous lattice strains, which also would contribute to the line broadening in the XRD spectra.

3.2. ION-IRRADIATED PERMALLOY FILMS

The RBS spectra and deduced Xe-implantation profiles in 72-nm thick permalloy films obtained for $(1-10) \times 10^{15}$ Xe-ions/cm^2 are illustrated in Figure 4. No

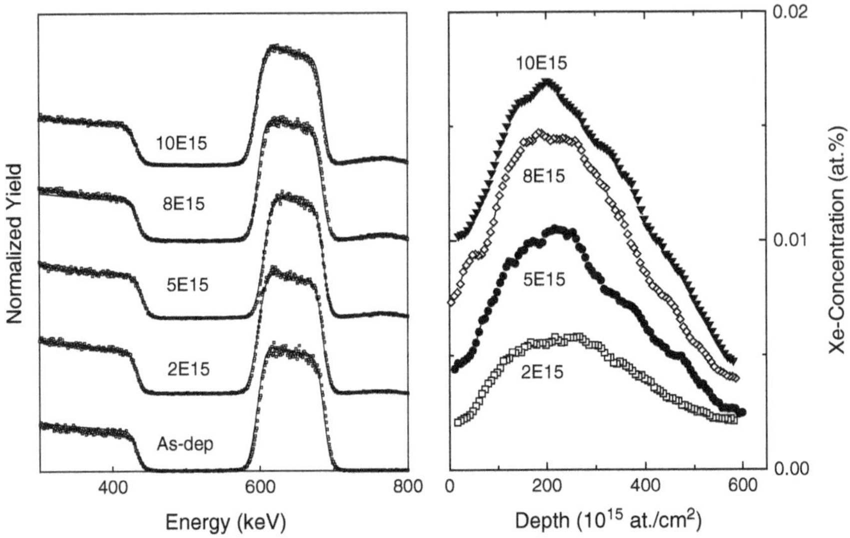

Figure 4. RBS spectra and simulations of as-deposited and Xe-ion irradiated permalloy films (*left*) and the deduced xenon depth profiles (*right*).

significant changes were seen in these spectra at different Xe fluences, indicating negligible interface mixing and ion-induced alloy decomposition. The observed small decrease in film thickness and the fact that centroid of the nearly Gaussian-shaped Xe-depth profiles shifted very little in its position, even for the highest fluence of 1×10^{16} ions/cm^2, were attributed to a weak sputtering effect, see left-hand side of Figure 4. The XRD patterns of an as-deposited and ion-irradiated film shown in Figure 5 confirmed the face-centred cubic lattice structure of the alloy. After irradiation, the diffraction peaks shifted slightly towards smaller angles, which is consistent with a small increase of the lattice constant; the line widths decreased.

The fluence-dependent coercivity along the hard axis after ion irradiation is shown in Figure 6 and exhibits a relatively strong increase for rising ion fluence, possibly due to domain-wall pinning. Due to the very narrow widths of the hysteresis loops of a few Oe as obtained from MOKE, we defined the hysteresis area function $A(\varphi) = \int M_r(\varphi)\, dH/M_s(\varphi)$ and considered its azimuthal dependence on the ion fluence. The examples shown in Figure 7 for $(5-10) \times 10^{15}$ Xe-ions/cm^2 revealed that the quantity $A(\varphi)$ became larger and more isotropic at the highest fluences. The disappearance of the uniaxial anisotropy is possibly associated with the in-plane demagnetization factor. Similar results have been reported for thin permalloy films with a surface roughness of about 1.4 nm [21]. The magnetic anisotropy attributed to surface roughness and called surface anisotropy originates from the dipolar and magneto-crystalline energy.

The coercivity value along the hard axis of permalloy is depicted in Figure 8 as a function of the Xe-ion fluence: it was rather constant up to 5×10^{15} ions/

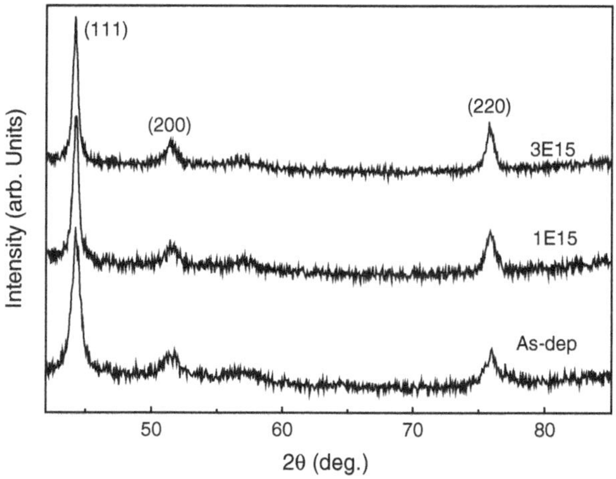

Figure 5. XRD patterns taken before and after Xe-ion irradiations of permalloy with fluences of 1×10^{15} ions/cm^2 and 3×10^{15} ions/cm^2.

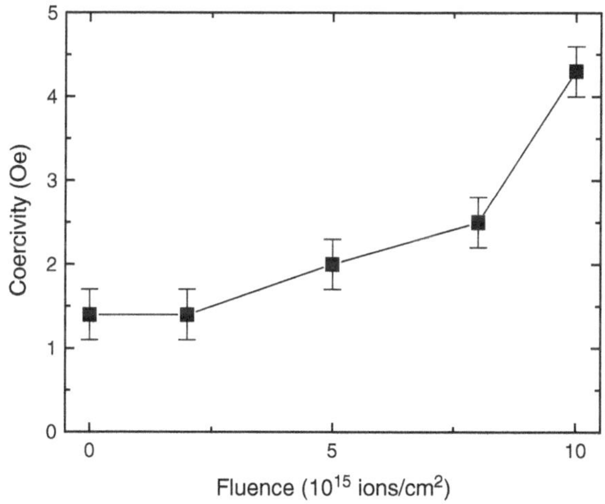

Figure 6. Coercivity H_C in permalloy films along the hard axis as a function of the Xe-ion fluence.

cm^2 and then increased. The in-plane anisotropy can be estimated from the angular dependence of the coercivity H_C. To induce the magnetization reversal, the sum of the external (MOKE) magnetic field H and magnetic anisotropy field $H_a \cos\varphi$ has to be equal to the domain-wall pinning coercivity H_{CW} [22]. The total coercivity along the easy axis can then be written as $H_C = H_a + H_{CW}$, while the coercivity of the hard axis is basically the value of H_{CW}. Then the value of H_a can be determined as the difference in the coercivities of the hard and easy axis. For instance, the coercivity of the hard axis increased from 0.3 Oe in the as-deposited specimen to 1.3 Oe after the irradiation with 8×10^{15} Xe ions/cm^2. The coercivity of the easy axis changed from 1.4 Oe to 2.5 Oe. These results

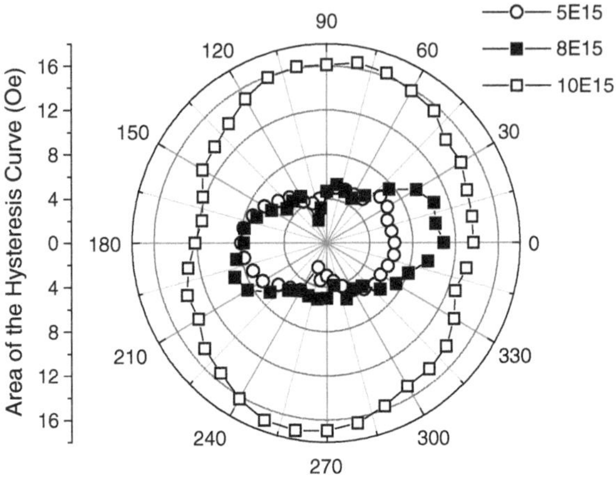

Figure 7. Hysteresis area $A(\varphi)$ in permalloy measured as function of the in-plane azimuthal angle φ for Xe fluences of 5×10^{15}, 8×10^{15} and 10×10^{15} ions/cm^2.

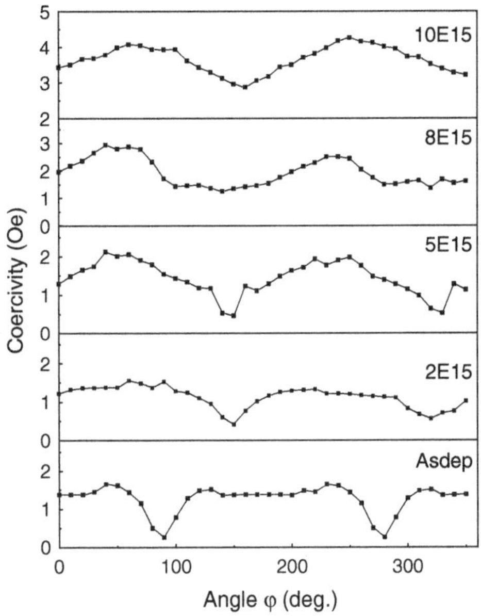

Figure 8. Angular dependence of the coercivity $H_C(\varphi)$ in permalloy: as-deposited sample, and after irradiations with the fluence of 2×10^{15}, 5×10^{15}, 8×10^{15} and 10×10^{15} ions/cm^2.

suggested that the anisotropy field $H_a \approx 1.0$ Oe remained quite constant during the irradiations, while the value of H_{CW} ($=H_C$ along the hard axis) increased due to the ion irradiation. The magnetization reversal thus proceeded not only by the coherent rotation of the magnetization, but also by the nucleation of magnetic domains and a rapid displacement of domain walls.

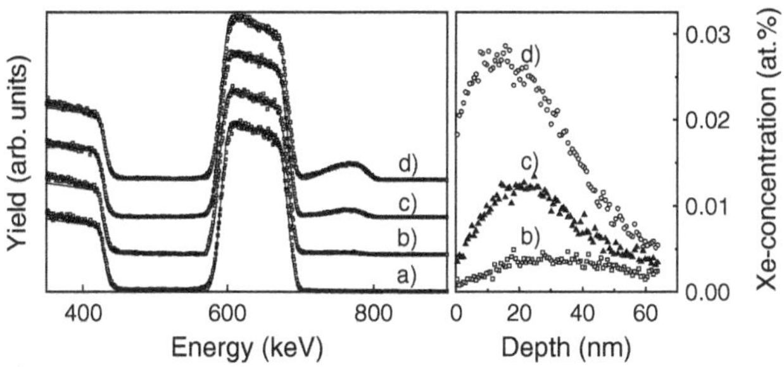

Figure 9. Left: experimental and simulated RBS spectra of FeCo films, a) as-deposited and at Xe-ion fluences of b) 1×10^{15} ions/cm^2, c) 5×10^{15} ions/cm^2 and d) 10×10^{15} ions/cm^2. The corresponding Xe depth profiles are shown at the right hand side.

The in-plane anisotropy can be estimated from the azimuthal dependence of the coercivity using the relation [23]

$$H_C(\varphi) = H_a \cos^2(\varphi - \varphi_0) + \left[H_{CW}^2 - H_a^2 \cos^2(\varphi - \varphi_0) \sin^2(\varphi - \varphi_0)\right]^{1/2},$$

where the angle φ_0 denotes the symmetry direction of the hard axis and φ the angle of the MOKE field relative to the long axis of the specimen. From the fits, we obtained the values H_{CW} = 3.1 Oe and H_a = 1.0 Oe. Upon irradiation, the orientation of the easy axis rotated by $\Delta\varphi_0$ = 80° with respect to the as-deposited film.

3.3. ION-IRRADIATED PERMENDUR FILMS

Figure 9 shows a set of RBS spectra taken for 73(1)-nm thick FeCo/Si bilayers before and after irradiation at different ion fluences and the deduced Xe-depth profiles. Even after the highest fluence of 1×10^{16} ions/cm^2, there were no visible changes in the spectra, apart from a small reduction in the film thickness, due to sputtering. From the glancing angle XRD pattern (not shown), we obtained the lattice constant and the average crystallite size using the Scherrer formula. The lattice constant increased monotonously as a function of ion fluence, see Figure 10(a), as did the crystallite size; at a fluence of 1×10^{16} ions/cm^2 the crystallite size was 18 nm. The width of the bcc-(110) reflex decreased as a function of the ion fluence, see Figure 10(a).

The variation of the coercivity along the hard axis obtained for different ion fluences is shown in Figure 10(b). Small fluences reduced the coercivity and a minimum was reached at 1×10^{15} ions/cm^2. For increasing fluence, the coercivity then increased steadily. Similar results had been obtained for Xe-irradiated polycrystalline Ni films of similar thickness [8, 9, 24]. The coercivity and remanence are extrinsic magnetic properties, which are very sensitive to local structural properties of the films such as defects, grain size, and strain. The

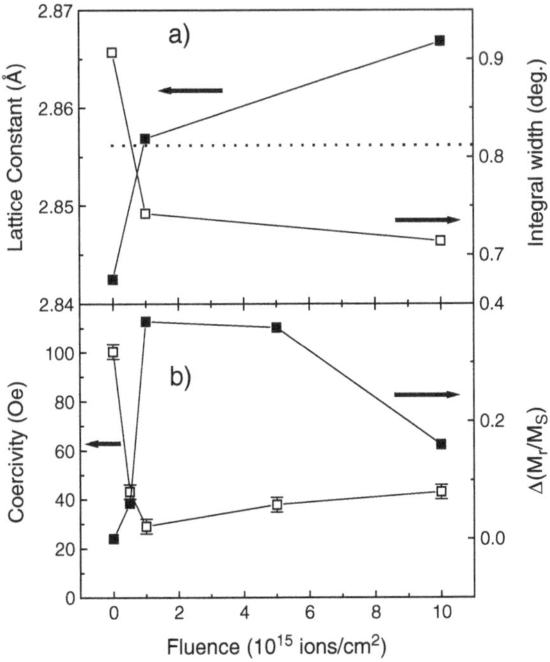

Figure 10. a) Lattice constant and integral line width of the bcc-(111) reflex in permendur as a function of the Xe-ion fluence. b) Fluence dependent coercivity of the easy axis (*right hand scale*) and total change in the remanence ratio, $\Delta(M_r/M_S)$, (*left hand scale*).

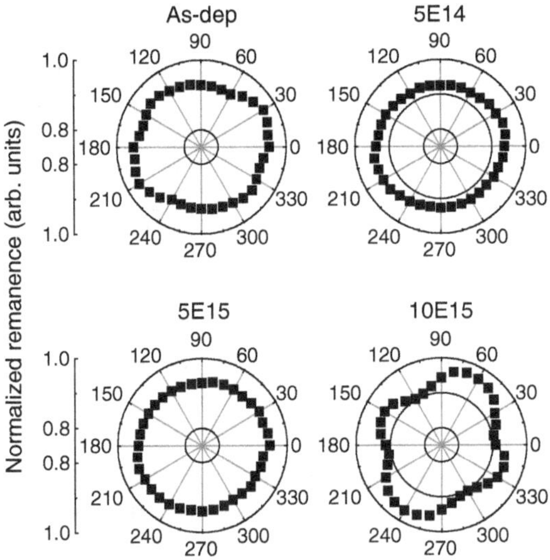

Figure 11. Polar MOKE plots taken before and after Xe-ion irradiation of a 72-nm permendur film.

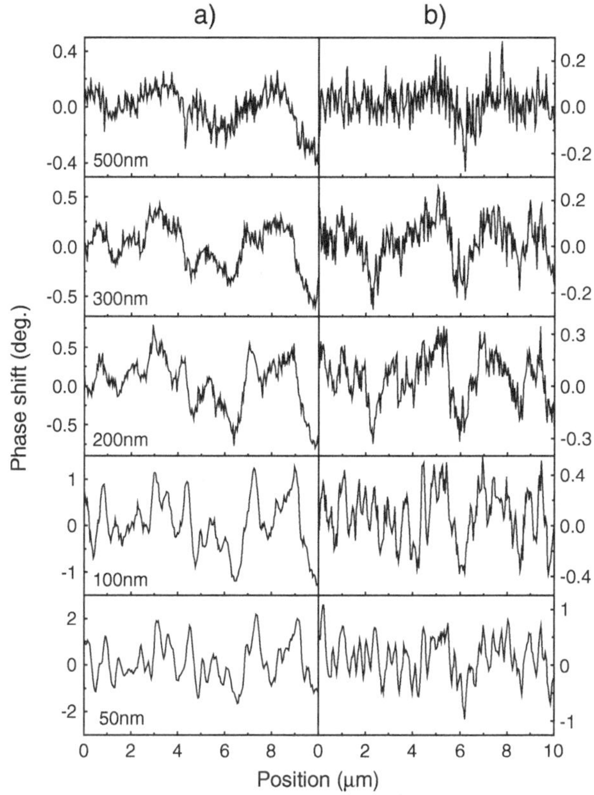

Figure 12. Phase shift across the length of the film for different tip-to-surface height Λ. *Left*: as-deposited sample; *right*: after implantation of 1×10^{16} Xe-ions/cm^2.

fast decrease of the coercivity at low fluences suggests that the implanted ions reduced the stress in the films [25], which is correlated to a sudden increase of the lattice constant as evidenced by XRD, see Figure 10(a). For larger fluences, the ions induced a higher density of defects, possibly leading to increased grain size and strain, generating pinning centers that interrupt magnetic switching and domain wall motion and in this way increase the coercivity. Besides this, one may also expect thermodynamically driven effects such as chemical ordering [26].

The measured angle-dependent MOKE hysteresis curves shown in Figure 11 clearly demonstrate that the easy axis was aligned parallel to the external magnetic field. For small fluences, a superposition of a fourfold-symmetric pattern, reflecting the magneto-crystalline anisotropy in the bcc-FeCo structure, and a uniaxial anisotropy was found. The quantity $\Delta(M_r/M_s)$, which is a suitable measure of the variation of the relative remanence M_r/M_s as function of φ, plotted in Figure 11. Clearly, the variations of H_C and $\Delta(M_r/M_s)$ are correlated with each other. After implanting 1×10^{15} or 5×10^{15} Xe-ions/cm^2 into the film,

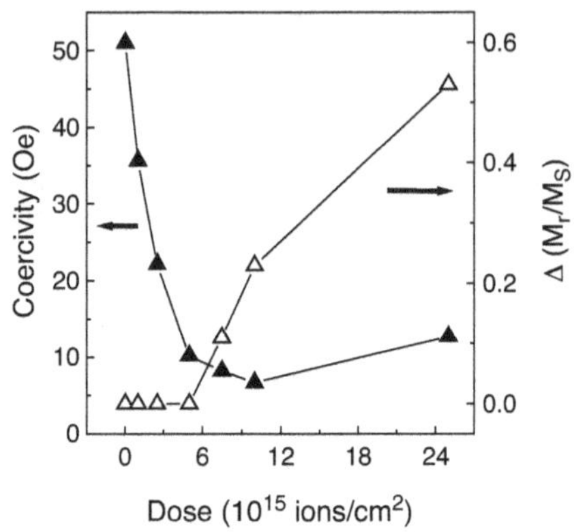

Figure 13. Fluence-dependent coercivity along the hard axis for a polycrystalline Fe film (*right hand side*) and total change in remanence ratio, $\Delta(M_r/M_S)$, (*left hand side*).

a pronounced four-fold symmetry was generated, leading to a rapid increase of $\Delta(M_r/M_s)$.

The magnetic microstructure was scanned with high resolution (~100 nm) by means of MFM in the remanent state. The micrographs of an as-deposited sample and one irradiated with 1×10^{16} ions/cm^2 are displayed in Figure 12. The method provides information on the internal structure of domain walls and some features of the domain pattern [27–30]. Figure 12 shows the phase shift δ as a function of tip-to-sample distance (scan height Λ = 50–500 nm). The scale of the abscissa indicates that the amplitude of the phase shift decreased as a function of Λ. It was highest for Λ = 50 nm and decreased for larger values of Λ suggesting that the MFM images are related to magnetic effects and not to artifacts [31]. To estimate the value of the magnetic moment, we choose the maximum phase shift for Λ = 50 nm and used the point-probe approximation of MFM. After the ion irradiation with 1×10^{16} Xe-ions, the value of the phase shift changed from 3.9 to 1.8, corresponding to a magnetic moment of 2.24×10^{-15} A/m^2.

4. Discussion and conclusions

Summarizing the changes of the microstructure in the permalloy and permendur films induced by the Xe-ion irradiations, we found in all the cases an increasing lattice constant (by some 0.010–0.015 Å at the highest ion fluence of typically 1×10^{16}/cm^2) and a decreasing line width of the bcc-(110) or fcc-(111) reflexes. Both findings suggest that defects due to film deposition were relaxed during the irradiations and that the degree of crystallinity improved. Similar trends had been observed for polycrystalline e-gun-deposited Fe films irradiated with Ne, Fe, Kr and Xe ions [15, 32]. Here, detailed XRD analyses had shown a decrease of the initial tensile stress of +3.8 GPa to a slight compressive stress of − 1.5 Gpa, and an increase of the lattice constant by 0.01–0.02 Å at about 1×10^{16} ions/cm^2. This variation was rather independent on the ion mass (Ne, Fe, Kr and Xe).

Evidently, the variation of the coercivity in the two ferromagnetic alloys showed marked differences. In the case of permalloy, H_C strongly increased (from 1.4 to 4.3 Oe, Figure 6), while in permendur it first decreased (from 100 to 28 Oe) and then increased slowly (to 42 Oe, Figure 10b). This difference obviously reflects the very different magnetostriction constants of the two materials, which almost vanishes in permalloy ($\lambda = 1.2 \times 10^{-6}$), but is very large in permendur ($\lambda = (6-7) \times 10^{-5}$). In the case of 75-nm thick polycrystalline Fe films prepared by e-gun deposition, Müller *et al.* [15,32] measured the variations of the quantities H_C and $\Delta(M_r/M_s)$ with the Xe-ion fluence shown in Figure 13. Iron having an average magnetostriction constant of $\lambda = -4.4 \times 10^{-6}$, which is intermediate between permalloy and permendur, shows also an intermediate behaviour of H_C: it decreases from 51 Oe in the as-deposited state, reaches a minimum of 7 Oe at a fluence 7×10^{15} Xe-ions/cm^2 and then increases slowly to 13 Oe at 2.5×10^{16}/cm^2. We interpret this behaviour as due to two processes, which come into play at different ion fluences. Low Xe-ion fluences relax the stress field in the as deposited films, possibly as a consequence of thermal spikes, which upon recondensation locally remove stress. Higher ion fluences, due to the accumulation in the film and the generation of extended radiation defects, pinning centers and/or grain growth, increase the structural disorder and therefore the coercivity, due to hindered domain wall motion [33].

Magnetic anisotropies depend fundamentally on the symmetry and the form of local interactions and may have magneto-crystalline and magneto-elastic contributions [33]. The surface roughness normally increases under ion impact, but the present RBS spectra showed little changes of the surface roughness, which would be visible as a less steep slope of the metal back edge or the Si front edge (see Figures 4 and 9). Thus, a decrease in $\Delta(M_r/M_s)$ for the higher Xe fluences may be due to an isotropic stress based on a higher density of pinning centers. Large magneto-elastic coupling comes from strong spin–orbit coupling plus a strong interaction between an ion charge cloud and displacements of its surrounding ions. Since the magneto-elastic coupling energy is the strain derivative of the anisotropy, it may be large even if the anisotropy itself is small. The strong dependence of $\Delta(M_r/M_s)$ on the ion fluence suggests an interplay between a cubic anisotropy, a uniaxial anisotropy and a magneto-strictive component. Furthermore, the different polar curves of the remanence ratio M_r/M_s demonstrate that the in-plane easy axis of the specimens is shifted to the direction of the external magnetic field during irradiation. This was also observed in the case of other ion species implantated into the alloys [31, 34].

Chang *et al.* [35, 36] observed that the magnetic moment of CoPt bulk material increased after ion beam mixing in the presence of magnetic field compared to mixing without magnetic field. These authors demonstrated that the metastable phase which has been formed during ion implantation has expanded unit cell and reduced co-ordination that greatly reduces the electronic interaction between atoms. A large shift in the direction of in-plane easy axis during magnetic-field-assisted ion-beam mixing as compared to mixing without magnetic field has been also observed in case of FePd alloys [37]. In this experiment a line-shaped domain structure was observed with the long axis along pointing into the induced magnetization direction.

In conclusion, magnetic and structural properties of thin polycrystalline permendur and permalloy films have been studied under the impact of 200-keV Xe ions in the presence of an external magnetic field of about 100 Oe. For increasing film thickness or after vacuum annealing, the coercivity in permendur measured by longitudinal MOKE increased due to increased grain size. The coercivity and the magnetization curves showed pronounced variations with the ion fluence. They were correlated with the relaxation of stress in the as-deposited films at low ion fluences and the build-up of pinning centers at higher ion fluences. The comparison with results obtained in polycrystalline Fe films implies the magnetostriction constant to be the decisive parameter. In the case of iron and permalloy, a two-fold uniaxial anisotropy was generated by the Xe-ions, while in the case of permendur, a four fold pattern appeared. During implantation, the in-plane easy axis aligned along the direction of the external magnetic field. Clearly, experiments are required which test this interpretation on a more microscopic level, for example by looking at the microstructure with transmission electron microscopy. The MFM pictures taken for permendur are a first step into this direction.

Acknowledgements

We are grateful to H.-U. Krebs and Y. Luo (University of Göttingen) for the preparation of some films, and to K.-H. Han (University of Leipzig) for carrying out the MFM measurements. It is a pleasure to thank D. Purschke for his assistance with the ion implantations and RBS analyses.

References

1. Neél L., *J. Physi. Radium* **15** (1954), 225.
2. Johnson M. T., Bloemen P. J. H., den Broeder F. J. A. and de Vries J. J., *Rep. Prog. Phys.* **59** (1996), 1409.
3. Chappert C., Bernas H., Ferre J., Kottler V., Jamet J.-P., Chen Y., Cambril E., Develoder T., Rousseaux F., Mathet V. and Launois H., *Science* **280** (1998), 1919.
4. Devolder T., Chappert C., Chen Y., Cambril E., Bernas H., Jamet J.-P. and Ferre J., *Appl. Phys. Lett.* **74** (1999), 3383.
5. Gupta A., Paul A., Gupta R., Avasthi, D. K. and Principi G., *J. Phys., C* **10** (1998), 9669.
6. Lewis W. A., Farle M., Clermens, B. M. and White R. C. J., *J. Appl. Phys.* **75** (1994), 5644.
7. Neubauer M., Reinecke N., Uhrmacher M., Lieb K. P., Münzenberg M. and Felsch W., *Nucl. Instrum. Methods, B* **139** (1998), 332.

8. Zhang K., Lieb K. P., Schaaf P., Uhrmacher M., Felsch W. and Münzenberg M., *Nucl. Instr. Methods, B* **161–163** (2000), 1020.
9. Lieb K. P., Zhang K., Müller G. A., Schaaf P., Uhrmacher M., Felsch W. and Münzenberg M., *Acta Phys. Pol., A* **100** (2001), 751.
10. Müller G. A., Gupta R., Lieb K. P. and Schaaf P., *Appl. Phys. Lett.* **82** (2003), 73.
11. Zhang K., Gupta R., Lieb K. P., Luo Y., Mueller G. A., Schaaf P. and Uhrmacher M., *Eur. Phys. Lett.* **64** (2003), 668.
12. Zhang K., Gupta R., Müller G. A., Schaaf P. and Lieb K. P., *Appl. Phys. Lett.* **84** (2004), 3915.
13. Gupta R., Müller G. A., Schaaf P., Zhang K. and Lieb K. P., *Nucl. Instrum. Methods, B* **216** (2004), 350.
14. Uhrmacher M., Pampus K., Bergmeister F. J., Purschke D. and Lieb K. P., *Nucl. Instrum. Methods, B* **9** (1985), 234.
15. Müller G. A., Doctoral thesis, Universität Göttingen (2003).
16. Han K.-H., Spemann D., Esquinazi P., Höhne R., Riede V. and Butz T., *Adv. Meth.* **15** (2003), 1719.
17. Han K.-H., Spemann D., Höhne R., Setzer A., Makarova T., Esquinazi P. and Butz T., *Carbon* **41** (2003), 785.
18. Smith D. O. J., *Appl. Phys.* **30** (1959), 2645; Smith D. O., Cohen M. S. and Weiss G. P., *J. Appl. Phys.* **31** (1960), 1755; Cohen M. S., *J. Appl. Phys.* **32** (1961), 875.
19. Valetta R. M., Guthmiller G. and Gorman G., *J. Vac. Sci. Technol.* **A9** (1991), 2093.
20. Freeland J. W., Bussmann K. and Idzerda Y. U., *Appl. Phys. Lett.* **76** (2000), 2603.
21. Choe G. and Steinback M., *J. Appl. Phys.* **85** (1999), 5777.
22. Li M., Wang G. C. and Min H. G., *J. Appl. Phys.* **83** (1998), 5313.
23. Li M., Zhao Y.-P., Wang G. C. and Min H. G., *J. Appl. Phys.* **83** (1998), 6287.
24. Zhang K., Lieb K. P., Müller G. A., Schaaf P., Uhrmacher M. and Münzenberg M., *Europ. J. Phys. B* **42** (2004), 193.
25. Devolder T., Chappert C., Chen Y., Cambril E., Jamet J.-P. and Ferré J., *Appl. Phys. Lett.* **74** (1999), 3383.
26. Ravelosona D., Chappert C., Mathet V. and Bernas H., *Appl. Phys. Lett.* **76** (2000), 236.
27. Lohau J., Kirsch S., Carl A. and Wassermann E. F., *Appl. Phys. Lett.* **76** (2000), 3094.
28. Martin Y. and Wickramansinghe H. K., *Appl. Phys. Lett.* **50** (1987), 1455.
29. Allenspach R., Salemik H., Bischof A. and Waibel E. Z., *Phys. B: Condens. Matter* **67** (1987), 125.
30. Martin Y., Rugar D. and Wickramansinghe H. K., *Appl. Phys. Lett.* **52** (1988), 244.
31. Gupta R., Han K.-H., Lieb K. P., Müller G. A., Schaaf P. and Zhang K., *J. Appl. Phys.* **97** (2005), 073911.
32. Müller G. A., Lieb K. P., Carpene E., Gupta R., Schaaf P. and Zhang K., *submitted to Europ. J. Phys. B.*
33. Hubert A. and Schaefer R., *Magnetic Domains, The Analysis of Magnetic Microstructures*, Springer, Heidelberg, Berlin, New York, 2000.
34. Gupta R., Klein H., Lieb K. P., Luo Y., Müller G. A., Schaaf P., Uhrmacher M. and Zhang K., to be published.
35. Chang G. S., Lee Y. P., Rhee J. Y., Lee J., Jeong K. and Whang C. N., *Phys. Rev. Lett.* **87** (2003), 067208.
36. Chang G. S., Callcott T. A., Zhang G. P., Woods G. T., Kim S. H., Shin S. W., Jeong K., Whang C. N. and Moewes A., *Appl. Phys. Lett.* **81** (2002), 3016.
37. Bernas H., Attane J.-P., Heinig K.-H., Halley D., Ravelosona D., Marty A., Auric P., Chappert C. and Samson Y., *Phys. Rev. Lett.* **91** (2003), 077203.

Hyperfine Interactions (2005) 160:123–142
DOI 10.1007/s10751-005-9144-x

Depth Resolved Structural Studies in Multilayers Using X-ray Standing Waves

AJAY GUPTA
UGC-DAE Consortium for Scientific Research University Campus, Khandwa Road, Indore 452017, India

Abstract. X-ray based characterization techniques are powerful tools for the study of atomic scale structure of materials. However, high penetrating power of X-rays make them less suitable for depth selective studies, as required in the characterization of multilayer structures. In the present work, it is shown that depth selectivity of the techniques like, X-ray fluorescence, X-ray absorption spectroscopy and nuclear resonance fluorescence can be greatly enhanced by generating X-ray standing waves inside the multilayer structure. The concentration profiles of various elements can be obtained with a depth resolution of the order of 0.1 nm. Depth dependent information about the local structure around a given atom can be obtained from XAFS under standing wave conditions. It is demonstrated that detection of nuclear resonance fluorescence by tuning the energy of the incident X-rays to a Mössbauer transition can yield depth profile of a particular isotope, and can be used for self-diffusion studies. The techniques of X-ray reflectivity and conversion electron Mössbauer spectroscopy are used to provide useful complementary information.

1. Introduction

Thin films and multilayers have emerged as an important class of nanostructured materials with immense possibilities of tailoring their properties in order to achieve the desired functionality. In multilayer structures, as the thickness of individual layers decrease and become comparable to the characteristic length scale of a given property (e.g., exchange length in case of magnetic properties, and wavelength of the electromagnetic radiation in the case of optical properties), that particular property can get modified drastically. This provides an unprecedented control over the properties of the multilayers through the control of their structure. Depending upon the application, layer thicknesses may vary from a fraction of a nanometer to a few tens of nanometers. Since in such multilayers, a large fraction of atoms reside at the surface/interfaces, the interfacial region plays a dominant role in determining their properties. Therefore, it is important to elucidate the interface structure in multilayers in order to both understand their novel properties as well as to tailor the same through tailoring the interface structure. A detailed characterization of the interface includes: (i) the topological structure of the interface which can be characterized, in general, in

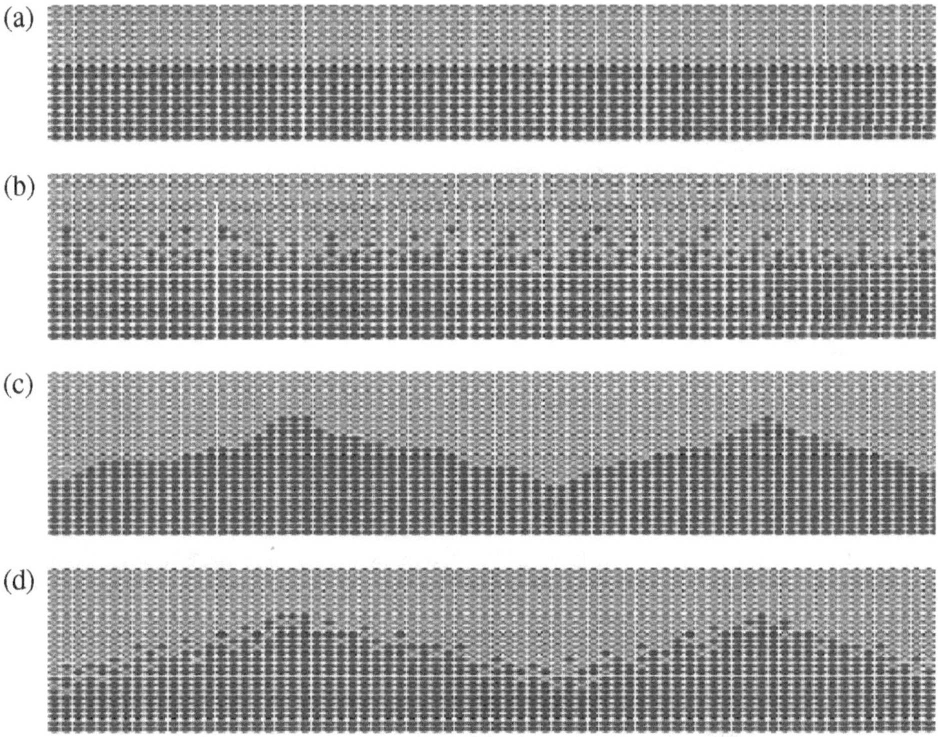

Figure 1. Schematic representation of various imperfections at the interface of two layers. Deviations from a perfectly sharp and smooth interface (a) can be represented in terms of some interdiffusion (b) or a finite roughness at the interface (c). In general, a real interface will be characterized by having both a finite interdiffusion as well as interface roughness (d).

terms of rms interface roughness σ, the correlation length of the interface-height in both film plane (ξ) as well as perpendicular to it (ζ), as well as interdiffusion at the interfaces (Figure 1) [1], and (ii) short range order in the interfacial region. Different physical properties of the multilayers are affected by different aspects of the interface structures: The performance of X-ray multilayer mirrors is affected only by topological structure of the interfaces [2]. On the other hand, in case of magnetic multilayers the relevant properties strongly depend also upon the short range order in the interfacial region [3–5]. For example, in Fe/Tb multilayers the perpendicular magnetic anisotropy is mainly determined by the anisotropy of Fe–Tb bonds in the interfacial region [3], while in GMR multilayers of Fe/Si, coupling between two Fe layers depends upon the structure of the phase formed in the interfacial region between Fe and Si layers [4, 5].

X-rays can be used to get a variety of information about the structure of the interfaces: X-ray reflectivity (XRR) and diffuse scattering (XDS) can be used to get information about the topological structure of the interfaces [6]. X-ray reflectivity in conjunction with conversion electron Mössbauer spectroscopy (CEMS) can be used to separate the interface roughness from the interdiffusion [7]. The

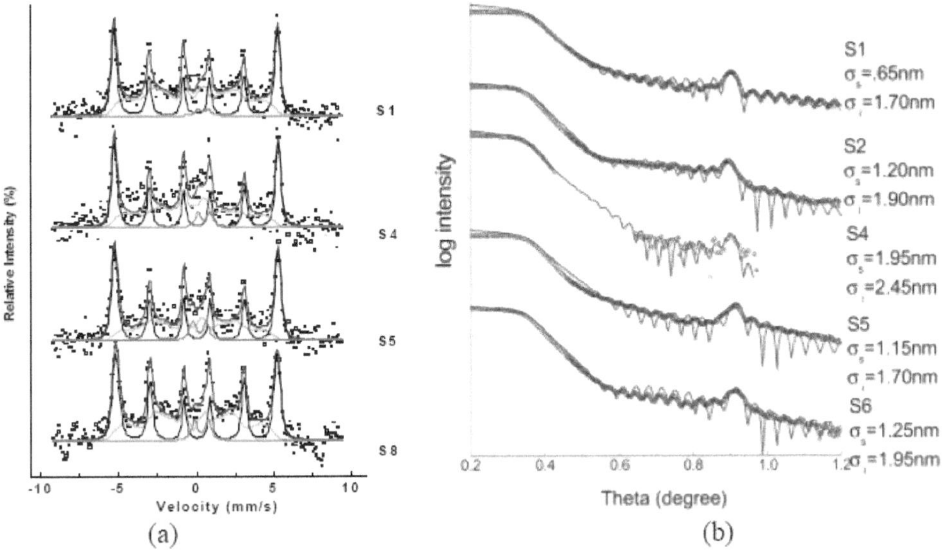

Figure 2. (a) CEMS of Fe/Tb multilayers for specimens S1, S4, S5, and S8. The continuous curves represent the best fit to the experimental data obtained by taking two independent hyperfine field distributions corresponding to i) a sharp sextet representing bulk αFe and, ii) the remaining broad component representing the interfacial region. (b) True x-ray reflectivity of specimens S1, S2, S4, S5 and S6 obtained by subtracting off-specular reflectivity from the specular reflectivity data. The continuous curves represent the best fit to the data. Values of substrate roughness σ_s and the interface roughness σ_i corresponding to the best fit of the experimental data are shown on the right hand side of the curves (taken from Ref. [7]).

information obtained from XRR pertains to the electron density variation across the interface and is not sensitive to the chemical nature of the species. Concentration of different constituent species can be obtained using X-ray fluorescence measurement, while X-ray absorption fine structure (XAFS) measurement can be used to get information about the short range order around a specific element, and nuclear resonance scattering (by tuning the X-rays from a synchrotron radiation source to Mössbauer transition) can give information about the local order as well as magnetic state of the relevant atoms [8].

It may be noted that X-rays are highly penetrating radiation and therefore, are in general not capable of providing localized or depth selective information. However, it has been demonstrated that X-rays can be localized in the medium by generating standing waves [9–11], and under such condition, the X-ray based techniques can be made depth selective. There are two different ways in which X-ray standing waves can be formed in thin film structures: It has been shown by the Bedzyk *et al.* that in case the thin film structure consists of a periodic multilayer with sufficient X-ray scattering contrast between successive layers, X-ray standing waves are formed within this structure for the incident angle of the X-rays around the Bragg angle [9, 10]. Later on it was demonstrated that the

Table I. Results of computer fitting of CEMS and XRR data of [Fe(3.0 nm)/Tb(2.0 nm)] × 20 multilayers on FG substrate with different etching times

Sample no.	Etching time (s)	CEMS Results				XRR Results		
		Parameters of the α-Fe component		Intermixed layer thickness (nm)	ϕ (degree)	σ_s (nm)	σ_i (nm)	$(\sigma_i^2-\sigma_s^2)^{1/2}$ (nm)
		<(B_{hf})> (T)	Aα (%)					
1	0	32.7 ± 0.2	44 ± 2	1.0 ± 0.3	42 ± 3	0.65 ± 0.05	1.70 ± 0.05	1.57 ± 0.07
2	15	–	–	–	–	1.20	1.90	1.47
3	30	–	–	–	–	1.45		
4	60	32.8	46	1.0	43	1.95	2.45	1.48
5	90	32.7	51	0.8	41	1.15	1.70	1.25
6	120	–	–	–	–	1.25	1.95	1.49
7	150	–	–	–	–	1.45		
8	180	32.3	48	0.9	50	3.35		

Aα represents the percentage area of the total spectrum in the α-Fe subspectra. σ_s and σ_i are the rms roughnesses of the substrate and interfaces respectively. The last column gives the uncorrelated part of the interface roughness.

standing waves in different films can also be generated using total external reflection of X-rays from a underlying layer of high Z material like Au or Pt [9, 11]. In the present work we demonstrate that the X-rays in conjunction with the hyperfine techniques can be used for a detailed characterization of the topological structure of the interfaces as well as for depth resolved information about the short range order in the interfacial region. In Section 2, a combination of XRR, XDS and CEMS is used to get a detailed information about the topological structure of the interface, while in Section 3, X-ray standing waves formed by total external reflection have been effectively used to get depth resolved elemental concentration as well as short range order in the system. In Section 4, it has been demonstrated that nuclear resonance fluorescence under standing wave conditions can be used to get depth profile of a particular isotope (Mössbauer active), and thus can be used to study self-diffusion in isotopically labeled homogeneous systems.

2. Characterization of the topological structure of the interfaces: Case of Fe/Tb multilayer

Fe/Tb multlayers with appropriate layer thicknesses exhibit perpendicular magnetic anisotropy (PMA) even at room temperature [3, 7, 12, 13]. The origin of PMA in these systems is not yet fully understood, however it is generally agreed that the dominant contribution to PMA comes from the single ion anisotropy of the rare-earth ions coupled with anisotropic distribution of TM–RE bonds at the interfaces [3]. The effect of interface structure on PMA in Fe/Tb multilayer having composition: substrate (float glass)/(Fe3 nm/Tb2 nm)$\times$20 has been studied [7]. The films were prepared by electron-beam evaporation in an ultra-high vacuum (UHV) chamber with a base pressure of 5×10^{-10} torr. The multilayers simultaneously deposited on float glass substrates with different surface roughnesses were characterised using X-ray reflectivity, X-ray diffuse scattering and Mössbauer spectroscopy. Figure 2a shows the Mössbauer spectra of the multilayers prepared on different substrates with varying surface roughness. These spectra consist of an overlap of a sharp sextet corresponding to α-Fe with a broad sextet having a lower hyperfine field, representing the Fe atoms interfaced with Tb layer. Table I gives the results of computer fitting of the Mössbauer spectra on the basis of the above model. It may be noted that about 56% of the area is under the broad sextet. The area under the broad sextet can be used to get information about the thickness of the intermixed layer at the interfaces: for perfectly sharp interfaces two monolayers of Fe will be interfaced with Tb atoms at each interface. This will constitute about 20% of the total atoms in a layer, considering that the total thickness of Fe layer is 20 monolayers. Therefore, 56 − 20 = 36% of Fe atoms exist in the intermixed layers at the two interfaces. Taking the composition of the intermixed layers as $Fe_{0.5}Tb_{0.5}$, the thickness of intermixed layer at each interface corresponds to about 1.0 nm [7, 13, 14]. No significant change in the thickness of the intermixed layer is observed in the samples with different substrate roughness. Figure 2b gives the true specular X-ray reflectivity of the multilayers. The X-ray reflectivity patterns were fitted using a computer program based on Parratt's formulism [15]. The multilayer was considered to be consisting of alternate layers of Fe and Tb with an intermixed layer of composition $Fe_{0.5}Tb_{0.5}$ at each interface. Thickness of the intermixed layer was taken from CEMS results. Table I also includes the average rms roughness of the interfaces as obtained from the fitting of the X-ray reflectivity data. For comparison, the surface roughness of the substrates as obtained from the X-ray reflectivity measurements before the deposition of the multilayers are also shown in the Table I. It may be noted that in multilayers, in general, the interface roughness consists of two parts: a correlated part of the roughness σ_c which is replicated from one interface to the other interface, and an uncorrelated part of the roughness σ_u. The total roughness can be written as $\sigma^2 = \sigma_c^2 + \sigma_u^2$ [16, 17]. If one assumes that the correlated part of the roughness is essentially coming from the roughness of the substrate itself, then the uncorrelated part of the roughness can be obtained using the above relation. Taking σ_c equal to the substrate roughness, σ_u as calculated using above relation is also shown in the last column of Table I. One may note that the uncorrelated part of the roughness is same in all the multilayers within experimental errors. This is as per expectation, since all of them are deposited simultaneously. The in-plane correlation length of the height variation was obtained by measuring diffuse

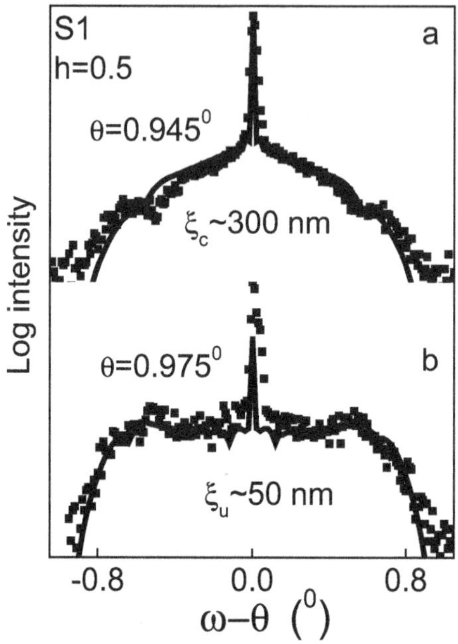

Figure 3. X-ray diffuse scattering data on the specimen S1 taken at two different values of the angle of incidence corresponding to the first Bragg peak in the reflectivity ($\theta = 0.945°$), and the successive minimum ($\theta = 0.975°$). Continuous curves represent fit to the experimental data. The obtained values of in-plane correlation length are also shown.

scattering at two distinct scattering angles corresponding to the first Bragg peak in the reflectivity ($2\theta = 1.89°$) and the consecutive minimum in the reflectivity ($2\theta = 1.95°$). The results of measurements along with the fitting of the data using the program TRDS developed by Stepanov [18] are shown in Figure 3. It may be noted that the diffusedly scattered X-rays from different interfaces due to correlated part of the roughness add-up coherently while that due to uncorrelated part of the roughness add incoherently [16–19]. Therefore, the diffuse scattering from the uncorrelated part of the roughness does not depend upon the scattering angle, while that due to correlated part of the roughness varies with the scattering angle, being maximum at the Bragg angle (due to constructive interface between various amplitudes). Thus the diffuse scattering measurements at Bragg peak is dominated by the correlated part of the roughness while that at scattering angle corresponding to a minimum in the reflectivity is dominated by the uncorrelated part of the roughness [17, 19]. Therefore, in Figure 3 curve (a) gives the in-plane correlation length of the correlated part of the roughness as 300 nm, while curve (b) gives that of the uncorrelated part of the roughness as 50 nm. According to the stochastic theory for the evolution of roughness in thin film deposition, the replication of roughness from one interface to the other is strongly frequency dependent; low-frequency components of roughness propagate with less attenuation between interfaces than high-frequency components [20]. There-

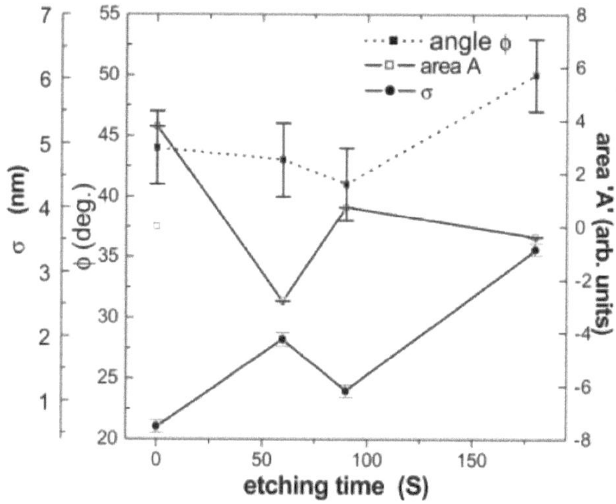

Figure 4. Dependence on the etching time of the float glass substrate of the substrate roughness σ, the spin canting angle ϕ, and the area enclosed between the magnetization profile of perpendicular and parallel measurements (taken from Ref. [7]).

fore, generally the high frequency part of the roughness is uncorrelated while the low frequency part constitute the correlated roughness. Present results are in good agreement with this theory, as the uncorrelated part of the roughness has a smaller in-plane correlation length and hence constitutes the high frequency component. The in-plane correlation length for the roughness of the bare substrate was found to be 450 nm which is comparable to that for the correlated part of the interface roughness. This again supports the conjecture that the correlated part of the roughness is essentially coming as a result of the replication of the substrate roughness.

The perpendicular magnetic anisotropy in the films was estimated both from the relative intensities of 2nd and 5th lines in conversion electron Mössbauer spectra at room temperature, as well as from SQUID measurements at 4.2 K. Figure 4 shows that the dependence of PMA on interface roughness is rather weak. This weak dependence is understandable since the films differ from each other only in the correlated part of the roughness that has an in-plane correlation length of about 300 nm. Thus, the slope of the interface at a given point, which may be approximately written as σ/ξ, is rather small and should not effect the anisotropy of the bonds significantly [7].

3. X-ray standing wave characterization of thin film structures

In the above section it has been shown that a combination of conversion electron Mössbauer spectroscopy in conjunction with X-ray reflectivity and X-ray diffuse scattering can be used to get a detailed information about the topological structure of the interfaces including the correlated and uncorrelated parts of the

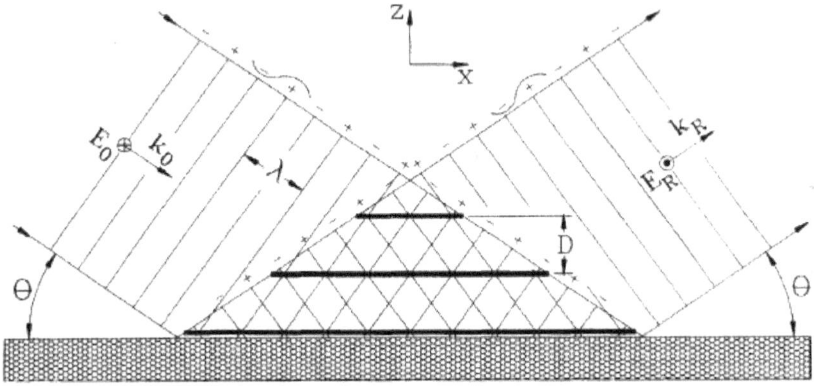

Figure 5. Schematic representation of the formation of X-ray standing wave as a result of total external reflection from a layer of high electron density (taken from Ref. [11]).

interface roughness, their in-plane correlation length, as well as the interdiffusion at the interfaces. However, in several situations it may be necessary to get a much more detailed information about the interface structure, including the depth distribution of various elements (and not just the variation of electron density), and of various phases formed at the interfaces. As discussed earlier, in general X-rays can be used to get information about elemental concentration (through X-ray fluorescence) and local structure (through XAFS), and the depth selectivity can be achieved using X-ray standing waves.

3.1. FORMATION OF X-RAY STANDING WAVES

X-ray standing waves can be generated by making use of total external reflection from a surface [9, 11]. It may be noted that the refractive index of a medium for X-rays can be written as $n = 1 - \delta - i\beta$. For angle of incidence θ less than $\theta_c = (2\delta)^{1/2}$, the X-rays undergo total external reflection (Figure 5). Taking the electric field vector of the incident traveling plane wave as:

$$\varepsilon_0(r,t) = E_0 \exp\{i[\omega t - 2\pi(k_x x - k_z z)]\}, \tag{1}$$

where z–x is the scattering plane as illustrated in Figure 5, the electric field vector of the reflected wave can be written as:

$$\varepsilon_R(r,t) = E_R \exp\{i[\omega t - 2\pi(k_x x + k_z z)]\}. \tag{2}$$

Here, $z = 0$ has been taken at the surface, and we consider the case of σ polarization. Ratio of the reflected to the incident amplitude can be described in terms of Fresnel coefficient of reflection as:

$$\frac{E_R}{E_0} = \frac{|E_R|}{|E_0|} e^{iv} = \frac{\theta - (\theta^2 - 2\delta - 2i\beta)^{1/2}}{\theta + (\theta^2 - 2\delta - 2i\beta)^{1/2}} \tag{3}$$

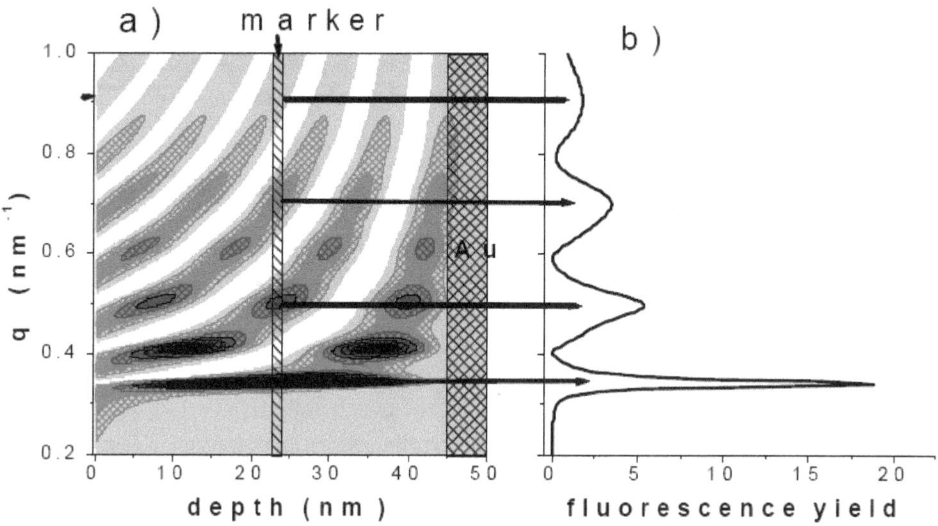

Figure 6. a) Simulated depth distribution of X-ray intensity inside a layer Au/Si (45 nm) as a function of the scattering vector q. Nodes and antinodes are formed inside the Si layer due to total reflection of X-rays from the surface of Au. b) Simulated fluorescence from a thin marker layer placed in the center of the Si layer.

Here ν is the phase difference between incident wave and the reflected wave at $z = 0$, and $R = |E_R/E_0|^2$ is the reflectivity, which is expected to be closed to unity below θ_c. The incident and reflected amplitudes interfere with each other to form standing waves above the surface, with planes of maximum intensity (antinodes) parallel to the surface with a period [9, 11]

$$D = \lambda/2\text{Sin}\theta = 2\pi/q, \tag{4}$$

q being the scattering vector. From Equation (3) one may note that for $\theta = 0$, the phase difference between the incident and the reflected waves at $z = 0$ is π, and therefore, there is a node at the surface, and the first antinode will be at ∞. At $\theta = \theta_c$, as $\nu = 0$, there is an antinode at the surface. Thus, with increasing angle of incidence the first antinode moves down from ∞ at $\theta = 0$ to the surface at $\theta = \theta_c$, at the same time the separation between the successive antinodes goes on decreasing as per the Equation (4). Since the intensity of the X-rays is strongly localized in the region of antinodes, the standing waves can be used as a spatially localized periodic probe with a spatial resolution (along Z direction) of $D = \lambda/(2 \text{ Sin } \theta)$ [21, 22]. In order to illustrate this aspect, Figure 6a shows the q dependence of the depth distribution of X-ray intensity above the reflecting surface. In case a marker layer of an element is placed at a distance L above the surface (using appropriate spacer layer) then the fluorescence from the marker layer will depend upon the X-ray intensity at that point. Thus, as the angle of incidence increases the antinodes will move down, and whenever an antinode passes across the marker layer the fluorescence from that layer will exhibit a

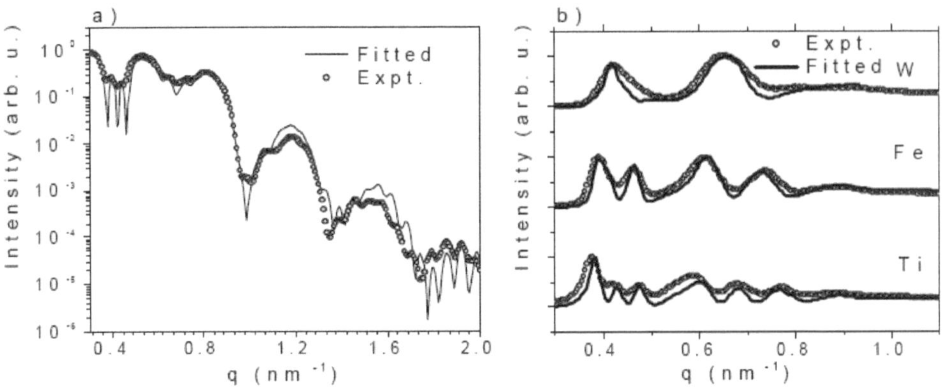

Figure 7. a) X-ray reflectivity of pristine specimen Au/Si/W/Si/Fe/Si/Ti/Si. The continuous curve represents theoretical fit to the data. b) The corresponding X-ray fluorescence yields from the W, Fe, Ti layers. The continuous curves represent fit to the data (taken from Ref. [25]).

peak, as shown in Figure 6b. Thus the q dependence of the fluorescence from a given element can be used to get the depth profile of the same. In the following use of this technique is demonstrated in the study of swift heavy ion induced intermixing between various metals and Si layers.

3.2. SWIFT HEAVY ION INDUCED MIXING IN METAL/SILICON SYSTEM

Metal silicides are important for metalisation applications in VLSI circuits because of their low metal like resistivity, and the high temperature stability and high electromigration resistence. In this context, formation of metal silicides using heavy ion induced mixing has attracted considerable attention over last couple of decades [23]. More recently, use of swift heavy ions (having energies in the range of MeV/nucleon) in producing metal silicides in-situ in thin film form has also been explored [24]. Since intermixing can be as small as a nm or less, such studies require depth profiling techniques with a higher depth resolution as compared to the conventional SIMS or RBS depth profiling. X-ray fluorescence measurements under standing wave conditions can yield concentration profile with a depth resolution of better than 1 nm. For this purpose a multilayer having the following nominal structure was prepared using electron-beam evaporation under UHV conditions: Substrate/Cr 20 nm/Au 70 nm/Si 12.5 nm/W 3 nm/Si 12.5 nm/Fe 3 nm/Si 12.5 nm/Ti 3 nm/Si 12.5 nm. The bottom Au layer is deposited in order to achieve the formation of X-ray standing waves in the multilayer structure. The multilayer was irradiated with 100 MeV Au ions to a fluence of 1×10^{13} ions/cm^2. The simultaneous X-ray reflectivity and fluorescence measurements were done on the samples using ID32 beamline of ESRF, Grenoble. Figure 7 gives the X-ray reflectivity as well as q dependence of fluo-

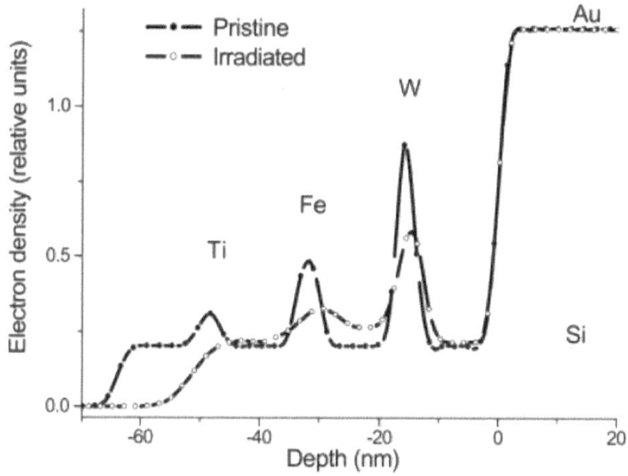

Figure 8. The electron density profile in the pristine as well as irradiated specimens of Au/Si/W/Si/Fe/Si/Ti/Si, as obtained from the combined fitting of the X-ray reflectivity and fluorescence yield data (taken from Ref. [25]).

rescence intensity of Ti, Fe and W layers [25]. Information about the depth profile of different elements can be obtained by comparing the simulated curves with the experimental data. A simultaneous fitting of both X-ray reflectivity and the X-ray fluorescence data can give a much more reliable information about the depth distribution of various elements. The continuous curves in Figure 7 represent such fittings to the data. The corresponding depth profiles of W, Fe and Ti before and after irradiation with swift heavy ions are shown in Figure 8. The results show that while W depth profile is only slightly broadened after irradiation, the Fe depth profile shows a significantly large broadening. On the other hand, Ti profile is broadened so much that no peak corresponding to the same is observed after irradiation. Thus, similar to their behaviour in bulk form [26], the sensitivity of W, Fe and Ti films to swift heavy ions irradiation varies in the order W < Fe < Ti. Further from Figure 8 one may note that irradiation results in some densification of Si layer (as indicated by the shifting of the peaks corresponding to W and Fe towards the Au surface). A large shift in the top surface may partly be attributed to some sputtering of the Si from the surface.

While measurements of X-ray fluorescence can give information about the depth distribution of different elements, the local structure around a given metal atom can be determined using X-ray absorption measurements. The depth resolution in X-ray absorption can again be achieved by properly choosing the q value so that the position of an antinode is made to coincide either with the center of a marker layer or with the metal/Si interface. In this way the XANES measurements were done in order to elucidate the structure of W and Fe layers before and after irradiation. Measurements were done in fluorescence made and an energy range of about 350 eV was scanned across the absorption edge of the corresponding element. The measurements were done in a constant q mode so

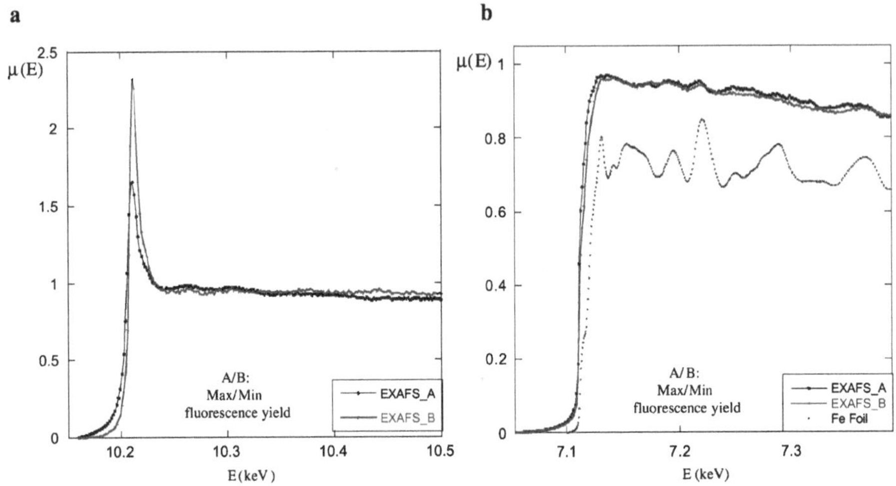

Figure 9. a) WL_3 edge X-ray absorption spectra and b) Fe K edge X-ray absorption spectra, of the pristine specimen Au/Si/W/Si/Fe/Si/Ti/Si. Spectrum A corresponds to the situation when the first antinode of the x-ray standing wave coincides with the center of the W (Fe) layer, while the spectrum B corresponds to the situation when the successive node coincides with the center of the W (Fe) layer (taken from Ref. [25]).

that the position of antinode does not shift as the energy of incident X-ray is varied. Figure 9a shows the XANES at W L_3-edge of the pristine specimen taken with q corresponding to the first maximum and the subsequent minimum in the fluorescence yield. While in the first case, the information is preferentially obtained from the center of the layer, in the second case information is obtained from the interfacial region. The spectrum A presents a weak while-line of W at the position of absorption edge corresponds to metallic W. In contrast to this, in spectrum B, the white-line has become very intense which is typical of W–Si compounds. The edge also exhibits a shift to higher energy, again indicating Si near-neighbour environment of W atoms. Irradiation causes a decrease in the intensity of white-line in both the A and B regions. However, the intensity ratio and the edge position in A and B regions remains similar. It may be noted that X-ray fluorescence analysis shows a small broadening of W profile after irradiation. This suggests that irradiation mainly provokes the structural disorder (decreasing of peak intensity and broadening of fluorescent peaks) but does not change the intermixing (no edge shift). Figure 9b gives Fe K-edge X-ray absorption spectra of pristine specimen. In this case the depth resolution is not sufficient to differentiate between the center of the layer and the interface region. The pristine sample shows features similar to metallic Fe evident in the region around 7.2 KeV. Irradiation in this case results in a drastic modification in XANES spectrum: features of metallic Fe are totally lost and XANES spectrum consisting of broad oscillations corresponding to highly disordered amorphous FeSi alloy.

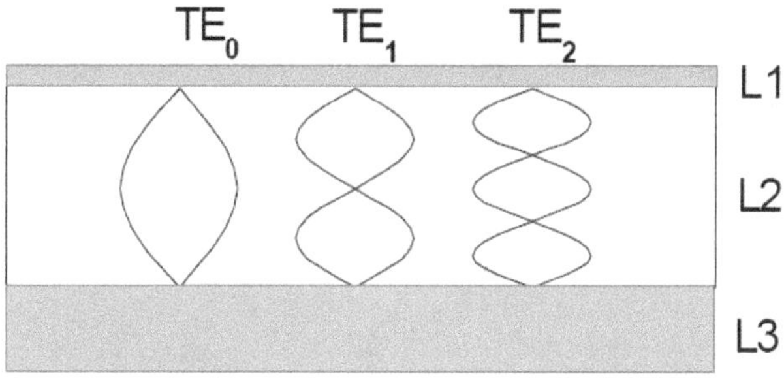

Figure 10. Schematic representation of waveguide structure and various resonance modes.

3.3. X-RAY WAVEGUIDES IN THIN FILM CHARACTERIZATION

X-ray waveguide structures consist of a low density guiding layer sandwiched in between layers of higher density. A trilayer structure shown in Figure 10 can act as a planar waveguide (WG), and electromagnetic modes can be excited within the layer 2 (cavity), provided the following condition is fulfilled for the refractive indices of its constituent: $n_2 > n_1 \geq n_3$. In a WG structure with thickness of cavity equal to D, only desecrate modes can be excited with the nodes of their intensity distribution coinciding with the interfaces. These modes are designated as TE_n modes. The condition for excitation of nth mode is given by:

$$2D\,\mathrm{Sin}\phi_{in} = n\lambda,$$

where ϕ_{in} is the angle at which the X-rays fall at the interfaces of the cavity. The intensity distribution inside the WG for TE_0, TE_1 and TE_2 modes is schematically shown in Figure 10. Excitation of X-ray waveguide modes in thin film structures was first demonstrated by Sinha and coworkers [27]. It has been shown that such waveguides can be used to study the structure of the waveguiding medium itself [28]. The q dependence of the field intensity inside the waveguide structure gets modified significantly as compared to the standing wave field considered in Section 3.1. In the following we demonstrate that this modified field distribution can be used to improve the depth resolution of the fluorescence and XAFS measurements.

Multilayer having a nominal structure: substrate/Cr 20 nm/Au 70 nm/Si 15 nm/Fe 4 nm/Si 9 nm/Au 2 nm, deposited using electron beam evaporation in a UHV environment has been studied [29]. The layer [Si 15 nm/Fe 4 nm/Si 9 nm] nm acts as the waveguide cavity. The 4 nm thick Fe layer is intentionally kept asymmetrically with respect to the center of the cavity. The X-ray reflectivity as well as fluorescence measurements were done using ID32 beamline of ESRF, Grenoble. Figure 11a gives the X-ray reflectivity of the waveguide structure taken at 12 keV X-ray energy. Three sharp dips observed in the reflectivity below the critical angle for Au indicate the resonance coupling of incident X-rays to the

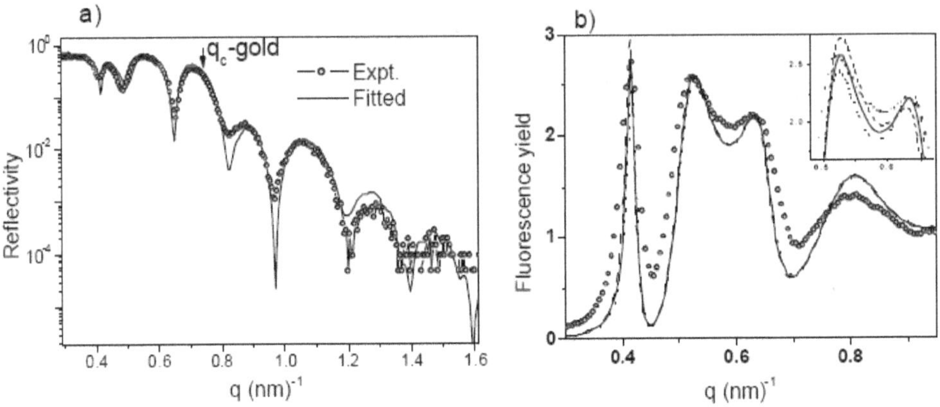

Figure 11. a) X-ray reflectivity from the waveguide structure Au/Si/Fe/Si/Au. Continuous curves represent theoretical fit to the data. b) Fluorescence yield from the Fe layer in the above structure as a function of q. The continuous curve represents the best fit to the data. The inset shows simulated curves obtained by shifting the Fe marker layer by +0.2 nm (– – –) or by −0.2 nm (...) from the best-fit value.

waveguide, corresponding to the modes TE_0, TE_1 and TE_2, respectively; when the scattering vector corresponds to one of the resonant modes of the waveguide, the absorption in the structure is greatly enhanced because of the prolonged duration of the X-rays inside the film (as a result of multiple reflections from the walls of the waveguide). Figure 11b gives the K_α fluorescence from the Fe marker layer as a function of the scattering vector q. The intensity of the fluorescence is normalized with Cr K_α-fluorescence from a control foil kept upstream of the sample, in order to correct for the detector deadtime. Three distinct peaks in the fluorescence at q values corresponding to the TE_0, TE_1 and TE_2 modes clearly evidence the excitation of waveguide modes. It may be noted that, had the Fe marker layer been placed at the center of the cavity, one would have observed peaks corresponding to TE_0, TE_2 modes only, since for TE_1 mode one of the nodes would have coincided with the marker layer. This point is clear from Figure 12 which gives contour plot of X-ray intensity inside the film as a function of q. One may note that the relative intensities of the fluorescence peaks corresponding to TE_1 and TE_2 modes would depend sensitively on the exact location of the marker layer, and therefore, by fitting the experimental data it should be possible to determine the position of the marker layer with high accuracy. The X-ray reflectivity and fluorescence data have been fitted simultaneously in order to obtain the structure of the multilayer. The results of fitting are given in Table II. In order to demonstrate the sensitivity of the fluorescence data to the exact location of the Fe marker layer, in Figure 11b the simulated fluorescence pattern for Fe layer displaced by ±0.2 nm from the best fit value are also shown as inset. The disagreement of these simulated curves with the experimental data is well above the experimental error. Thus, the accuracy with

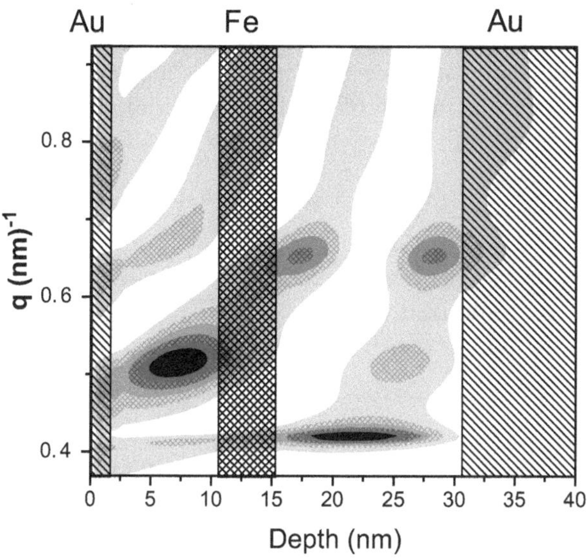

Figure 12. Simulated depth distribution of X-ray intensity inside a waveguide structure: Au/Si (15 nm)/Fe (4 nm)/Si (9 nm)/Au (2 nm), as a function of the scattering vector q.

Table II. Results of the combined fitting of the x-ray reflectivity and fluorescence yield data of the waveguide structure: Substrate/Cr/Au/Si/Fe/Si/Au

Layer	Thickness (nm)	Roughness (nm)
Au	1.6	1.6
Si	8.9	1.6
Fe	4.3	1.6
Si	15.5	1.6
Au	70	1.6
Cr	20	1.6
Substrate	∞	0.5

which proposition of the Fe marker layer inside the cavity can be determined using this technique is better than 0.2 nm.

4. Nuclear resonance fluorescence studies for isotopic depth profiling

The technique of X-ray standing wave can be made isotope-sensitive if instead of detecting fluorescence from the atomic levels one detects the nuclear resonance fluorescence by appropriatly tuning the energy of incident X-rays (e.g., from a synchrotron radiation source) to a nuclear transition (Mössbauer transition).

In the presence of nuclear resonance scattering the total scattering amplitude of an atom can be written as a sum of the scattering amplitudes due to electronic scattering and nuclear resonance scattering. As a result, at grazing incidence geometry the index of refraction of the layer material can be written as:

$$n_{ab} = \sqrt{1 + \frac{\lambda_0^2}{\pi} \sum_i \sigma_i (f_i)_{ab}} \approx 1 + \frac{\lambda_0^2}{2\pi} \sum_i \sigma_i \left(\left(f_i^e \right)_{ab} + \left(f_i^n \right)_{ab} \right) \quad (5)$$

Where $(f_i^e)_{ab}$ and $(f_i^n)_{ab}$ are the electronic and nuclear scattering amplitudes for scattering $\vec{\varepsilon}_b$ – polarized radiation into $\vec{\varepsilon}_a$ – polarized radiation, respectively and σ_i is the atomic density of the species i. The nuclear scattering amplitude is given by [30]

$$\left(f^n\right)_{ab} = \frac{2\lambda_0 P f_{LM}}{2j_0 + 1} \left[\frac{\Gamma_\gamma}{\Gamma} \right] \sum_{M=-L}^{L} \sum_{m_0=-j_0}^{j_0} C^2(j_0 L j_1; m_0 M) \quad (6)$$

$$\hat{\varepsilon}_a^* \cdot Y_{LM}^\lambda \left(\hat{k}_0 \right) \left[Y_{LM}^{(\lambda)} \left(\hat{k}_0 \right) \right]^* \cdot \hat{\varepsilon}_b / [x(m_0 M) - i]. \quad (7)$$

Where P is the enrichment of resonant nuclei, f_{LM} is the Lamb–Mossbauer factor, Γ_γ is the radiative and Γ the total width (due to radiative, internal conversion, and inhomogeneous broadening), j_0 and j_1 are the spin quantum numbers of ground and excited states respectively, λ designates the multipolarity of the radiation, $Y_{LM}^{(\lambda)}\left(\hat{k}_0\right)$ is the vector spherical harmonic, and

$$x(m_0 M) = 2[E(j_1, m_0 + M) - E(j_0, m_0) - \hbar\omega]/\Gamma, M = \Delta j_z = m_1 - m_0. \quad (8)$$

For the Mossbauer transition of ^{57}Fe nuclei (M1 transition), in absence of a magnetic splitting, and for a polycrystalline sample, the nuclear forward scattering amplitude can be written as:

$$\left(f^n\right)_{ab} = \delta_{ab} \frac{\lambda_0}{4\pi} \frac{f_{LM}}{1+\alpha} \frac{2j_1 + 1}{2j_0 + 1} \frac{A}{x - i} \quad (9)$$

Where $x = (\Delta E - \hbar\omega)/\Gamma_0$, A denotes the inhomogeneous broadening, ΔE is the quadrupole splitting and Γ_0 the natural line width. The nuclear part of the refractive index shows a resonance behaviour and is significant only within a narrow energy range of the order of μeV around the nuclear transition energy. However, in this energy range the Mössbauer isotope presents a strong scattering contrast, and the information about the depth distribution of this isotope can be obtained both from the q-dependence of the nuclear resonance reflectivity as well as nuclear fluorescence.

The multilayer used for studying the nuclear resonance fluorescence has the following structure: substrate/Pt(70 nm)/$Fe_{60}Zr_{40}$(12.5 nm)/$^{57}Fe_{60}Zr_{40}$(5 nm)/

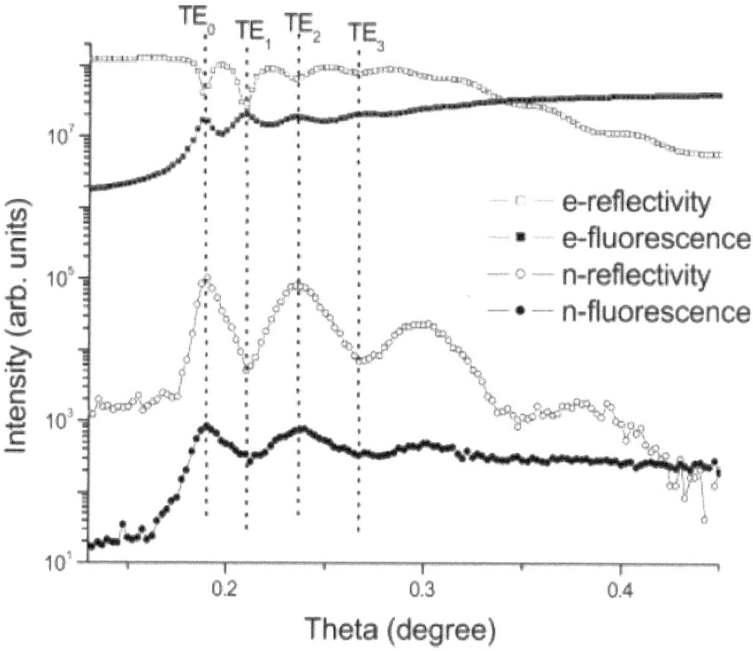

Figure 13. The electronic as well as nuclear reflectivity and fluorescence yield data for the multilayer Pt/FeZr/^{57}FeZr/FeZr.

$Fe_{60}Zr_{40}$(12.5 nm) [31], i.e., as per the chemical composition it consists of a single 30 nm thick layer of $Fe_{60}Zr_{40}$ on Pt, in the middle of which a 5 nm thick layer is enriched in ^{57}Fe to more than 90%. The film was deposited on a float glass substrate using magnetron sputtering. Measurements were done at ID22N beamline of ESRF, Grenoble [32]. The incident X-ray beam was monochromatised down to a width of 4 meV and the energy was tuned to the 14.4 keV Mössbauer transition for ^{57}Fe. The q dependence of reflectivity as well as the fluorescence measurements were done simultaneously using two different sets of avalanche photo diodes. The nuclear and electronic parts of reflectivity as well as fluorescence were separated by making use of the fact that nuclear transitions are delayed in time due to finite life time of the Mössbauer excited state (140 ns in case of ^{57}Fe isotope) [32]. Figure 13 gives the electronic and nuclear parts of the reflectivity as well as fluorescence from the multilayer. The electronic fluorescence includes both the K-edge fluorescence from Fe as well as L-edge fluorescence from the bottom Pt layer (Zr L-edge fluorescence being low in energy, will not contribute significantly). However, since the fluorescence from Pt layer is expected to increase monotonically with the angle of incidence upto θ_c, the observed features in the electronic fluorescence are mainly due to that from Fe atoms. Since the X-ray scattering contrast at air/FeZr and FeZr/Pt interfaces is of the same order, the above structure acts as an asymmetric waveguide. Therefore, field inside the FeZr layer will get enhanced whenever the condition $2D\sin\theta = n\lambda$

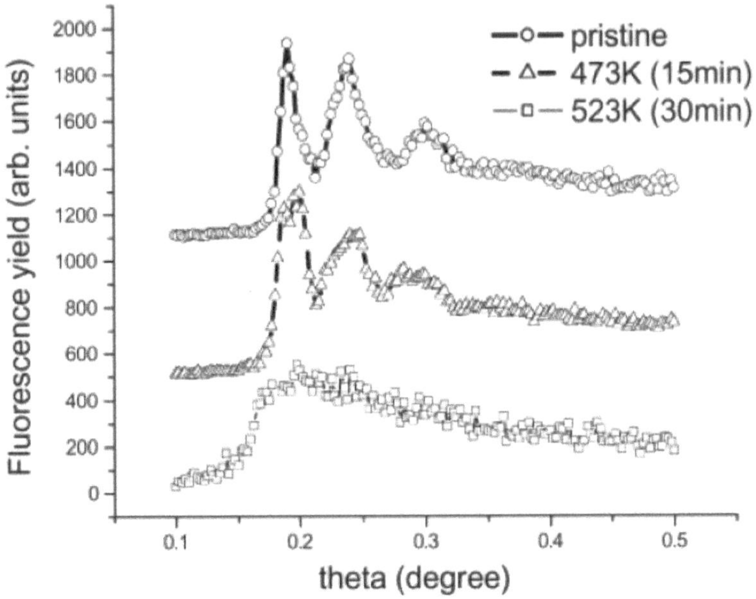

Figure 14. Nuclear fluorescence yield from the multilayer Pt/FeZr/^{57}FeZr/FeZr after different annealing treatments.

is satisfied. This aspect is reflected in peaks in electronic fluorescence at θ = 0.187, 0.21, 0.235, 0.267, corresponding to TE_0, TE_1, TE_2 and TE_3 modes respectively. It may be noted that for θ corresponding to TE_1 and TE_3 modes, although the average field inside FeZr layer gets enhanced, at the center of the layer there is a node (Figure 10). Therefore, the nuclear resonance fluorescence from the ^{57}FeZr marker layer which lies in the center of the FeZr layer, would exhibit a dip at this q value. The experimentally observed nuclear resonance fluorescence does exhibit such behaviour: it shows peaks corresponding to TE_0 and TE_2 modes only. Figure 14 shows the nuclear fluorescence from the layer after annealings at 473 K and 523 K. Thermal annealing is expected to result in broadening of the depth profile of ^{57}FeZr marker layer as a result of self-diffusion of Fe. This broadening of the depth profile is reflected in the broadening of the peaks in the nuclear resonance fluorescence pattern. These results demonstrate that nuclear resonance fluorescence under X-ray standing wave condition can be used for depth profiling of a specific Mössbauer isotope.

5. Conclusions

In conclusion, it is shown that X-ray based characterization techniques can be made depth selective by generating standing waves inside the multilayer to be studied. The specific case of standing waves generated as a result of total reflection from an underlying layer of high electron density is considered. X-ray reflectivity in conjunction with conversion electron Mössbauer spectroscopy has been used to get a detailed information about the topological structure of the interfaces in multilayers. X-ray reflectivity combined with X-ray fluorescence measurements under standing wave conditions can be used for elemental depth profiling with an accuracy of the order of 0.1 nm. On the other hand, depth resolved structural information can be obtained using X-ray absorption spectroscopy under standing wave conditions. The depth resolution of the technique is determined by the separation between successive antinodes and nodes which in turn depends upon the scattering vector q through the relation $D = \lambda/2\sin\theta = 2\pi/q$. Use of X-ray waveguide structure can result in a stronger localization of the intensity, and thus can be used to improve the depth resolution of the technique. By tuning the energy of the incident X-rays to a given Mössbauer transition the technique can be made isotope sensitive and under this condition it can be used to measure self-diffusion of an element. However, such measurements are limited only to Mössbauer isotopes suitable for study using synchrotron radiation source.

References

1. Sinha K., Sirota E. B., Garoff S. and Stanley H. B., *Phys. Rev. B* **38** (1988), 2297; Holý V., Kube¡na J., Ohlýdal I., Lischka K. and Plotz W., *Phys. Rev. B* **47** (1993), 15896.
2. Kingetsu T. and Yamamoto M., *Surf. Sci. Rep.* **45** (2002), 79; Kortright J. B., Joksch S. and Ziegler E., *J. Appl. Phys.* **69** (1991), 168.
3. Shan Z. S., Sellmyer D. J., Jaswal S. S., Wang Y. J. and Shen J. X., *Phys. Rev. B* **42** (1990), 10446; Sato N., *J. Appl. Phys.* **59** (1986), 2514.
4. de Vries J. J., Kohlhepp J., den Broeder F. J. A., Coehoorn R., Jungblut R., Reinders A. and de Jonge W. J. M., *Phys. Rev. Lett.* **78** (1997), 3023; Pruneda J. M., Robles R., Bouarab S., Ferrer J. and Vega A., *Phys. Rev. B* **45** (2001), 024440.
5. Telling N. D., Faunce C. A., Bonder M. J., Grundy P. J., Lord D. G. and Van den Berg J. A., *J. Appl. Phys.* **89** (2001), 7074.
6. Chason E. and Mayer T. M., *Crit. Rev. Solid State Mater. Sci.* **22**(1) (1997); Stepanov S. A., Kondrashkina E. A., Schmidbauer M., Kohler R., Pfeiffer J. U., Jach T. and Souvorov A. Y., *Phys. Rev. B* **54** (1996), 8150; Holý V. and Baumbach T., *Phys. Rev. B* **49** (1994), 10668.
7. Paul A., Gupta A., Shah P., Kawaguchi K. and Principi G., *Hyperfine Interact.* **139/140** (2002), 205; Paul A., Ph.D. thesis, Devi Ahilya University, Indore, India, 2001, unpublished.
8. Gerdau E., Rüffer R., Winkler H., Tolksdorf W., Klages C. P. and Hannon J. P., *Phys. Rev. Lett.* **54** (1985) 835; Gerdau E., van Bürck U. and Rüffer R., *Hyperfine Interact.* **123–125** (2000), 3.
9. Bedzyk M. J. and Cheng L., In: Fenter P. A., Rivers M. L., Sturchio N. C. and Sutton S. R. (eds.), *Reviews in Mineralogy and Geochemistry*, Vol. 49, Mineralogical Society of America, Washington, D.C., 2002, pp. 221–266.
10. Bedzyk M. J., Bilderback D., Bommarito G. M., Caffrey M. and Schildkraut J. S., *Science* **241** (1988), 1788.
11. Bedzyk M. J., Bommarito G. M. and Schildkraut J. S., *Phys. Rev. Lett.* **62** (1989), 1376; Bedzyk M. J., Bommarito G. M., Caffrey T. A. and Penner M., *Science* **248** (1990), 52.
12. Richomme F., Teillet J., Fnidiki A. and Keune W., *Phys. Rev. B* **64** (2001), 094415; Findiki A., Juraszek J., Teillet J., Richomme F. and Lebertois J. P., *J. Magn. Magn. Mater.* **165** (1997), 405.

13. Amitesh P. and Gupta, A., *J. Alloys Compd.* **326** (2001), 246.
14. Gupta A., Paul A., Gupta R. and Principi G., *J. Phys., Condens. Matter* **10** (1998), 9669.
15. Parratt L. G., *Phys. Rev.* **95** (1954), 359.
16. Savage D. E., Kleiner J., Schimke N., Phang Y.-H., Jankowski T., Jacobs J., Kariotis R. and Lagally M. G., *J. Appl. Phys.* **69** (1991), 1411.
17. Salditt T., Metzger T. H., Brandt C., Klemradt U. and Peisl J., *Phys. Rev. B* **51** (1995), 5617.
18. TRDS-Stepanov.
19. Payne A. P. and Clemens B. M., *Phys. Rev. B* **47** (1993) 228.
20. Stearns D. G. and Gullikson E. M., *Physica B* **283** (2000), 84.
21. Ghose S. K., Dev B. N. and Gupta A., *Phys. Rev. B* **64** (2001), 233403; Pelka J. B., Cedola A., Lagomarsino S., di Fonzo S., Jark W. and Soullie G., *J. Alloys Compd.* **286** (1999), 313.
22. Zheludeva S., Kovalchuk M. and Novikova N., *Spectrochim. Acta B* **56** (2001), 2019; Zheludeva S. I., Novikova N. N., Konovalov O. V., Kovalchuk M. V., Stepina N. D. and Tereschenko E. Y., *Mater. Sci. Eng. C* **23** (2003), 567; Woodruff D. P., *Surf. Sci.* **482–485** (2001), 49.
23. Ayache R., Richter E. and Bouabellou A., *Nucl. Instrum. Methods, B* **216** (2004), 137; Walterfang M., Kruijer S., Keune W., Dobler M. and Reuther H., *Appl. Phys. Lett.* **76** (2000), 1413.
24. Dhuri P., Gupta A., Chaudhari S. M., Phase D. M. and Avasthi D. K., *Nucl. Instrum. Methods B* **156** (1999), 148.
25. Gupta A., Meneghini C., Saraiya A., Principi G. and Avasthi D. K., *Nucl. Instrum. Methods B* **212** (2003), 458.
26. Dunlop A., Lesueur D., Legrand P., Dammak H. and Dural J., *Nucl. Instrum. Methods B* **90** (1994), 330; Dunlop A. and Lesueur D., *Radiat. Eff. Defects Solids* **126** (1993), 123.
27. Feng Y. P., Sinha S. K., Deckmann H. W., Hastings J. B. and Siddons D. P., *Phys. Rev. Lett.* **71** (1993), 537.
28. Salditt T., Pfeiffer F., Perzl H., Vix A., Mennicke U., Jarre A., Mazuelas A. and Metzger T. H., *Physica B* **336** (2003), 181.
29. Gupta A., Meneghini C., Rajput P. and Principi G. to be published.
30. Hannon J. P., Hung N. V., Trammell G. T., Gerdau E., Mueller M., Rüffer R. and Winkler H., *Phys. Rev. B* **32** (1985), 6363.
31. Gupta A., Rajput P., Gupta M. and Rüffer R. to be published.
32. Rüffer R. and Chumakov A. I., *Hyperfine Interact.* **97/98** (1996), 589.

Hyperfine Interactions (2005) 160:143–156
DOI 10.1007/s10751-005-9145-9

Magnetic Study of Nanocrystalline Ferrites and the Effect of Swift Heavy Ion Irradiation

RAVI KUMAR[1,*], S. K. SHARMA[2], ANJANA DOGRA[1,2], V. V. SIVA KUMAR[1], S. N. DOLIA[3], A. GUPTA[4], M. KNOBEL[5] and M. SINGH[2]
[1]*Materials Science Division, Aruna Asaf Ali Marg, Nuclear Science Centre, Post Box 10502, New Delhi 110 067, India; e-mail: ranade@nsc.ernet.in*
[2]*Department of Physics, H.P. University, Shimla 171005, India*
[3]*Department of Physics, University of Rajasthan, Jaipur 302004, India*
[4]*University Campus, IUC-DAEF, Khandwa Road, Indore 452017, India*
[5]*Instituto de Fisica, UNICAMP, Campinas SP, Brazil*

Abstract. 100 MeV Si^{+7} irradiation induced modifications in the structural and magnetic properties of $Mg_{0.95}Mn_{0.05}Fe_2O_4$ nanoparticles have been studied by using X-ray diffraction, Mössbauer spectroscopy and a SQUID magnetometer. The X-ray diffraction patterns indicate the presence of single-phase cubic spinel structure of the samples. The particle size was estimated from the broadened (311) X-ray diffraction peak using the well-known Scherrer equation. The milling process reduced the average particle size to the nanometer range. After irradiation a slight increase in the particle size was observed. With the room temperature Mössbauer spectroscopy, superparamagnetic relaxation effects were observed in the pristine as well as in the irradiated samples. No appreciable changes were observed in the room temperature Mössbauer spectra after ion irradiation. Mössbauer spectroscopy performed on a 12 h milled pristine sample (6 nm) confirmed the transition to a magnetically ordered state for temperatures less than 140 K. All the samples showed well-defined magnetic ordering at 5 K, whereas, at room temperature they were in a superparamagnetic state. From the magnetization studies performed on the irradiated samples, it was concluded that the saturation magnetization was enhanced. This was explained on the basis of SHI irradiation induced modifications in surface states of the nanoparticles.

Key Words: irradiation, nanocrystalline, Mössbauer spectroscopy, superparamagnetism, magnetization.

PACS: 61.80.Jh, 61.82.Rx, 76.80.+y, 75.20.−g, 75.60.E.

1. Introduction

Swift heavy ion (SHI) irradiation is known to generate controlled defects of various types such as point/cluster and columnar defects in the materials [1] and

* Author for correspondence.

has been studied extensively for magnetic oxides and ferrites for the last two decades [2–5]. SHI irradiation provides several interesting and unique aspects in understanding damage structures and materials modifications. The swift heavy ions during their passage through the materials loose their energy by two processes; by inelastic collisions with the electrons (electronic energy loss S_e) and by elastic collisions with the nuclei (nuclear energy loss S_n). For heavy ions in the megaelectron volt range, the electronic energy loss dominates over the nuclear energy loss and is able to generate various types of defects such as point/clusters and columnar defects depending upon the magnitude of S_e. To create columnar defects, certain threshold of S_{eth} is required. If S_e is less than the S_{eth} it will create only point or clusters of defects. The SHI-induced defects are known to create structural strain and disorder in oxide materials such as in CMR, high-Tc and magnetic oxides [6–9] and are responsible to modify their physical properties.

Spinel ferrites have been a subject of research because of their remarkable magnetic and dielectric properties. Among the various variety of spinel ferrites Mg–Mn is especially interesting due to its rectangular hysteresis loop characteristic, suitable for the applications in memory and switching circuits of digital computers [10, 11]. In recent years, many efforts have also been made in the research of nanoparticles of spinel ferrites due to their wide applications to information storage, ferrofluid technology, and magnetic diagnostics [12–16]. The nanoparticles of ferrites are being fabricated by various techniques such as sol–gel [17], co-precipitation [18], hydrothermal [19], etc. High-energy ball milling (HEBM) is also extensively employed as an alternative route to obtain the nanoparticles in large amounts for the synthesis of novel materials through solid-state reaction methods. Reduction in particle size and possible chemical transformations of the system during milling are believed to be influenced by different factors such as ball to powder mass ratio, milling time, milling environment, milling speed, as well as the type of ball mill itself.

In recent years, a lot of attention has been devoted towards the SHI irradiation induced defect creation and modifications in the bulk and thin films of ferrites, but no detailed study is available on the effect of SHI irradiation of nanoparticles ferrite except that of Shinde *et al.* [20]. In the present paper, we are presenting a detailed study on the structural and magnetic properties of $Mg_{0.95}Mn_{0.05}Fe_2O_4$ ferrite nanoparticles synthesized by high-energy ball milling and irradiated with 100 MeV Si^{+7} ions.

2. Experimental

The mixed ferrite of composition $Mg_{0.95}Mn_{0.05}Fe_2O_4$ was synthesized using standard solid-state reaction technique. Stoichiometric amounts of MgO, MnO and Fe_2O_3 were mixed thoroughly and pre-heated at 1000°C for 12 h for calcination. The calcinated powder was pressed into pellets and sintered at

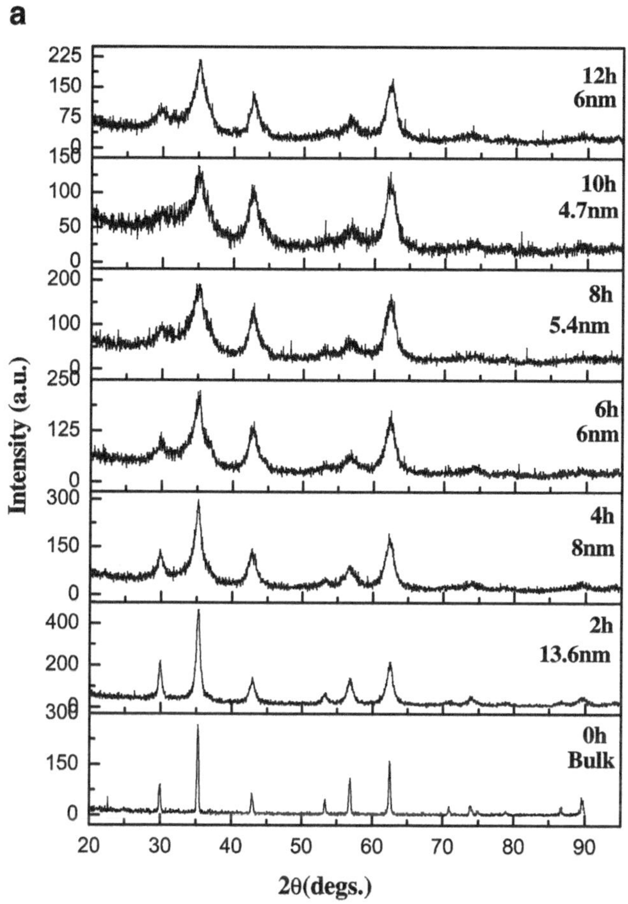

Figure 1. **a**) X-ray diffraction patterns for $Mg_{0.95}Mn_{0.05}Fe_2O_4$ bulk material and the samples milled for different duration of times. **b**) X-ray diffraction patterns for bulk material (a) and after irradiations with 100 MeV Si^{+7} ions at fluences of (b) 1×10^{12}, (c) 1×10^{13} and (d) 5×10^{13} ions/cm^2. **c**) X-ray diffraction patterns for the 12 h milled sample irradiated with 100 MeV Si^{+7} ions; (a) pristine, (b) 1×10^{13}, and (c) 5×10^{13} ions/cm^2.

1300°C for 24 h, followed by slow cooling to room temperature. To prepare nanoparticles, 5 g of the final material was used for the milling in a high-energy ball mill (SPEX 8000D). Two hardened stainless steel vials each of 72 cm^3 volume and charged with five hardened steel balls (of 12 mm diameter) and three (of 6 mm diameter) were used for milling. The ball to powder mass ratio was kept at 10:1. The powder was milled in air at room temperature without any additive under closed milling condition. The $Mg_{0.95}Mn_{0.05}Fe_2O_4$ material was milled for different duration.

The particle size and structure was studied using powder X-ray diffraction (XRD) with the Cu K_α radiation and a Siemens X-ray diffractometer (D5000). The slow scan was performed in the 2θ range from 20°–95°. The well-

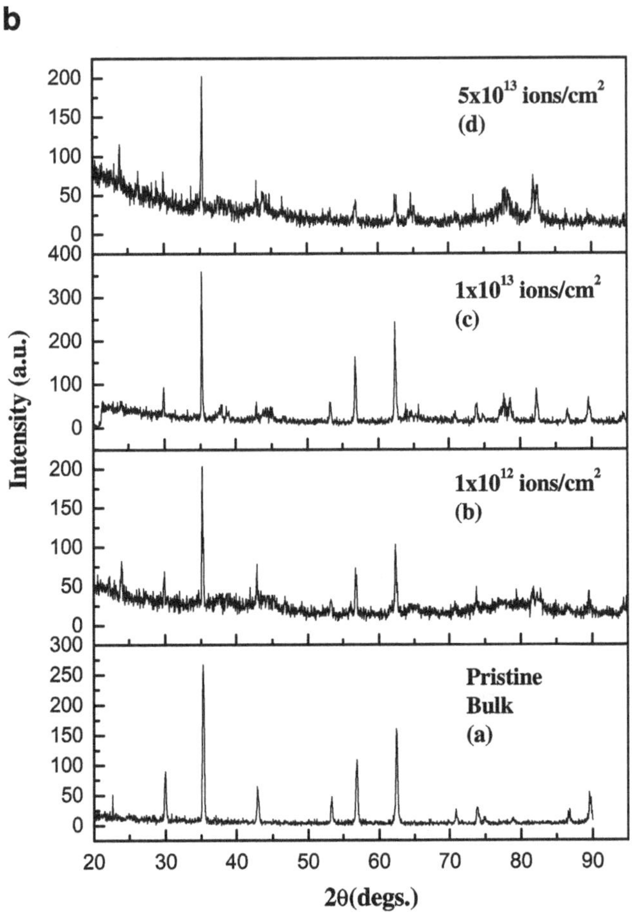

Figure 1. (Continued.)

characterized samples were irradiated with 100 MeV Si^{+7} ions at 1×10^{12}, 1×10^{13} and 5×10^{13} ions/cm^2 using the 15UD Tandem Pelletron accelerator at NSC, New Delhi. The Mössbauer spectroscopy measurements for pristine and irradiated nanoparticles were performed using a conventional constant acceleration spectrometer in transmission geometry with a source of about 25 mCi ^{57}Co in an Rh matrix. The spectra were analyzed using least square fitting program NORMOS (SITE/DIST) [21]. The isothermal dc magnetization measurements of the nanoparticles were performed using a Quantum design SQUID magnetometer.

3. Results and discussion

The X-ray diffraction patterns were recorded to ensure the single-phase nature of the nanocrystalline $Mg_{0.95}Mn_{0.05}Fe_2O_4$ ferrite material and to study the structural

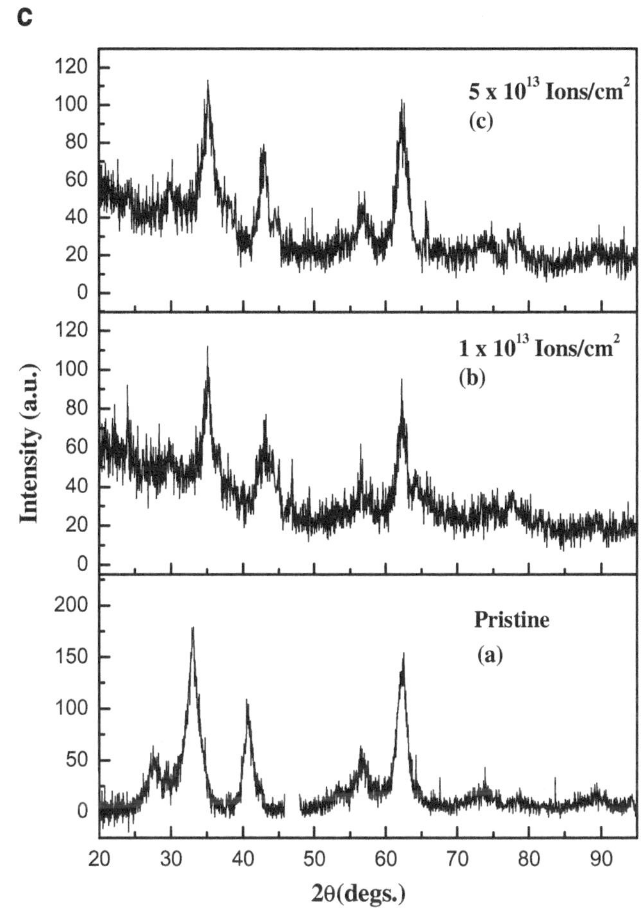

Figure 1. (Continued.)

changes induced by ion irradiation. Figure 1(a) shows the powder XRD patterns of bulk and the samples milled for different durations. Clearly the bulk as well as milled samples exhibit spinel structure. However, in the case of the milled samples the XRD peaks are getting broadened which indicates a reduction in particle size. The maximum intensity peak (311) of milled sample was fitted with a Gaussian shape to determine the exact peak position as well as the full width half the maximum (FWHM). The average particle size was calculated using Scherrer's equation [22],

$$D = k\lambda/B \cos\theta \tag{1}$$

where $B^2 = B_M^2 - B_S^2$, B_M and B_S are the measured peak broadening and instrumental broadening in radians, D is the particle size in Å, k is the shape factor (usually taken as 0.9), λ is the X-ray wavelength (1.54 Å), and θ is Bragg's angle in degrees. Using Equation (1), the calculated average particle

a

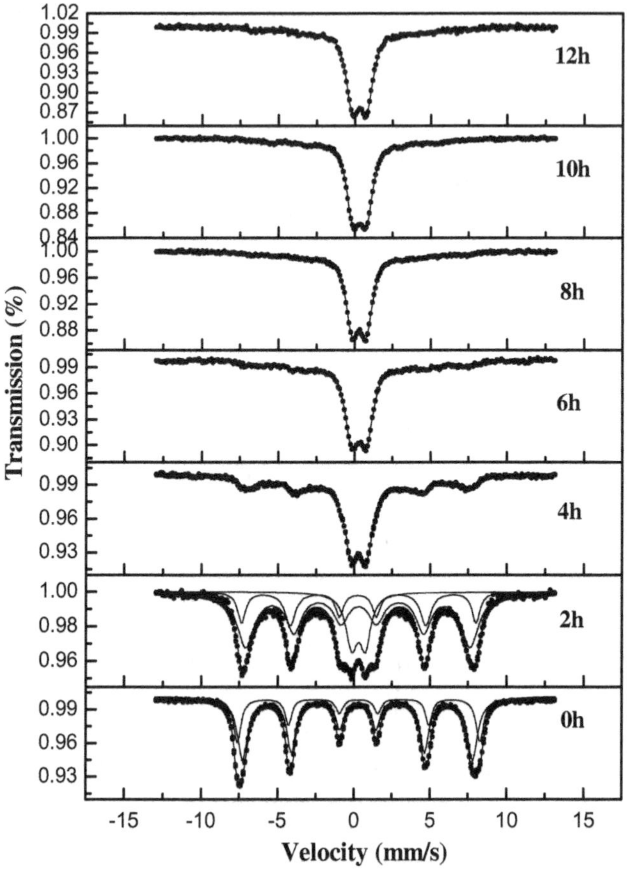

Figure 2. **a**) Mössbauer spectra of bulk $Mg_{0.95}Mn_{0.05}Fe_2O_4$ ferrite and nanoparticles milled for various times at room temperature. **b**) Mössbauer spectra of the 12 h milled sample at temperatures from 20 to 220 K. **c**) Mössbauer spectra of bulk ferrite irradiated with100 MeV Si^{+7} ions: (a) bulk pristine, (b) 1×10^{12}, (c) 1×10^{13} and (d) 5×10^{13} ions/cm^2, at room temperature. **d**) Mössbauer spectra of 12 h milled ferrite nanoparticles irradiated with 100 MeV Si^{+7} ions: (a) pristine, (b) 1×10^{12}, (c) 1×10^{13} and (d 5×10^{13} ions/cm2, at room temperature.

size were found to decrease with the milling times up to 10 h and further increase in the time increase the size. The well-characterized nanoparticles of various sizes were irradiated with 100 MeV Si^{+7} ions at various fluences. The electronic energy loss, nuclear energy loss and projectile range for the present system are 3.2 keV/nm, 2.6 eV/nm and 27 μm, respectively, calculated using the SRIM code. Clearly S_e is about three orders of magnitude larger than S_n, therefore the dominant process is the electronic energy loss process in the present system. However, the threshold to create columnar defects i.e., $S_{eth}\sim13$ keV/nm, indicates that there is no possibility to create columnar defects, but only point defects or defect clusters. We would like to mention that these calculations are

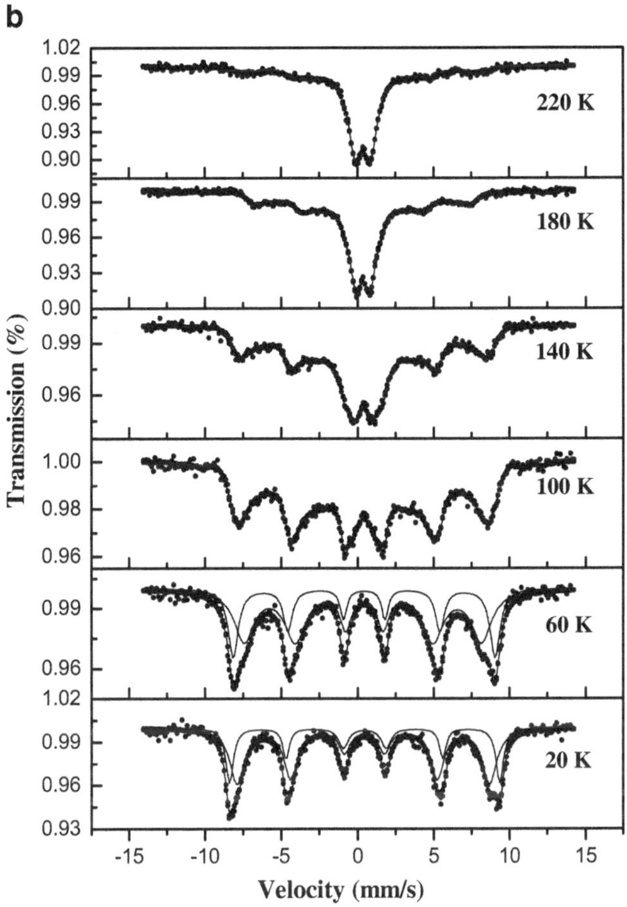

Figure 2. (Continued.)

for bulk and thin film samples, not for nanoparticles. In the case of nanoparticles, the surface to volume ratio is high; the surface energy plays a major role in their physical properties and may be modified by the SHI irradiation. We have investigated these effects by measuring the powder X-ray diffraction after irradiation. Figure 1(b,c) show the XRD patterns for the bulk and for 12 h milled (6 nm) nanoparticles irradiated at various fluences. It is observed that the basic crystal structure remains the same after irradiation. However, the peak intensities decreased and the widths of the peaks increased slightly. From these observations, it is clear that the SHI irradiation has generated some defect states in the system. Also the peak positions are shifted to lower 2θ values, indicating a lattice expansion. One can say that with irradiation, the surface strains were released. Similar types of observations were also reported in thin films of oxide materials [23–25]. From the XRD pattern, the calculated average particle size for the 12 h irradiated nanoparticles increased from 6 to 6.8 nm for 1×10^{13} ions/cm^2 and to 7.4 nm for 5×10^{13} ions/cm^2, respectively. However, the SHI irradiation of

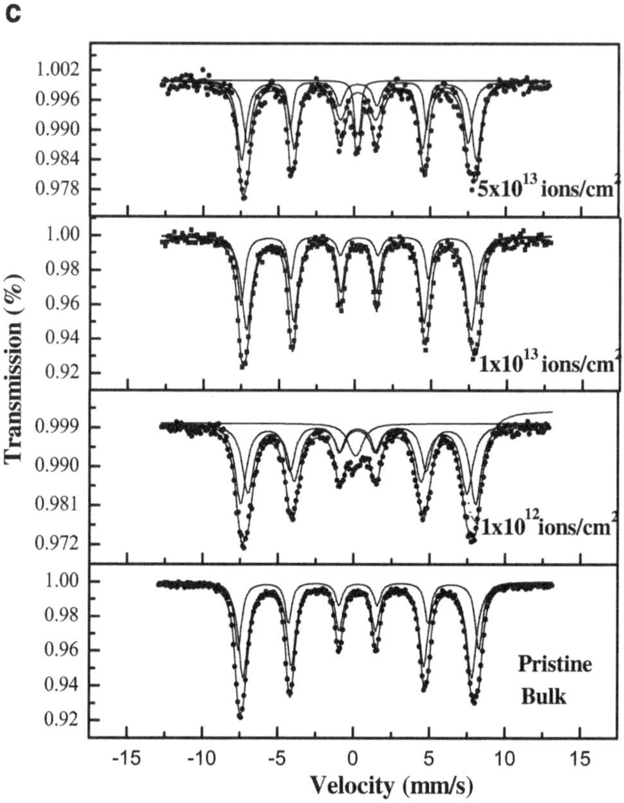

Figure 2. (Continued.)

the bulk sample amorphized it, as is clearly indicated by the decrease in line intensity and the increase in background.

Figure 2(a) shows the Mössbauer spectra of bulk and nanoparticles of different size at room temperature. The dots represent the data and the solid line through the data the least square fit. In the bulk sample, two sextets were observed, which represent two sites of Fe^{+3}, i.e., tetrahedral (A-site) and octahedral (B-site). Evidently, the material is magnetically ordered. With the milling, the sextet in the spectrum disappeared and was replaced with a central doublet, maybe due to superparamagnetism in nanosized ferrite. The spectrum of 6 nm-sized particles showed a broad doublet, which indicates that the long range magnetic ordering has vanished and the particles behave like superparamagnetic single domains. This superparamagnetic relaxation can also be seen when reducing the sample temperature. Figure 2(b) shows the Mössbauer spectra of the 6 nm-sized nanoparticles at 20–220 K. The estimated blocking temperature for the sample is found to be between 190 and 220 K. Figure 2(c,d) show the room temperature Mössbauer spectra for the bulk sample and 6 nm-sized nanoparticles at three different fluences 1×10^{12}, 1×10^{13}, and 5×10^{13} ions/cm^2, respectively.

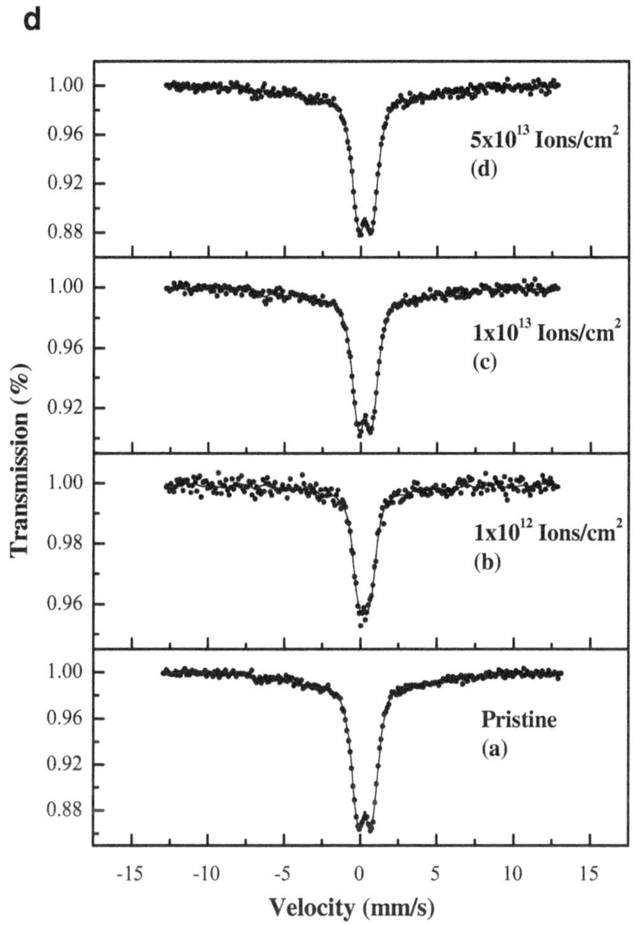

Figure 2. (Continued.)

After the ion irradiations, a small paramagnetic doublet appeared in the bulk sample, which increased with the ion fluence. These observations are in good agreement with the XRD results which indicate that after SHI irradiation the bulk samples are getting amorphized. On the other hand, no qualitative changes were observed in the room temperature Mössbauer spectra in case of nanoparticles (see doublet in Figure 2(d)). But the width of the doublet is decreasing with the ion fluence, which indicates that the surface strains or domain pinning may be reduced after irradiation (also observed in the XRD data). The Mössbauer studies at room temperature showed that after irradiation the nanoparticles remained in the single-domain superparamagnetic state.

Further, we studied the magnetic properties using dc magnetization measurements. Figure 3 shows a plot of magnetization vs temperature (M–T) recorded in zero-field cooling (ZFC) and field cooling (FC) modes in an external magnetic field of 20 Oe for the samples milled for 4, 8 and 12 h. The ZFC magnetization

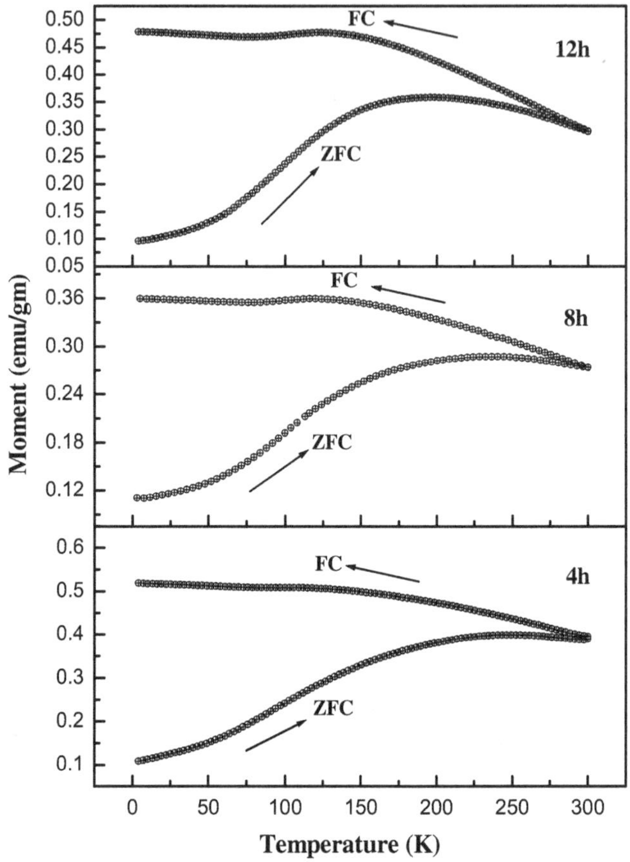

Figure 3. ZFC and FC curves for (a) 4, (b) 8, and 12 h milled samples in a 20 Oe external field *versus* temperature.

was recorded by first cooling the sample from 300 to 20 K in zero magnetic field, then applying the magnetic field and warming the sample up to 300 K in the presence of the field and recording the moment during this warming cycle. Field-cooled patterns were obtained by first cooling the sample from 300 K down to 20 K in the external field and then warming it up to 300 K and recording the moment. The ZFC curve exhibits a peak at ~195 K (for 12 h milled) which is the mean superparamagnetic blocking temperature. The departures of the FC curves from the ZFC magnetization curves suggest a temporal relaxation. However, the peak in ZFC curves is rather broad which indicates a broad particle distribution in our samples. The blocking temperature estimated for the 12 h milled sample from the Mössbauer spectra is not consistent with that from the temperature-dependence of magnetization since the superparamagnetic relaxation is a dynamic process and the measuring times of both methods are very different. A similar observation was made for some magnetic nanoparticles, where the

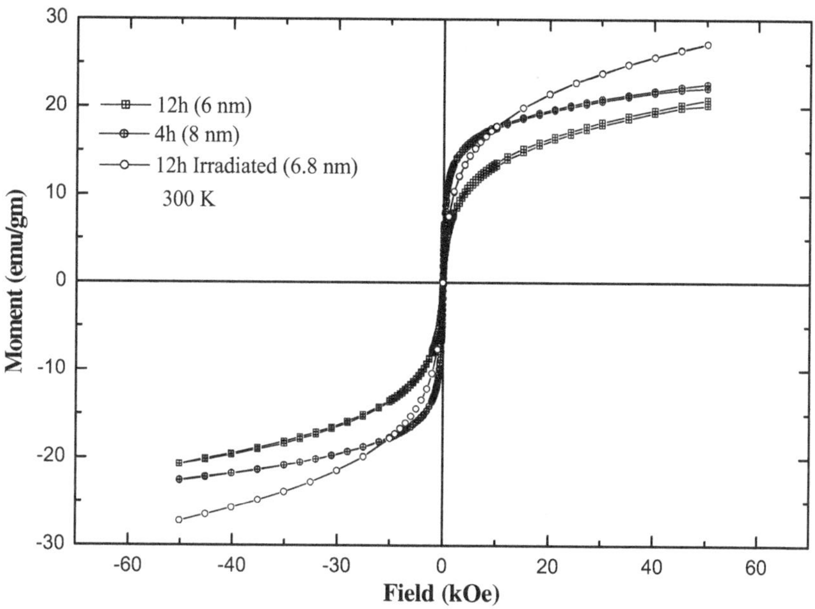

Figure 4. $M(H)$ hysteresis curves for the samples milled for 4 and 12 h and irradiated with 100 MeV Si^{+7} ions measured at 300 K.

estimated blocking temperature deduced from temperature-dependent magnetization was lower than that from Mössbauer data [26, 27].

However, we observed that the blocking temperature in these samples measured by both techniques are very close. This clearly proves that the 12 h milled sample shows single-domain superparamagnetic behavior with the blocking temperature of around 190 K.

Finally, we have studied the isothermal magnetization hysteresis of nanoparticles of different sizes. Figure 4 shows the isothermal dc magnetization hysteresis of 4 h, 12 h milled pristine nanoparticles and 100 MeV Si^{+7} ion irradiated at 1×10^{13} ions/cm^2 at 300 K. From Figure 5, no magnetic hysteresis is observed and the curves passed through the origin (H = 0). Also there is no saturation of magnetization up to the available magnetic field (H = 5.5 T). The samples show zero coercivity and retentivity at this temperature, which are the characteristics of single-domain superparamagnetic particles. These observations confirm that both pristine and irradiated nanoparticles show superparamagnetic behavior. The saturation magnetization value obtained by plotting the M vs. $1/H$ and extrapolating the data, we have observed that the saturation magnetization of the irradiated sample of 6nm size (12 h milled) is ~27 emu/gm, which is more than two times than that of pristine nanoparticles sample. However at 5 K, all the nanoparticles samples are magnetically ordered and show the magnetization hysteresis (see Figure 5). Here again the parameters for irradiated nanoparticles are modified. The coercivity, H_c, of the irradiated sample is found to be 1253 Oe, which is greater than that of the pristine nanoparticles having a value of 827 Oe.

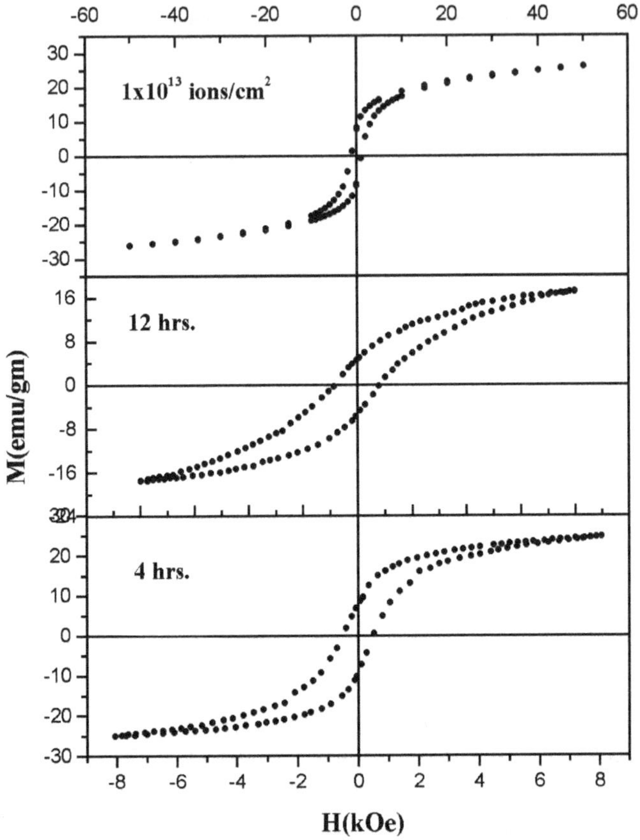

Figure 5. Same as Figure 4, but measured at 5 K.

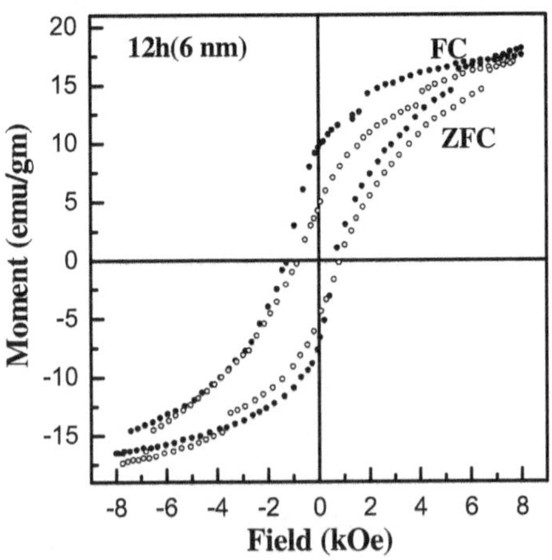

Figure 6. *M–H* loops measured in ZFC–FC mode for the 12 h milled pristine nano-particles.

As particle size increased, the coercivity increased, which is the nature of single-domain particles. This increase in the coercivity may be due to the controlled growth of single-domain particles in which the absence of domain walls makes the magnetization reversal difficult [29]. It is interesting to note here that the surface state pinning of domains is dominating as the size of the particles is decreasing. In order to test that we have also measured magnetization hysteresis (*M–H* curves) in two different modes i.e., ZFC and FC modes for the 12 h milled pristine sample at 20 K (Figure 6). It is clearly evident from Figure 6 that the *M–H* loops are shifted, which suggest that magnetic moments in the surface layer, outside a core of ordered moments, are in a state of frozen disorder [28]. On irradiation the surface state pinning of domains are release and the magnetization of the nanoparticles is increasing.

4. Summary

To summarize we have synthesized $Mg_{0.95}Mn_{0.05}Fe_2O_4$ ferrite nanoparticles of different size using a solid-state reaction technique followed by high-energy ball milling. We have studied the effects of 100 MeV Si ion rradiation on their structural and magnetic properties. X-ray diffraction revealed that the basic spinel structure remained unchanged. Only a slight increase in the peak widths and decrease in the peak intensities were observed after the SHI irradiation. The Mössbauer study indicated that after irradiation the nanoparticles remained in the single-domain superparamagnetic state. This was confirmed by the magnetization studies in which the blocking temperature was found to vary as a function of particle size. The coercivity of the irradiated nanoparticles increased at low temperature, which is consistent with the XRD results increase of particle size with irradiation. The saturation magnetization after irradiation was observed to increase by factor of 2 and was explained by SHI irradiation-induced modifications of surface disorder.

Acknowledgements

The authors are thankful to the Nuclear Science Centre, New Delhi (India), in providing financial support to this work under the project UFUP scheme and in providing experimental facilities in the Material Science Beam line to perform the experiments.

References

1. Thompson M. W., *Defects and Radiation Damage in Metals*, Cambridge University Press, Cambridge, 1969.
2. Studer F., Pascard H., Groult D., Houpert C., Nguyen N. and Toulemonde M., *Nucl. Instrum. Methods, B* **32** (1989), 389.

3. Meftah A., Merrien N., Nguyen N., Studer F., Pascard H. and Toulemonde M., *Nucl. Instrum. Methods, B* **59–60** (1991), 605.
4. Studer F. and Toulemonde M., *Nucl. Instrum. Methods, B* **65** (1992), 560.
5. Pascard H. and Studer F., *J. Phys.* **IV** (1997), C1-211.
6. Studer F. and Toulmonde M., *Nucl. Instrum. Methods, B* **65** (1992), 560.
7. Houpert C., Studer F., Groult D. and Toulmonde M., *Nucl. Instrum. Methods, B* **39** (1989), 720–723.
8. Kumar R., Samantra S. B., Arora S. K., Gupta A., Kanjilal D., Pinto R. and Narlikar A. V., *Solid State Commun.* **106**(12) (1998), 805–810.
9. Kumar R., Arora S. K., Kanjilal D., Mehta G. K., Bache R., Date S. K., Shinde S. R., Saraf L. V., Ogale S. B. and Patil S. I., *Radiat. Eff. Defects Solids* **147** (1999), 187.
10. Papian W. N., *Proceeding of Metal Powder Association*, London, 2, 1995, p. 183.
11. Heck C., *Magnetic Materials and their Applications*, Butterworth, London, 1974.
12. Dorman J. L. and Fiorani D. (eds.), *Magnetic Properties of Fine Particles*, North-Holland, Amsterdam, 1992.
13. Lu L., Sui M. L. and Lu K., *Science* **287** (2000), 1463.
14. Raj K., Moskowitz R. and Casciari R., *J. Magn. Magn. Mater.* **149** (1995), 174.
15. Ozaki M., *Mater. Res. Bull.* **XIV** (1989), 35.
16. Gleiter H., *Nanostruct. Mater.* **1** (1992), 1.
17. Verma A., Goel T. C. and Mendiratta R. G., *Mater. Sci. Technol.* **16** (2000), 712.
18. Kim Y. I., Kim D. and Lee C. S., *Physica B* **337** (2003), 42–51.
19. Rath C., Mishra N. C., Anand S., Das R. P., Sahu K. K., Upadhyay C. and Verma H. C., *Appl. Phys. Lett.* **76**(4) (2000), 475.
20. Shinde S. R., Bhagwat A., Patil S. I., Ogale S. B., Mehta G. K., Date S. K. and Marest G., *J. Magn. Mater.* **186** (1998), 342.
21. Brand R. A., Lauer J. and Herlach D. M., *J. Phys.,* **F14** (1984), 55.
22. Cullity B. D., *Elements of X-ray Diffraction*, Addison-Wesely, Reading, Massachusetts, 1978.
23. Deepthy A., Rao K. S. R. K., Bhat H. L., Kumar R. and Ashoken K., *J. Appl. Phys.* **89** (2001), 6560.
24. Basavraj A., Jali V. M., Lagare M. T., Kini N. S., Umarji M., Kumar R., Arora S. K. and Kanjilal D., *Nucl. Instrum. Methods, B* **187** (2002), 87.
25. Bathe R., Date S. K., Shinde S. R., Saraf L. V., Ogale S. B., Patil S. I., Kumar R., Arora S. K. and Mehta G. K., *J. Appl. Phys.* **83** (1998), 7174.
26. Rondinone A. J., Samia A. C. S. and Zhang Z. J., *J. Phys. Chem., B* **105** (2001), 7967.
27. Rondinone A. J., Samia A. C. S. and Zhang Z. J., *Appl. Phys. Lett.* **76** (2000), 3624.
28. Kodama R. H. and Berkowitz A. E., *Phys. Rev.,* **B59** (1999), 6321.
29. Smit and Wijn H. P. J., *Ferrites*, Gloeilampenfabriken, Eindhoven, Holland, 1959.

Hyperfine Interactions (2005) 160:157–163
DOI 10.1007/s10751-005-9154-8

Preparation of Fe/Pt Films with Perpendicular Magnetic Anisotropy

S. KAVITA[1], V. RAGHAVENDRA REDDY[1,*], AJAY GUPTA[1] and MUKUL GUPTA[2]
[1]*Inter-University Consortium for DAEF, University Campus, Khandwa Road, Indore-452 017, India; e-mail: vrreddy@csr.ernet.in*
[2]*Laboratory for Neutron Scattering, 5232 Villigen, PSI Switzerland*

Abstract. We have investigated the microstructures and magnetic properties of $L1_0$ ordered equiatomic FePt thin films prepared by ion beam sputtering and subsequent annealing. It is observed from X-ray reflectivity and X-ray diffraction measurements that the mixing at Fe/Pt interfaces starts to occur with annealing and leading to the FePt alloy phase formation. The rapid increase in the coercivity values above 275°C, obtained from vibrating sample magnetometer (VSM) measurements, confirms the formation of the ordered $L1_0$ FCT FePt phase.

1. Introduction

Magnetic layers with a high perpendicular magnetic anisotropy appears as an attractive solution to increase the information density in magneto optic recording media. Currently longitudinal magnetic recording systems reach densities of 10 Gbits/in.^{2}, while computer simulations based on the PMR mode anticipate densities of 300 Gbits/in^2 for the 21st century. Such high density recording requires a bit dimension of only a few nanometers in diameter, and thus precise control of media microstructure, especially grain size, grain size dispersion and grain isolation to break exchange coupling is necessary. In order to overcome the thermal instability of magnetic recording expected from such nanoscale ferromagnetic particles, it is required that the value of K_u (the magnetocrystalline anisotropy constant) should be high. Thus the tetragonal intermetallic alloys with the $L1_0$ ordered structure such as FePt and CoPt, which have 20–40 times higher K_u values than the present hexagonal Co-alloy based media are regarded as the best candidates for the next generation high density recording media [1].

The transformation of the disordered face centered cubic (fcc) phase to the ordered face centered tetragonal (FCT) $L1_0$ phase has been attributed to be the reason for the high magnetic anisotropy. The desired FCT phase is achieved by post annealing of the as-prepared samples. In granular films in which the fer-

* Author for correspondence.

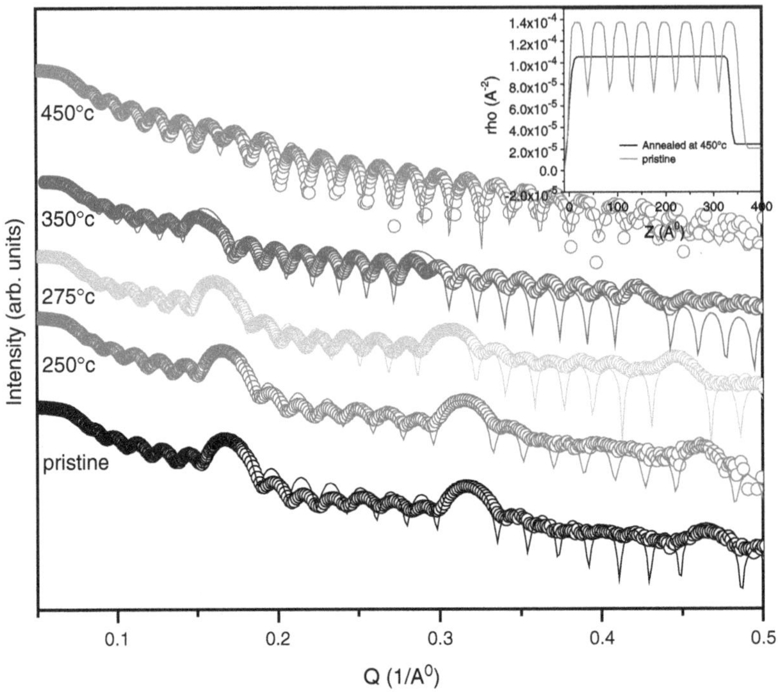

Figure 1. XRR patterns of Si(sub)/[Fe (19 Å)/Pt (25 Å)]$_{\times 8}$ multilayer. The *solid line* represents the best fit to the data. The *inset* shows the EDP of the pristine and 450°C-annealed sample.

romagnetic particles are embedded in non-magnetic medium the phase transition temperature is reported to be above 650°C [2]. In the case of FePt sputtered films, the ordering temperature is above 500°C [3], which makes the film unsuitable for recording media. Many attempts were made to obtain the $L1_0$ FePt phase at lower temperatures, e.g., by addition of ternary element [4], deposition on heated substrate, monolayer deposition using molecular beam epitaxy (MBE) [5].

The focus of this paper is on the efforts of lowering the ordering temperature of $L1_0$ phase with vacuum annealing of the multilayer films. The thickness of bilayers is in the proportion of equi-atomic stoichiometry.

2. Experimental

[Fe (19 Å)/Pt (25 Å)] multilayer with eight bilayers were deposited on silicon substrate using ion beam sputtering (IBS). A beam of argon ions was used to sputter Fe and Pt targets using Kaufman type hot-cathode ion source. A base vacuum of 1×10^{-7} Torr was achieved before deposition. The targets were mounted on rotary motion feed through to switch over from Fe to Pt in order to deposit alternate layers. The as-deposited samples were annealed at different temperatures for 1 h in a vacuum of 1×10^{-6} Torr. X-ray reflectivity (XRR) and

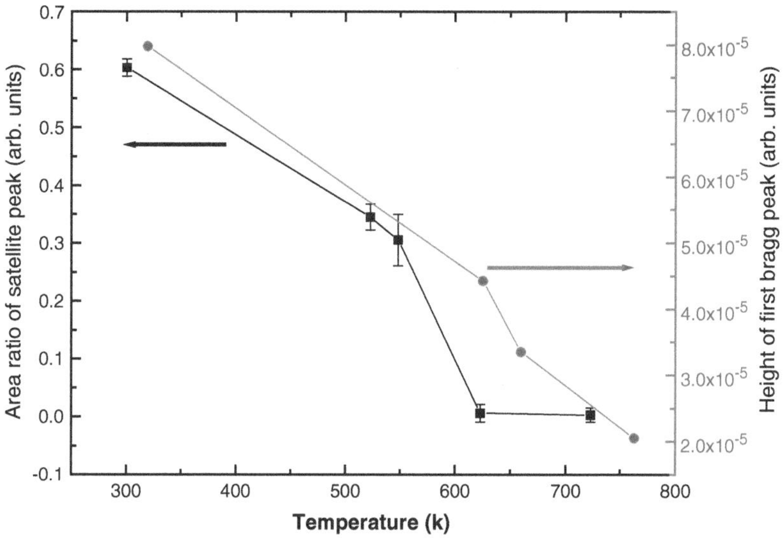

Figure 2. Variation of the first Bragg peak height and the area ratio of the satellite peak from XRD of the as-deposited multilayer with annealing.

X-ray diffraction (XRD) measurements were carried out using Siemens D5000 diffractometer with CuKα radiation. The magnetic measurements were carried out using VSM.

3. Results and discussion

Figure 1 shows the XRR patterns of the as-deposited multilayer and at different annealing temperatures. The as-deposited sample shows Bragg peaks up to third order indicating clear stratification. The reflectivity pattern is fitted with Parratt formalism [6]. The obtained bilayer periodicity is 43 ± 1 nm. The solid line represents the best fit to the data in Figure 1. The reflectivity pattern of pristine sample shows that the mixing at the interfaces is negligible. It is clearly observed that the second and third order reflections disappear with annealing. At 450°C all the Bragg peaks disappeared indicating the mixing in the multilayer structure and formation of FePt alloy because of negative heat of mixing. The electron density profile (EDP) which is shown in the inset of Figure 1 of the as-deposited and annealed at 450°C clearly indicates the formation of FePt alloy. Further, the height of the first order Bragg peak, after removing the q^{-4} dependence, is estimated [6]. The height of the first-order Bragg peak systematically decreases with the annealing as shown in Figure 2. This reduction in the height of the first Bragg peak is due to the diffusion of atoms across the interfaces.

Figure 3 shows the XRD of the as-deposited multilayers with the annealing. The XRD shows that the layers are polycrystalline with strong (111) texture. The coherence length calculated from the width of the Pt(111) peak using Scherrer

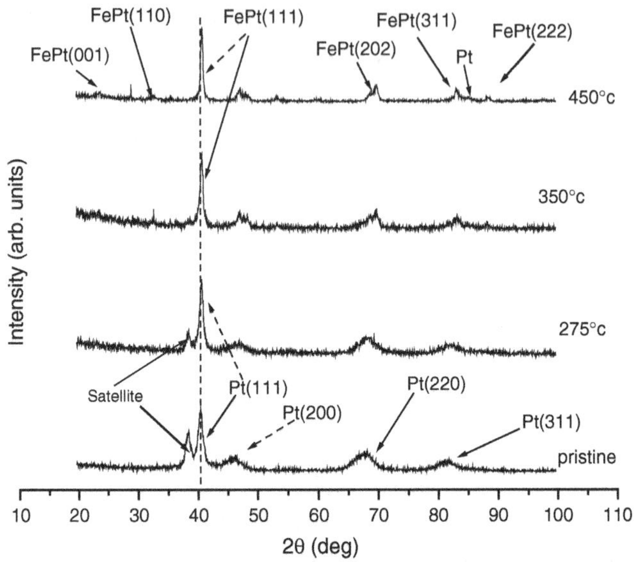

Figure 3. XRD patterns of Si(sub)/[Fe (19 Å)/Pt (25 Å)]$_{\times 8}$ at different annealing temperatures.

formula is 8.94 nm, which is greater than the bilayer periodicity. Fullerton *et al.* [7] have also observed similar longer coherence lengths in IBS-sputtered antiferromagnetically coupled Fe/Si films and inferred that the spacer Si layer must be crystalline. Similarly, in the present study also the intermediate Fe layer must be responsible for increasing the coherence length more than that of bilayer periodicity. The position of the first-order satellite peak of the multilayer calculated using the formula $\sin^2\theta = \left(\frac{n\lambda}{2\Lambda}\right)^2 + 2\delta_s$, where λ is the wavelengths, n is the order of reflection, δ_s is the real part of average refractive index of the super-lattice [8], around the Bragg peak of Pt(111) matches with the observed peak position at 38.24°. It may be noted that the coherence length calculated from the other peak position of Pt matches well with the thickness of the Pt layer. As the annealing temperature increases, the satellite peak is observed to disappear. Peaks corresponding to FePt(001), FePt(110), FePt(111) and FePt(222) can be observed at 2θ values 23.44°, 32.5°, 40.6° and 88.31°, respectively. The (111) reflection of both the disordered and the ordered phases appear within the same angular region and overlap one another. One interesting feature of the XRD pattern is the gradual changes that occurred in the position, width and asymmetry of this composite (111) reflection with the annealing. It is observed that with the annealing the peak corresponding to (111) reflection shifts to higher angle and the width reduces, as shown in Figure 4, which is due to the diminishing intensity of fcc and growing of FCT phases. [9]. The area ratio of the satellite peak calculated from the XRD is as shown in Figure 2. The systematic reduction of this ratio indicates the distortion of the layered structure due to the diffusion across the interfaces.

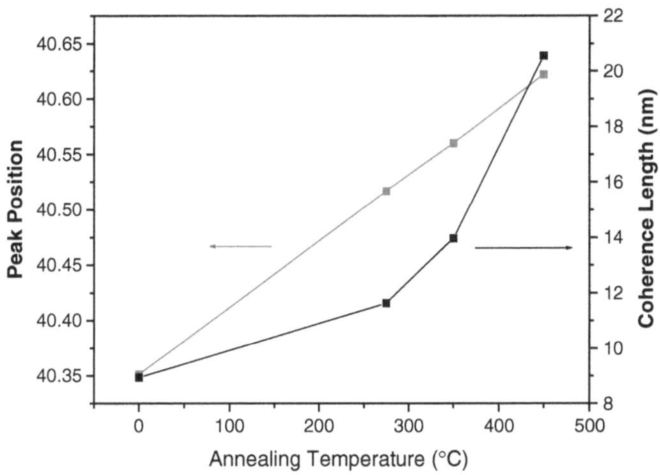

Figure 4. Variation of Pt(111) and FCT FePt(111) peak position and coherence length with annealing.

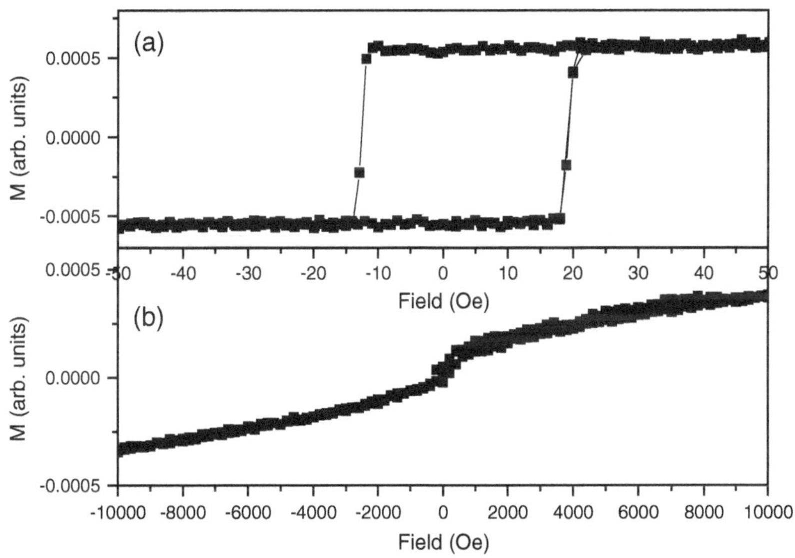

Figure 5. Magnetization measurements of the as-deposited films. (a) When the field is applied parallel to the film plane and (b) represents when the field is applied perpendicular to the film plane.

Figure 5 shows the magnetization measurements of the pristine sample. Figure 5(a) represents the magnetization when the field is applied parallel to the film plane and (b) represents when the field is applied perpendicular to the film plane. Both the loops clearly indicate that the magnetization of the as-deposited film lies in-plane. Figure 6 represents the magnetization measurements of the 450°C-annealed sample, which indicates that the magnetization has rotated from the in-plane to the out-of-plane. The variation of the coercivity (H_c) with annealing temperature is shown in Figure 7. The sharp increase in the H_c after

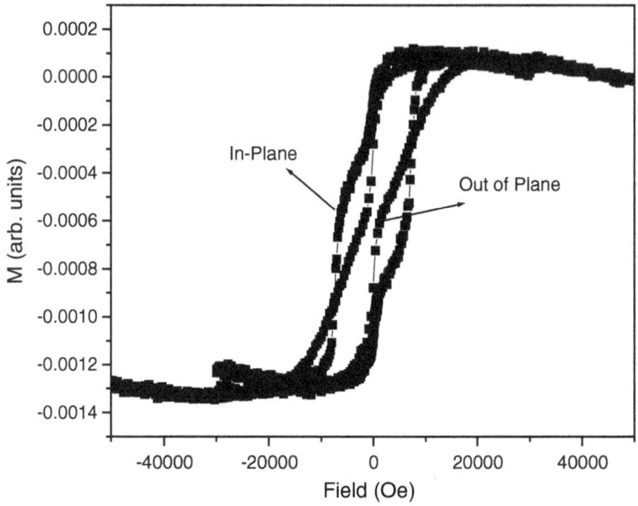

Figure 6. Magnetization measurements of the film annealed at 450°C.

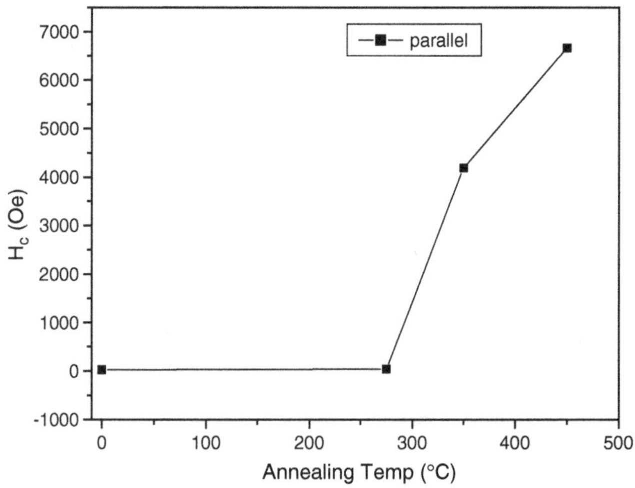

Figure 7. Variation of H_c obtained from magnetization measurements with the field applied parallel to the film plane.

275°C indicates that the FCT FePt formation is taking place at considerably lower temperature as compared to that of FePt alloy films. From the magnetization measurements of the film annealed at 450°C, it appears that the film still consists of two phases, i.e., fcc and FCT FePt.

4. Conclusions

The magnetic and structural characterisation of equi-atomic Fe/Pt multilayer as a function of annealing temperature is carried out. The intermixing of the as-deposited film increases with annealing. The obtained H_c values also increase rapidly with annealing above 275°C suggesting that the ordered FCT phase formation is taking place with annealing at considerably lower temperatures as that of the sputtered FePt films.

References

1. Ovanov O. A., Solina L. V. and Demshina V. A., *Phys. Met. Metallogr.* **35** (1973), 81.
2. Luo C. P., Liou S. H. and Sellmyer D. J., *J. Appl. Phys.* **87** (2000), 6941.
3. Visokay M. R. and Sinclair R., *App. Phys. Lett.* **66** (1999), 1692.
4. Maeda T., Kai T., Kikitsu A., Nagase T. and Akiyama J., *Appl. Phys. Lett.* **80** (2002), 2147.
5. Shima T., Moriguchi T., Mitani S. and Takanashi T., *Appl. Phys. Lett.* **80** (2002), 288.
6. Parratt L. G., *Phys. Rev., A* **41** (1954), 359.
7. Fullerton E. E. *et al.*, *J. Appl. Phys.* **73** (1993), 6335.
8. Fullerton E. E. *et al.*, *Phys. Rev., B* **45** (1992), 9292.
9. Spada F. E., Parker F. T., Platt C. L. and Howard J. K., *J. Appl. Phys.* **94** (2003), 5123.

Hyperfine Interactions (2005) 160:165–172
DOI 10.1007/s10751-005-9160-x

Effects of Interface Roughness on Interlayer Coupling in Fe/Cr/Fe Structure

DILEEP KUMAR and AJAY GUPTA*
Inter-University Consortium for DAE Facilities, University Campus, Khandwa Road, Indore, 452 017, India; e-mail: agupta@csr.ernet.in

Abstract. Effect of interface roughness on antiferromagnetic coupling between Fe layers in a Fe/Cr/Fe trilayer, with Cr layer having a wedge form has been studied. All the samples have been deposited simultaneously on substrates having different roughness, thus it is being considered that there is no variation in the morphological features like grain size and grain texture of the films. Measurements have been done as a function of Cr spacer layer thickness and the peak value of antiferromagnetic coupling strength is compared among different trilayers, thus any influence of spacer layer thickness fluctuation from sample to sample has also been avoided. The samples are characterized by X-ray reflectivity (XRR) and magneto-optic Kerr effect (MOKE). XRR results show that the roughness of the substrate is not replicated at the successive interfaces. Antiferromagnetic coupling between Fe layers decreases with the increase of roughness of Fe/Cr/Fe interfaces.

Key Words: interface roughness, interlayer coupling, magnetic films, MOKE.

1. Introduction

Since the discovery of the antiferromagnetic interlayer exchange coupling in the Fe/Cr/Fe layered system by Grünberg *et al.* [1], this effect has been widely investigated both theoretically and experimentally. The coupling was shown to oscillate between antiferromagnetic and ferromagnetic one, with the increase in Cr spacer layer thickness [2–6]. The short and long oscillation period of interlayer coupling in epitaxial Fe/Cr/Fe sandwiches was found to be about two monolayers and 10–12 monolayers of Cr, respectively [3], but in polycrystalline Fe/Cr/Fe multilayers, only long oscillation period about 12 Å has been observed [2, 4–6].

A number of experiments aiming to correlate the interlayer coupling with interface roughness in magnetic multilayers have been performed. In these experiments the interface roughness was varied through changing the sputtering gas pressure/sputtering power [7], varying the substrate temperature during deposition [8, 9] or post-deposition annealing at different temperatures [5]. How-

* Author for correspondence.

ever such variation in deposition conditions or post-deposition treatments may also induce changes in the morphological features of films like grain size and grain texture. Therefore, in such experiments, it is not possible to separate the effect of interface roughness from those of changes in morphological features. In order to avoid the above mentioned ambiguity, in the present work variation in interface roughness is achieved by simultaneous deposition of films on a number of substrates having different roughness. Identical deposition conditions ensure similar morphological features. Furthermore, it is known that the interlayer coupling sensitively depends on the thickness of spacer layer and even a variation of 1 to 2 Å in the thickness of spacer layer can significantly affect the interlayer coupling. Therefore, it is important to compare the interlayer coupling in films with different interface roughness but identical spacer layer thickness. In order to avoid any effect of fluctuation of spacer layer thickness due to a possible variation of source to target distance, films have been prepared with the spacer layer being in the form of a wedge. Systematic measurements have been done as a function of spacer layer thickness, and comparison among different films has been made at the peak of the antiferromagnetic coupling.

2. Experimental

Films of following structure: substrate/Cr (140 Å)/Fe (30 Å)/Cr_{wedge}/Fe (30 Å)/Au (20 Å) have been prepared with thickness of the Cr wedge ranging from 6 Å to 20 Å. The films were deposited on a set of float glass substrates with different surface roughness of 5 Å, 19 Å and 28 Å (designated as F5, F19 and F28). Simultaneous deposition on all the three substrates was done in a UHV chamber at a base pressure of 5×10^{-9} mbar using electron beam evaporation at the rate of 0.1 Å/s. Cr wedge was prepared by moving masks across the length of the specimen at a speed of about 0.5 mm/sec in order to achieve a thickness gradient of about 0.2 Å/mm. 140 Å Cr buffer layer has been deposited in order to improve the film quality [4, 10], while 20 Å Au capping layer has been deposited in order to avoid the oxidation of the top Fe layer. All the samples have been deposited simultaneously thus it is expected that there are no variations in the morphological features like grain size and grain texture of the specimens. In an earlier study it has been shown that films deposited simultaneously on substrates of different roughness have similar morphological feature like grain size and grain texture [4]. Therefore these films are expected to differ only in their surface and interface roughness. (as a part of the substrate roughness is expected to be replicated at successive interfaces). It may be noted that, in order to elucidate the effect of interface roughness on interlayer coupling, it is important to compare films having similar grain size and grain texture, as their morphological features may also affect the interlayer coupling [11, 12]. Substrates with varying surface roughness were prepared by etching the float glass (FG) substrates in dilute HF

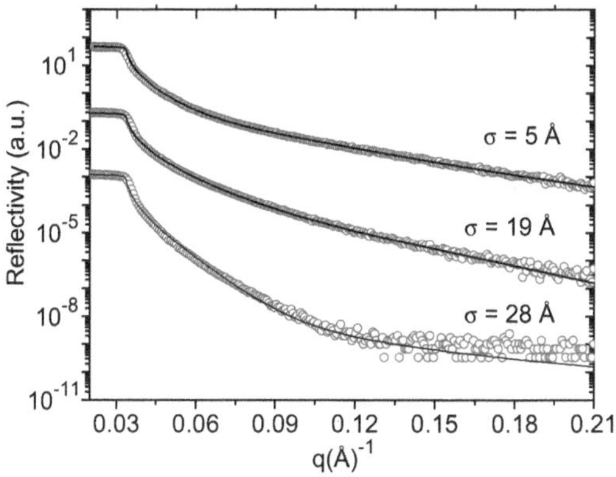

Figure 1. Fitted reflectivity data of the different roughness float glass substrates, F5, F19 and F28.

for different periods of time (0, 30 and 60 s). Cr spacer layer thickness range of 6 Å to 20 Å covers the long period peak of antiferromagnetic coupling between Fe layers. X-ray Diffractometer model D5000 of Siemens with $Cu - K_\alpha$ radiation was used to measure the specular X-ray reflectivity (XRR), which gives information about the layer thicknesses and root mean square (rms) interface roughness. Antiferromagnetic coupling between Fe layers through Cr spacer layer was characterized using the magneto-optic Kerr effect (MOKE) with magnetic field applied in the film plane. Magnetic remanence obtained from the MOKE hysteresis curve gives information about the type and the strength of magnetic coupling.

3. Results and discussion

Figure 1 shows X-ray reflectivity spectra of the bare substrates fitted using Parratt's formalism [13], which gives the roughness of the surface of the substrates as 5 Å, 19 Å and 28 Å, respectively. The XRR on two trilayers F5 and F19 are presented in Figure 2(a,b), respectively. The electron density in Fe and Cr layers are very close to each other, therefore, the contrast between these layers is very poor, thus the Kiessig fringes seen in the reflectivity spectra come from the total thickness of all the layers (Cr/Fe/Cr/Fe) modulated by the Au capping layer. Therefore fittings of XRR provide the total thickness of Cr/Fe/Cr/Fe layers. Thickness of Cr wedge layer at a given point along the length of the sample was determined by subtracting the thickness of buffer layer as well as the two iron layers, as determined through *in-situ* quartz crystal thickness monitor from the total thickness of Cr/Fe/Cr/Fe layers. The quartz crystal thickness monitor was pre-calibrated by comparing with the results of thickness measurements of single Fe and Cr layers using XRR. Fitting parameters of XRR spectra are given in Table I. It may be noted that the best-fit value of substrate roughness in the multilayer is different from the value determined by fitting the reflectivity of bare substrate. This may be due to some surface layer formed during etching,

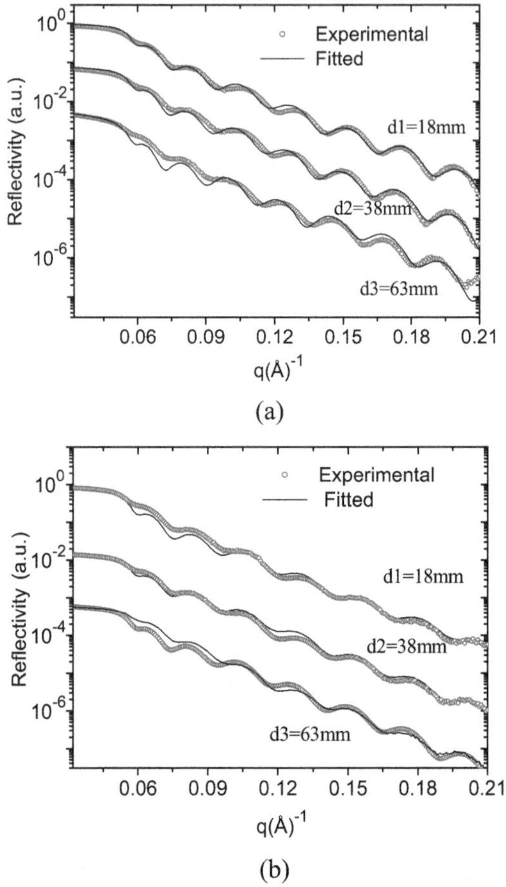

Figure 2. The X-ray reflectivity spectra of the samples (a) F5 and (b) F19, respectively at three different points, at distances *d*1, *d*2 and *d*3 from one end of the film.

Table I. The result of the fitting of reflectivity data of the samples F5, F19 and F28

	F5				F19				F28			
	d_{Cr} (Å)	σ_{in} (Å)	σ_{top} (Å)	σ_{sub} (Å)	d_{Cr} (Å)	σ_{in} (Å)	σ_{top} (Å)	σ_{sub} (Å)	d_{Cr} (Å)	σ_{in} (Å)	σ_{top} (Å)	σ_{sub} (Å)
*d*1	9.8	7.2	7.5	7.0	9.6	10.0	7.5	11.0	9.0	14.2	12.9	18.0
*d*2	13.6	7.3	7.5	7.0	14.2	10.0	7.8	10.0	13.5	14.0	13.0	18.5
*d*3	18.3	7.1	7.5	7.0	18.5	10.0	8.1	10.0	18.7	14.0	12.5	18.9

*d*1, *d*2 and *d*3 are the points on the samples at distances of 18 mm, 38 mm and 63 mm, from one edge. d_{Cr} is the thickness of the Cr spacer layer; σ_{in}, σ_{top} and σ_{sub} are the interface roughness, top surface roughness, and substrate roughness after deposition, respectively. Errors in layer thickness as well as roughnesses are ±0.5 Å.

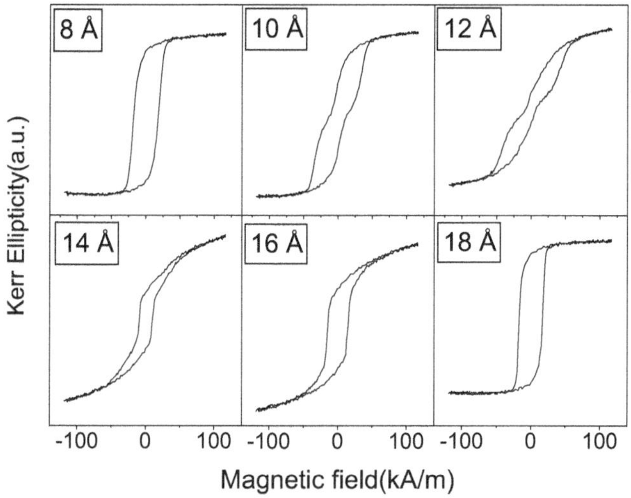

Figure 3. MOKE hysteresis loops of Fe/Cr wedge samples on float glass with surface roughness of 28 Å with different Cr thicknesses.

which gets removed in the vacuum during thin film deposition. The Cr layer thickness at a few selected points along the length of the sample obtained from Table I confirms that the thickness of the Cr spacer layer increases with a gradient of 0.2 Å/mm. This data has been used to determine the thickness of Cr layer at each position along the length of the sample, taking a linear variation of the Cr layer thickness as a function of thickness (which is the case, since mask was moved with a constant speed). It is interesting to note that while in samples F5 and F19 the interface roughness remains comparable to the substrate roughness, in sample F28 the interface roughness is lower than that of the substrate roughness. Thus in case of highly rough surface the deposition of the 140 Å thick Cr buffer layer, has some smoothening effect. The results are in accordance with some earlier studies on the effect of substrate roughness on W/C multilayer mirrors [14].

In order to see the variation in antiferromagnetic coupling with increasing Cr thickness, MOKE measurements have been done on all the samples at different points. Some of the representative MOKE loops are given in Figure 3. The information about the strength of antiferromagnetic coupling between two Fe layers is obtained from the remanence fraction (M_R) defined as,

$$M_R = M_r / M_s$$

where M_r is the remanence magnetization and M_s is the saturation magnetization. Since in the absence of antiferromagnetic coupling between Fe layers the remanence fraction should be close to unity, the quantity ($1\text{-}M_R$) is a measure of the strength of antiferromagnetic coupling. It may be noted that for spacer layer thickness around 12 Å the magnetization does not reach the saturation and therefore M_r/M_s can not be calculated from such a hysteresis curve. However,

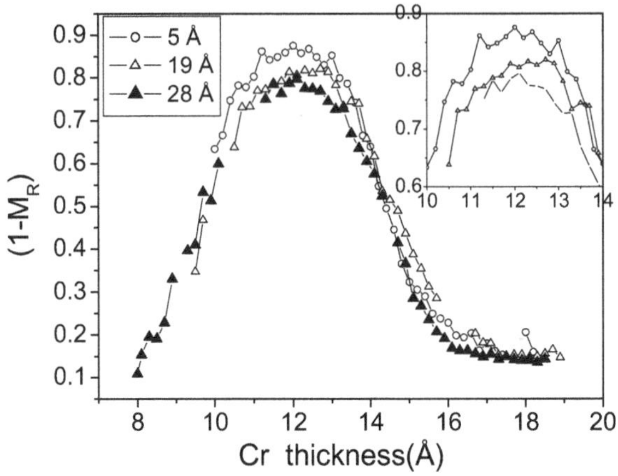

Figure 4. Variation of (1-M_R), which is the measure of the antiferromagnetic coupling as a function of thickness of Cr layer.

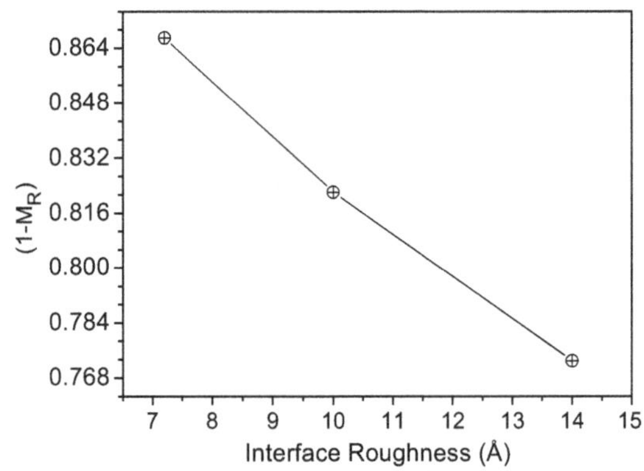

Figure 5. 1-M_R as function of interface roughness of the samples F5, F19 and F28, determined at Cr spacer layer thickness of 12 Å, which corresponds to the maximum of the antiferromagnetic coupling in all the samples.

since the thickness of Fe layers are the same at all the positions along the length of the wedge corresponding to different Cr layers, the value of saturation magnetization is expected to be the same. Therefore the value of M_s has been taken from the hysteresis curve at the position corresponding to Cr layer thickness of 8 Å at which the magnetization already saturates at the applied field. The variation of (1-M_R) with Cr thickness in all the three samples taken at different points along the length of the sample (with a separation of 1 mm) is shown in Figure 4. It is seen that the antiferromagnetic coupling is maximum at 12 Å thickness of the Cr layer.

Figure 5 gives the strength of antiferromagnetic coupling as a function of interface roughness. One may note that with increasing interface roughness the strength of antiferromagnetic coupling decreases almost linearly. Since in the present case all the films are deposited simultaneously, they are expected to differ from each other only in the correlated part of the interface roughness. Therefore the present results pertain to the effect of correlated roughness on the antiferromagnetic coupling.

4. Conclusions

In conclusion, an attempt has been made to determine unambiguously the effect of interface roughness on antiferromagnetic coupling between the Fe layers separated by Cr spacer layer. The effect of any morphological variation has been avoided by the deposition of all the films simultaneously under identical conditions. Effect of possible fluctuations in the thickness of Cr spacer layer has been eliminated by depositing Cr layer in the form of a wedge and comparing different interface roughness films at the peak of the antiferromagnetic coupling. XRR results show that in the case of high substrate roughness film, buffer layer partially screens out the substrate roughness. The maximum of the antiferromagnetic coupling occurs around 12 Å of the Cr layer thickness in all the three samples. The strength of antiferromagnetic coupling decreases almost linearly with increasing interface roughness.

Acknowledgements

One of the authors Dileep Kumar is thankful to Council for scientific and Industrial research, New Delhi, India for financial support in the form of junior research fellowship.

References

1. Grünberg P., Schreiber R., Pang Y., Brodsky M. B. and Sowers H., *Phys. Rev. Lett.* **5** (1986), 2442.
2. Parkin S. S. P., More N. and Roche K. P., *Phys. Rev. Lett.* **64** (1990), 2304.
3. Unguris J., Celotta R. J. and Pierce D. T., *Phys. Rev. Lett.* **67** (1991), 140.
4. Gupta A., Paul A., Chaudhary S. M. and Phase D. M., *J. Phys. Soc. Jpn.* **69** (2000), 2182.
5. Fullerton E. F., Kelly D. M., Guimpel J., Schuller I. K. and Bruynseraede Y., *Phys. Rev. Lett.* **68** (1992), 859.
6. Fullerton E. E., Conover M. J., Mattson J. E., Sowers C. H. and Bader S. D., *Appl. Phys. Lett.* **63** (1993), 1699.
7. Schad R., Belien P., Verbanck G., Moshchalkov V. V., Bruynseraede Y., Fischer E., Lefebvre S. and Bessiere B., *Phys. Rev. B* **59** (1999), 1242.
8. Belien P., Schad R., Potter C. D., Verbanck G., Moshchalkov V. V. and Bruynseraede Y., *Phys. Rev. B* **50** (1994) 9957.
9. Schad R., Barnas J., Belien P., Verbanck G., Potter C. D., Fisher H., Lefebvre S., Bessiere M., Moshchalkov V. V. and Bruynserade Y., *J. Magn. Magn. Mater.* **158** (1996), 39.
10. Paul A., Gupta A., Chaudhari S. M. and Phase D. M., *Vacuum* **60** (2001), 401
11. Modak A. R., Smith D. J. and Parkin S. S. P., *Phys. Rev. B* **50** (1994), 4232.

12. Fullerton E. E., Conover M. J., Mattson J. E., Sowers C. H. and Bader S. D., *Phys. Rev. B* **48** (1993), 15755.
13. Parratt L. G., *Phys. Rev.* **95** (1954), 359
14. Houdy P., Boher P., Schiller C., Luzeau P., Barchewitz R., Alehyane N. and Ouahabi M., *SPIE* **984** (1988), 95.

Hyperfine Interactions (2005) 160:173–180
DOI 10.1007/s10751-005-9161-9

Effect of Addition of Process Control Agent (PCA) on the Nanocrystalline Behavior of Elemental Silver during High Energy Milling

P. B. JOSHI[1], G. R. MARATHE[1], ARUN PRATAP[2,*] and VINOD KURUP[3]
[1]*Department of Metallurgical Engineering, M. S. University of Baroda, Vadodara 390 001, India*
[2]*Department of Applied Physics, M. S. University of Baroda, Vadodara 390 001, India;*
e-mail: apratapmsu@yahoo.com
[3]*Electrical Research and Development Association, Vadodara 390 010, India*

Abstract. Mechanical Alloying (MA) or High Energy Milling has been a subject of great interest for last few decades. However, in the majority of the cases the investigations are confined to areas like alloying in binary or multi-component systems from premixed powders. Very little work has been reported on high-energy milling of pure metals. There are some reports on mechanical alloying of pure metals that undergo polymorphic transformation on milling, but relatively few papers have been reported in the literature pertaining to attrition milling of pure metals, which do not fall under this category. One such attempt has been made in this investigation by subjecting a noble metal like silver with fcc crystal structure to attrition milling. The present work deals with the investigation of the effect of addition of a process control agent (PCA) on the nanocrystalline behavior of elemental silver powder subjected to high energy milling in an attritor. Elemental silver powder was subjected to attrition milling with and without addition of stearic acid as PCA. The powder samples drawn at periodic intervals during the course of milling were subjected to characterization using techniques like XRD, SEM and DSC. The variation in particle shape morphology, crystallite size and lattice strain as a function of PCA was studied.

Key Words: effect of PCA, mechanical alloying, silver powder.

1. Introduction

Nano-crystals are ultrafine single or multiphase mono/polycrystalline materials with nanometric grain size (typically <20 nm). Nano-crystals are subject of intense research due to many attractive mechanical (hardness, elastic modulus, etc.) and functional (heat capacity, magnetic, catalytic, chemical reactivity and electrical resistivity) properties and associated applications of them [1–3]. Among the possible routes of synthesis of nanocrystalline metallic and ceramic materials, mechanical alloying/milling offers an easy, flexible and inexpensive option capable of producing materials with interesting microstructure in large

* Author for correspondence.

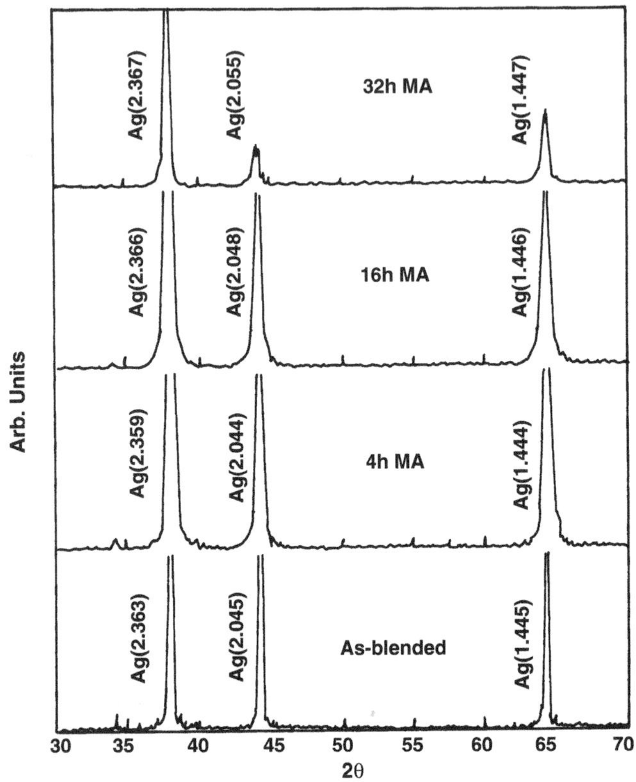

Figure 1. Variation of the peak intensity of the silver powder containing PCA for various milling times.

quantity. Mechanical alloying/milling mostly involves multi-component powder blends, which leads to intermixing of two atomic species on an atomic scale with a large negative heat of mixing [4–6]. Very little work has been reported on high energy milling of pure metals. There are some reports on mechanical alloying of pure metals which undergo polymorphic transformation on milling revealing some fundamental aspects of phase transformation [7].

In the present work, high energy milling of elemental silver powder (fcc) was taken up. Silver nano powders are on the edge of becoming commercial filler materials for electrically conducting polymer matrix composites. The morphology of the powder as such contributes to the electrical and mechanical properties [8, 9]. The study presents the effect of PCA additions on the behavior of elemental silver powder during milling.

2. Experimental

Pure silver powder (less than 5 micron particle size and 99 percent purity) was milled in an indigenously fabricated high energy attrition mill. Two independent

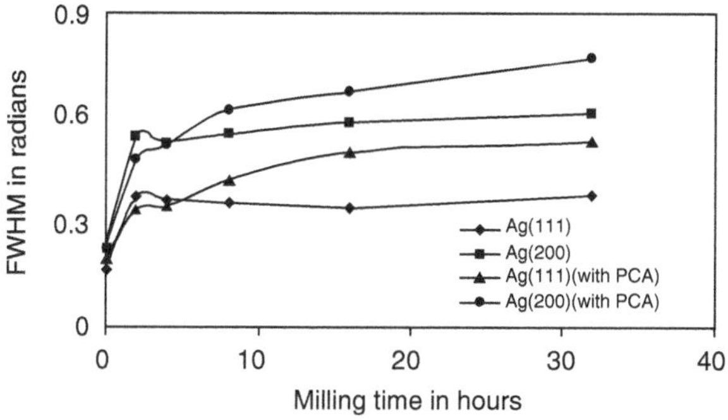

Figure 2. Variation of full width at half maximum intensity for the silver peaks with respect to milling time.

batches of 30 g each of silver powder, one with stearic acid as process control agent (PCA) and the other without PCA, were run. The ball to charge ratio was 20:1. Approximately 3 g of sample was drawn at definite time intervals of 2, 4, 8, 16, 32 and 48 hours during the course of milling.

The milled samples with and without PCA were characterized by X-ray diffraction (XRD) and scanning electron microscopy (SEM) in secondary electron imaging (SEI) mode. The change in thermal behavior of powder samples due to milling was investigated using SHIMADZU DSC−50 (Differential Scanning Calorimetry) at a heating rate of 10°C/min.

3. Results

The X-ray diffraction patterns of the samples milled with and without PCA and drawn at different time intervals show a broadening of peaks with increasing milling time. Representative multiple XRD plots for powders milled with PCA are shown in Figure 1. The variation of the full width at half maximum intensity for the silver peaks with different milling duration is shown in Figure 2.

This is a general trend with regard to the broadening of diffraction peaks that follows due to mechanical disordering as well as particle refinement, but the degree of broadening is larger in the case of samples milled with PCA. The crystallite size was calculated from the broadening of the X-ray peaks using the Hall–Williamson method. The crystallite size variation with milling duration is shown in Figure 3. The crystallite size decreases sharply to around 30 nm during the initial period of milling (4 hrs) and later on it stabilizes between 20 to 30 nm and remains constant as milling proceeds. This phenomenon is observed in samples with and without PCA; however the crystallite size is finer in samples milled with PCA.

Figure 4 shows the variation of micro-strain with respect to milling time. As apparent from the figure, the lattice strain increases sharply up to 4 hours and

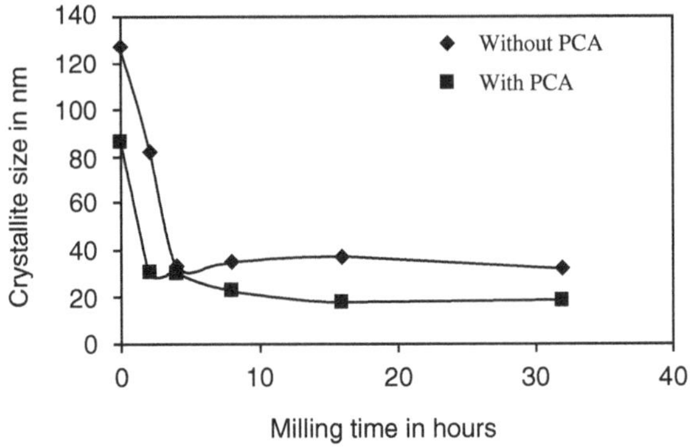

Figure 3. Variation of crystallite size of silver nano particles with respect to milling time.

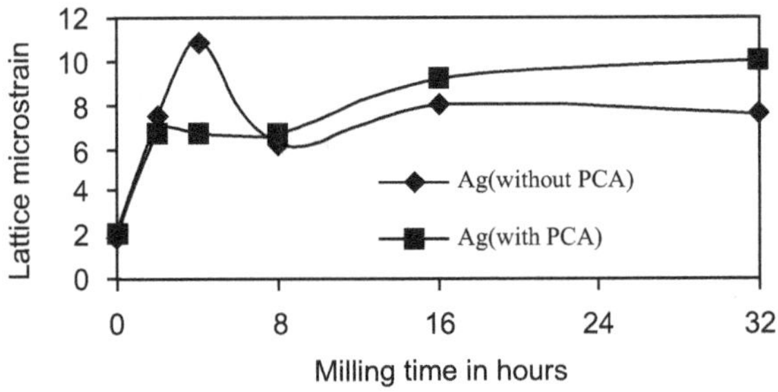

Figure 4. Variation in lattice micro strain of silver nano particles with respect to milling time.

then decreases gradually and finally becomes constant. This is attributed to the sudden increase in density of defect structures such as dislocations and grain boundaries [10]. Prolonged milling leads to an increase in temperature which results in re-crystallization and annealing. Thus the decrease in lattice strain can be said to be due to the re-crystallization and annealing phenomenon.

The change in particle morphology of Ag powder for two different conditions (i.e., with and without PCA) was evaluated by subjecting the representative powder samples to scanning electron microscopy (SEM). Figure 5 indicates the tendency towards coarsening of particles with progressive milling. At intermediate stages of milling excessive flattening of silver particles is observed, this happens more in case of powder without PCA. The initially round particles having an agglomerated chain like structure change their morphology to a flat flaky type with increasing milling time.

In view of noticeable changes in crystallite size and lattice strain, the starting powder and sample drawn after 4 hours of milling were subjected to DSC scans shown in Figure 6. The DSC plot in Figure 6(a) shows two small exothermic humps corresponding to the relaxation of the small amount of strain present in the starting material. However, a predominant exothermic peak at 626°K is ob-

(a) Elemental silver

(b) Silver-2hrs milled (without PCA)

(e) Silver-2hrs milled (with PCA)

(c) Silver-16hrs milled (without PCA)

(f) Silver-16hrs milled (with PCA)

(d) Silver-32hrs milled (without PCA)

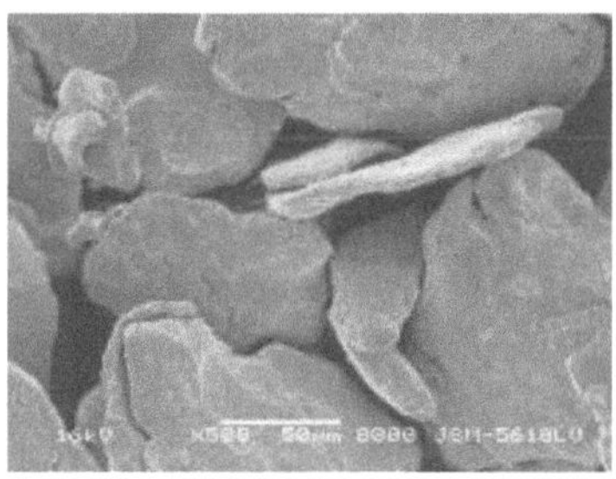

(g) Silver-32hrs milled (with PCA)

Figure 5. SEM Photomicrographs for (a) elemental silver powder, (b–d) Mechanically Alloyed powders without PCA and (e–g) Mechanically Alloyed powders with PCA.

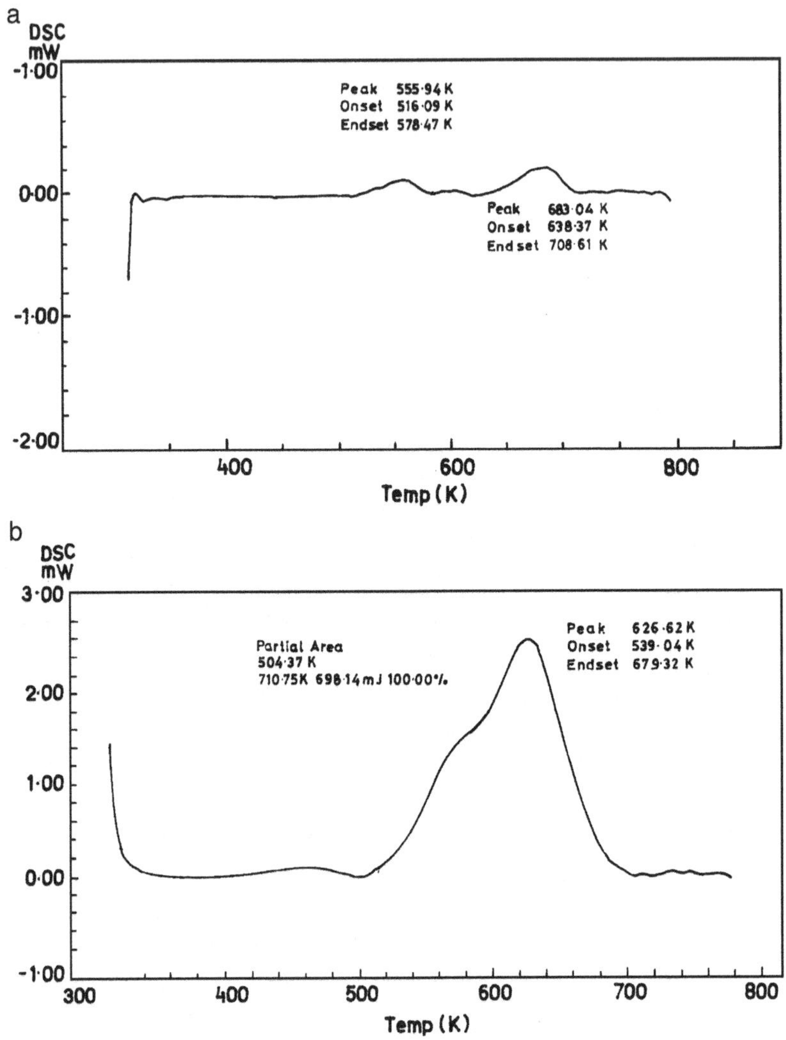

Figure 6. (a) DSC thermogram of initial silver powder before milling, (b) DSC scan for the same silver powder after 4 hrs of milling.

served for 4 hrs milled sample (Figure 6(b)), which may be due to recovery, i.e., grain growth and strain relaxation.

4. Discussion

The peak broadening at half maximum intensity of the X-ray diffraction lines is due to a reduction in crystallite size, flattening and micro-strains within the diffracting domains. The broadening is larger in samples milled with PCA, which indicates crystallite size reduction and increase in lattice strain as shown in Figures 3 and 4. In general, plastic deformation proceeds by slip and twinning at low and moderate strain rates, whereas at high strain rates the formation of a dense network of dislocations becomes the dominant mechanism [11, 12]. Plastic deformation by ball milling occurs via shear bands. In early stages of milling, the average strain increases due to an increase in dislocation density. At a certain dislocation density within the heavily strained region, the crystal disintegrates into sub-grains. This leads to a decrease in lattice strain after 5 hours of milling. With further processing, deformation occurs in shear bands located in previously unstrained parts of the material. The size of sub-grains in existing bands is further reduced reaching a steady condition as shown in Figure 3 at a later stage. The PCA on being absorbed on the surface of particles helps in inhibiting excessive cold welding and thereby agglomeration, by lowering the surface tension of the solid. Because the energy required for milling is a function of plastic deformation of powder particles and new surface area generated times the surface tension, a reduction in surface tension by absorption of PCA results in a finer crystallite size.

The SEM micrographs indicate a trend towards coarsening along with a tendency to convolution as milling progresses. This change in particle morphology is as expected due to the high ductility of silver [13]. The effect of PCA in reducing the cold welding tendency can be clearly observed. If we critically compare the SEM micrographs for 16 hours milled samples (with and without PCA), we observe large flattened particles with smooth surfaces due to excessive cold welding in case of the sample without PCA where the particle diameter tends to increase in one dimension. However, for the sample milled with PCA cold welding is suppressed, therefore we are able to attain a finer particle size. Cold welding and fracturing processes help the powder particles to be in contact with each other with atomically clean surfaces. The degree of cold welding depends on the ductility of the powder. PCA impedes the clean metal-to-metal contact by being absorbed on the surface of the particles and helps to inhibit cold welding. However, as the milling is continued, the particle sizes become larger as seen in samples milled for 32 hours. This is due to the decomposition of the process control agent (PCA) leading to diminishing effect of the same. Thus in the 32 hrs milled sample we get a flat and flaky morphology with excessive cold welding.

The thermal behavior of the samples was investigated with the help of DSC. DSC plot for starting material exhibits two small exothermic humps shown in Figure 6(a). These may be attributed to (i) relaxation phenomena of small amount of pre-existing microstrains associated with the starting material; and/or (ii) tarnishing of silver. This is also evidenced by Figure 4 for lattice microstrain vs. milling time wherein some amount of microstrain is clearly indicated for unmilled sample. Such small initial microstrain in the starting sample seems to be a result of the specific manufacturing practice for the silver powder used as starting material in this investigation. It is due to this that the heats of enthalpy for these two small exothermic events are 28.23 mJ and 59.16 mJ, respectively.

The DSC thermogram of the elemental silver after 4-hrs of milling shows that higher heat of enthalpy is associated (698.14 mJ) and the broad exothermic peak at ~626°K is due to grain growth and strain relaxation, which is confirmed by XRD peaks after isothermal annealing heat treatment.

5. Conclusion

The high energy milling of elemental silver does not involve any polymorphic transformation even after prolonged milling, as indicated by the X-ray diffraction. The broadening of diffraction peaks and reduction in peak intensities is attributed to crystallite size reduction and an increase in the micro-strain level. The addition of PCA helps in retarding the excessive tendency to cold welding and thus is able to offer finer crystallite sizes. The SEM studies bring out the tendency of silver powder to extensive flattening/leafing with prolonged milling. This has a direct influence on the ratio of surface area to volume of the powder produced.

References

1. Gleiter H., *Prog. Mat. Sci.* **33** (1989), 223.
2. Siegel R. W., In: Cahn R. W., Haasen P. and Kramer E. J. (eds.), *Processing of Metals and Alloys*, Vol 15, VCH, Weinheim, 1991, p. 583.
3. Manna I., Chattopadhyay P. P., Banhart F. and Fetch H. J., *Appl. Phys. Lett.* **81** (2002), 22.
4. Murthy B. S. and Ranganathan S., *Int. Mater. Rev.* **43** (1998), 101.
5. Lu L. and Lai M. O., *Mater. Des.* **16** (1995), 34.
6. Schaffer G. B. and McCormick P. G., *Review Article on Mechanical Alloying* **16** (1992), 91.
7. Chattopadhyay P. P., Nambissan P. M. G., Pabi S. K. and Manna I., *Phys. Rev. B.* **63** (2001), 054107.
8. Gunter B. H., *Int. J. Pow. Met.* **35** (1999), 53.
9. Busmann H. G., Gunther B. and Mayer U., *Nanostructured Mat.* **12** (1999), 130.
10. Hatherly M. and Malin A. S., *Scripta. Metall.* **18** (1984), 449.
11. Clifton R. J., Duffy J., Harley K. A. and Shawki T. G., *Scripta. Metall.* **18** (1984), 443.
12. Gust W., Predel B. and Stenzel K. J., *Z. Metallkd* **69** (1978), 721.
13. Lu L., Lai M. O. and Zhang S., *Mater. Des.* **15** (1994), 79.

Hyperfine Interactions (2005) 160:181–187
DOI 10.1007/s10751-005-9162-8

Electron- and Hole-Doping Effects in Manganites Studied by X-Ray Absorption Spectroscopy

K. ASOKAN[1,*,†], J. C. JAN[1], K. V. R. RAO[1,‡], J. W. CHIOU[1], H. M. TSAI[1], W. F. PONG[1], M.-H. TSAI[2] and RAVI KUMAR[3]
[1]*Department of Physics, Tamkang University, Tamsui 251, Taiwan; e-mail: asokan@iuac.res.in*
[2]*Department of Physics, National Sun Yat-Sen University, Kaohsiung 804, Taiwan*
[3]*Nuclear Science Centre, Aruna Asaf Ali Marg, New Delhi 110 067, India*

Abstract. We investigate the electronic structures of hole-doped, $La_{0.7}Ca_{0.3}MnO_3$, and electron-doped, $La_{0.7}Ce_{0.3}MnO_3$, manganites by x-ray absorption near edge structure (XANES) spectroscopy at the O and Mn *K*-edges. While, the O K-edge XANES results indicate that Ca and Ce doping induce holes in O 2*p* derived states, the Mn *K*-edge XANES do not give any evidence for creation of the Mn^{4+} (or Mn^{2+}) ions by Ca (or Ce) dopants. Such results further questions the validity of double exchange mechanism in understanding the anomalous properties of manganites.

Key Words: electron-doping, electronic structures, hole-doping, manganites, XANES.

PACS: 78.70.Dm, 71.20.Be, 71.28.+d, 75.30.Vn.

1. Introduction

Manganites exhibits extraordinary phase transitions and various anomalous properties such as colossal magnetoresistance, charge ordering, and magnetization [1]. Fundamental origin of such anomalous properties remains a puzzle. However, the interplay between spin, orbital, and charge degrees of freedom was found to be crucial and the double exchange (DE) model was usually used for a basic understanding of these properties [1, 2]. Extensive investigations were performed for the $La_{1-x}Ca_xMnO_3$ compounds, which was called hole-doped manganites because Ca ions generated holes in $LaMnO_3$ (hereafter referred to as LM) [1, 2]. In Ce doped LM, i.e., $La_{1-x}Ce_xMnO_3$, Ce ions donate electrons and are called electron-doped manganites [3, 4]. The DE model was not satisfactory to explain metal-insulator and ferromagnetic transitions, and the concomitant colossal magnetoresistance phenomenon in the manganites [1, 2].

---

* Author for correspondence.
† Present address: Nuclear Science Centre, Aruna Asaf Ali Marg, New Delhi-110 067, India.
‡ Present address: Department of Physics, University of Rajasthan, Jaipur, India.

X-ray absorption near edge structure (XANES) spectroscopy is a sensitive probe for the electronic properties, which can provide information about the valency, the unoccupied electronic states, and the effective charge of the absorber atom in a solid [5]. Asokan *et al.* [6] and Mitra *et al.* [7], based on XANES study mainly at Mn $L_{3.2}$ and Ce $M_{5.4}$-edges, found electron doping in Ce doped LM. In this study, we report XANES measurements at the O and Mn *K*-edges to understand the effects of hole- and electron-doping on the electronic property of the manganite, especially the valency of the Mn ions. We have chosen $La_{0.7}Ca_{0.3}MnO_3$ (Ca-LM) and $La_{0.7}Ce_{0.3}MnO_3$ (Ce-LM) to represent hole- and electron-doped manganites, respectively.

2. Experimental details

Samples were synthesized by the standard solid-state reaction route and characterized by x-ray diffraction and magnetoresistivity measurements. The Ca-LM and Ce-LM samples have magnetic transition (Curie) temperatures of 245 K and 274 K and orthorhombic structures [6–8]. XANES measurements at the O *K*-edges were carried out using high-energy spherical grating monochromator (HSGM) beamline at the National Synchrotron Radiation Research Center (NSRRC), Hsinchu, Taiwan, operating at 1.5 GeV with a maximum stored current of 200 mA. The spectra were obtained using the sample drain current mode at room temperature and the vacuum in the experimental chamber was in the low range of 10^{-9} Torr. The typical resolution of the HSGM beamline was better than ~0.2 eV. The wiggler beamline BL17C of NSRRC was also used for the XANES measurements at the Mn *K*-edge in the fluorescence mode and the resolution was about 0.5 eV.

3. Results and discussion

For manganites the electronic structure close to the Fermi level (E_F) is dominated by Mn-3*d* and O-2*p* states [1, 2, 9]. The O 2*p* derived states can be seen in the normalized XANES spectra of the Ca-LM, Ce-LM, and LM compounds at the O *K*-edge as shown in Figure 1. In this figure the spectra of Mn_2O_3 and MnO compounds are also given for reference. All spectra were normalized in the energy range between 550 and 560 eV (not fully shown in the figure) after subtracting the background. Four main features in the spectra are labeled as A_1 to D_1 in the energy range from 530 to 545 eV. The features A_1 and B_1 were attributed to O-2*p* and Mn t_{2g} and e_g hybridized states, respectively, because the crystal-field in the MnO_6 octahedron splits the Mn 3*d* band into t_{2g} and higher-energy e_g subbands. If one assigns the spectral features considering manganites as one of perovskites, above procedure is consistent [10–14]. The splitting between A_1 and B_1 is about 2.7 eV for the three Ca-LM, Ce-LM and LM spectra.

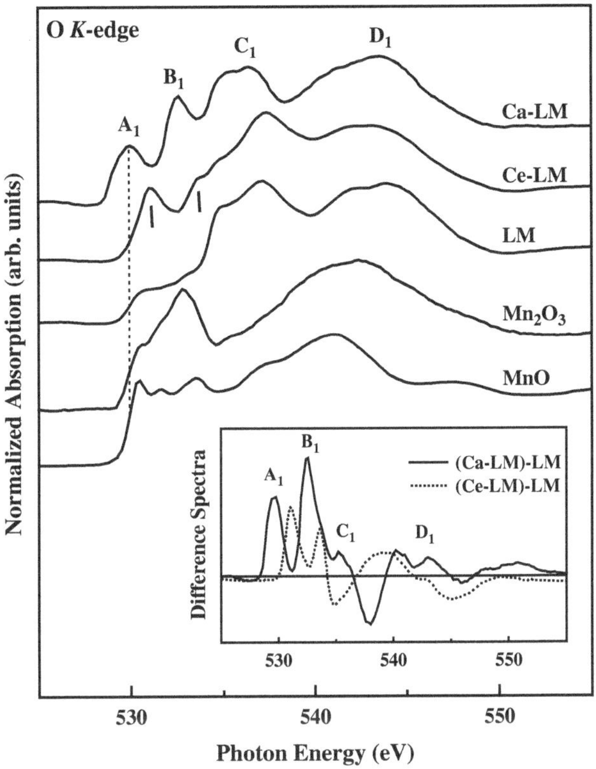

Figure 1. Normalized O *K*-edge XANES spectra of manganites and reference compounds, MnO and Mn_2O_3. *Dotted line* indicates the shift in the energy position with respect to that of Ca-LM. The *inset* shows difference spectra obtained after subtracting the spectrum of LM, which reveals major differences between these manganites.

Features A_1 and B_1 in the spectrum of Ca-LM are located at 529.9 eV and 532.6 eV with a chemical shift of ~ -0.5 eV relative to those of LM (shown in the figure by vertical lines). Both features A_1 and B_1 are much more prominent than those in the spectrum of LM. The enhancement of features A_1 and B_1 in the Ca-LM spectrum can be clearly seen in the difference spectra obtained by subtracting the spectrum of LM shown in the inset of Figure 1, which was referred to as an electron jump from the ligand [15]. The enhancement of features A_1 and B_1 and the -0.5 eV chemical shift consistently show that the negative charge on the O ion is reduced. Features A_1 and B_1 in the spectrum of Ce-LM are located at 531.2 and 533.8 eV (also shown in the figure by vertical lines) with a chemical shift of $\sim +1$ eV relative to those of Ca-LM shows a increase of the negative charge on the O ion. This chemical shift was recognized to be due to the more extended nature of the unoccupied O 2*p* derived states that are influenced by the reduced positive electrostatic potential at the Mn ions, which gain electrons from Ce dopants [15]. Peak C_1 (534–538 eV) in the spectrum of Ca-LM was assigned to the La 5*d*-Ca 3*d* and O 2*p* hybridized states [13]. The more prominent peak

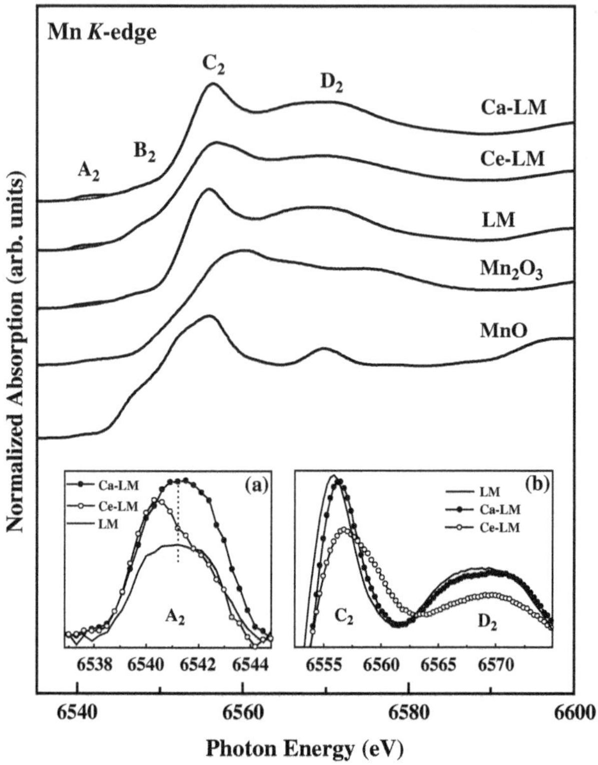

Figure 2. Normalized Mn *K*-edge XANES spectra of manganites and reference compounds, MnO and Mn_2O_3, measured in fluorescence mode. The main spectral features are labeled as A_2 to D_2. Inset (a) is a magnified view of the pre-edge feature, A_2, after a Gaussian type background was subtracted. Notably, the center of the pre-edge peak is shifted in the case of Ce-LM by ~1 eV. Inset (b) plots the differences between features C_2 and D_2.

C_1 for Ce-LM than for Ca-LM was interpreted as an evidence of the involvement of La/Ce 5*d*/4*f* states [6, 13]. It could be explained by the strong hybridization between the Ce (4*f*/5*d*) and Mn (3*d*) orbitals because Ca does not have near edge *f* states. Based on the spectra of reference compounds, MnO and Mn_2O_3, the energy region between 538 to 548 eV, marked by D_1, was assigned to Mn 4*sp*-O 2*p* hybridized states [6, 13].

Figure 2 displays the normalized XANES spectra at the Mn *K*-edges of the manganites and the reference compounds Mn_2O_3 and MnO. The insets show the differences between the spectral shapes of the pre-edge and main features upon doping Ca and Ce ions. The pre-edge peak A_2 was assigned to the direct quadrupole transition from 1*s* to the empty Mn 3*d* states [16–18]. Inset (a) in the figure presents a magnified view of pre-edge A_2 after subtracting a Gaussian type background as shown by the dashed curve. This feature is relatively broad but noticeable in the spectrum of Ca-LM, indicating a significant probability of transition from the Mn 1*s* electron to the empty t_{2g}- and e_g-O 2*p* hybridized states under the influence of the octahedral ligand field [16, 17]. The mixing of the Mn oxidation states was argued to lead to a mixture of Mn 1*s* $\rightarrow$ 3*d* transition energies, which consequently broadened the pre-edge feature [16]. Beside differences in the intensities, the centers of the pre-edge feature A_2 in the Ce-LM (6540 eV) and Ca-LM (6541 eV) spectra differ by ~ -1 eV. This shows consistency in O and Mn *K*-edge measurements. The very shallow feature B_2 is only noticeable in the Ce-LM spectrum, which was attributed to an increased *p*-character in the lowest empty Mn 3*d*-like electronic band [19]. Inset (b) in Figure 2 shows the comparison of features C_2 and D_2 in the Ca-LM, Ce-LM and LM Mn *K*-edge spectra. Features C_2 and D_2 for Ca-LM are essentially the same as for LM except a small shift to higher energy. The intensities of features C_2 and D_2 for Ce-LM, however, are substantially reduced relative to those for LM, which suggests that Ce doping enhance the occupation of Mn *p* orbitals. This result shows absence of features associated with Mn^{4+} and Mn^{2+} in these spectra.

The DE model has been usually used to explain the normal state properties including the CMR phenomenon of manganites [1, 2, 7]. However, some recent studies showed that the DE mechanism alone could not explain all experimental results [1, 2]. The DE model is based on the assumptions that first the e_g orbitals are sufficiently delocalized and the electrons occupying these orbitals are able to hop between adjacent Mn sites and give rise to metallic (strictly speaking semimetallic) property. It is understood that the occupation of e_g orbitals depends on the dopant and its concentration and these orbitals can be either localized or itinerant [2]. When the La^{3+} ions are partially substituted by Ca^{2+} ions, a corresponding number of Mn ions changes to Mn^{4+} ions. This permits hopping of itinerant e_g electrons between spin-aligned Mn^{3+} and Mn^{4+} ions and causes a DE interaction that results in an effective ferromagnetic interaction between Mn^{3+} and Mn^{4+} ions. Electron-doped manganites are obtained by doping Ce^{4+} ions, in which similar DE interaction is expected between Mn^{3+} and Mn^{2+} ions [3, 4, 6, 7]. According to these arguments, mixed-valence Mn charge states are responsible for correlation between magnetism and conductivity [1]. This mixed-valent picture ignores the electrostatic potential difference between Mn^{3+} and Mn^{4+} (or Mn^{3+} and Mn^{2+}) ions, which is about 8 eV without the screening effect. Ca-LM ($La_{0.7}Ca_{0.3}MnO_3$) is not a good conductor because it has a resistivity of an order of 10^{-2} Ω-cm at the room temperature, so that the screening effect, though will reduce the potential difference, will not be sufficient to suppress chemical shift difference between Mn^{3+} and Mn^{4+}. Thus both photoemission and XANES spectra will show two distinctive sets of Mn 3*d*-derived features in the spectra of Ca doped manganites. However, the extra set of Mn 3*d* derived features were absent in our XANES spectra and those obtained previously [16–18] and the photoemission spectra of Park *et al.* [9]. Thus above results question the DE model used to understand why LM is an antiferromagnetic insulator while Ca-LM and Ce-LM are ferromagnetic semimetal.

4. Conclusions

In conclusion, spectroscopic data obtained using XANES measurements at the O, and Mn *K*-edges for the hole-doped, $La_{0.7}Ca_{0.3}MnO_3$, electron-doped, $La_{0.7}Ce_{0.3}MnO_3$ and undoped $LaMnO_3$ manganites indicate that Ca and Ce doping induces holes in O 2*p* derived states. These results do not give evidence of the creation of the Mn^{4+} (or Mn^{2+}) state by Ca (or Ce) doping. Such results further questions the validity of double exchange mechanism in understanding the anomalous properties of manganites.

Acknowledgements

The authors are grateful to Dr. P. A. Joy (National Chemical Laboratory, Pune), Mr. S. Husain and Prof. J. P. Srivastava (Department of Physics, AMU, Aligarh, India) for providing samples. The author (K. A.) is thankful to the director of the Nuclear Science Center for granting them leave and encouragement. The authors (K. A. and W. F. P.) would like to thank the National Science Council of the Republic of China, for financially supporting this research under Contract no. NSC91-2112-M-032-015. The NSRRC staff is also appreciated for their technical support.

References

1. For review and references: Tokura Y. (ed.), Colossal Magnetoresistive Oxides, Gordon and Breach Science Publishers, The Netherlands, 2000.
2. Edwards D. M., *Adv. Phys.* **51** (2002), 1259.
3. Das S. and Mandal P., *Indian J. Phys.* **A71A** (1997), 231.
4. Kang J.-S., Kim Y. J., Lee W. B., Olson C. G. and Min B. I., *J. Phys., Condens. Matter* **13** (2001), 3779.
5. Stöhr J., *NEXAFS Spectroscopy*, Springer, Berlin Heidelberg New York, 1992.
6. Asokan K. *et al.*, *Surf. Rev. Lett.* **9** (2002), 1053.
7. Mitra C. *et al.*, *Phys. Rev.,* **B67** (2003), 092404, and reference therein.
8. Anil Kumar P. S., Joy P. A. and Date S. K., *J. Phys., Condens. Matter* **10** (1998), L269.
9. Park J.-H., Kimura T. and Tokura Y., *Phys. Rev.,* **B58** (1998), R13330; Park J.-H. *et al., Phys. Rev.,* **B58** (1998), R13330.
10. van Aken P. A., Liebscher B. and Styrsa V. J., *Phys. Chem. Miner.* **25** (1998), 494.
11. Asokan K. *et al.*, *J. Phys., Condens. Matter* **13** (2001), 11087.
12. Sarma D. D. *et al.*, *Phys. Rev.,* **B49** (1994), 14238.
13. Abbate M., Cruz D. Z. N., Zampieri G., Briatico J., Causa M. T., Tovar M., Caneiro A., Alascio B. and Morikawa E., *Solid State Commun.* **103** (1997), 9.
14. Solovyev V. *et al.*, preprint (arxiv:cond-mat/0303668 dated 29th May, 2003; Localized states in transition metal oxides, L. Hozoi, PhD thesis, Rijksuniversiteit Groningen (RUG), The Netherlands, 2003.
15. Glaser T., Hedman B., Hodgson K. O. and Solomon E. I., *Acc. Chem. Res.* **33** (2000), 859.
16. Ignatov A. Y., Ali N. and Khalid S., *Phys. Rev., B* **64** (2001), 14413, and references therein.

17. Bridges F., Booth C. H., Kwei G. H., Neumeiere J. J. and Sawatzky G. A., *Phys. Rev., B* **61** (2000), R9237; Zeng Z., Greenblatt M. and Croft M., *Phys. Rev., B* **63** (2001), 224410.
18. Croft M., Sills D., Greenblatt M., Lee C., Cheong S.-W., Ramanujachary K. V. and Tran D., *Phys. Rev., B* **55** (1997), 8726.
19. Franke R., Rothe J., Becker R., Pollmann J., Hormes J., Bonnemann H., Brijoux W. and Kopper R., *Adv. Mater.* **10** (1998), 126.

Hyperfine Interactions (2005) 160:189–192
DOI 10.1007/s10751-005-9163-7

Nanoencapsulation of Fullerenes in Organic Structures with Nonpolar Cavities

C. N. MURTHY
Applied Chemistry Department, Faculty of Technology and Engineering, M.S. University of Baroda, P. O. Box 51, Kalabhavan, Vadodara 390 001, India

Abstract. The formation of supramolecular structures, assemblies, and arrays held together by weak intermolecular interactions and non-covalent binding mimicking natural processes has been used in applications being anticipated in nanotechnology, biotechnology and the emerging field of nanomedicine. Encapsulation of C_{60} fullerene by cyclic molecules like cyclodextrins and calixarenes has potential for a number of applications. Similarly, biomolecules like lysozyme also have been shown to encapsulate C_{60} fullerene. This poster article reports the recent trends and the results obtained in the nanoencapsulation of fullerenes by biomolecules containing nonpolar cavities. Lysozyme was chosen as the model biomolecule and it was observed that there is no covalent bond formed between the bimolecule and the C_{60} fullerene. This was confirmed from fluorescence energy transfer studies. UV–Vis studies further supported this observation that it is possible to selectively remove the C_{60} fullerene from the nonpolar cavity. This behavior has potential in biomedical applications.

Key Words: calixarenes, cyclodextrins, fullerenes, lysozyme, noncovalent interactions.

1. Introduction

The concept of 'Chemistry beyond the molecule' has been gaining importance for the last few years [1] and has resulted in excellent reviews [2, 3]. The formation of supramolecular structures, assemblies, and arrays held together by weak intermolecular interactions and non-covalent binding has been used in applications being anticipated in both nanotechnology, biotechnology and the emerging field of nanomedicine [4]. Among the supramolecular structures with noncovalent binding, the encapsulation of fullerenes, specifically C_{60} fullerene has been a topic of interest for almost a decade. The interest in C_{60} fullerene chemistry is due to its unique physical and chemical properties. These include (1) its optical limiting property due to its electronic absorption in the entire UV–visible range, (2) efficient singlet oxygen generating ability, (3) radical scavenging character and (4) superconductivity on doping with alkali metals. While the properties (1) and (4) are useful in materials research, the properties (2) and (3) have applications in the biomedical area. C_{60} fullerene has been made preparatively accessible about a decade ago and recent research has shown the biological activity of C_{60} fullerene and its derivatives such as its DNA-cleaving ability and anti-HIV activity [8]. However, as C_{60} fullerene is a nonpolar molecule, it is insoluble in polar solvents and soluble in only a few solvents [9]. Encapsulation of C_{60} fullerene in unique molecules like cyclodextrins and calixarenes is one way of enhancing solubility in both polar as well as nonpolar solvents and has immense application potential in materials research and medicine.

Cyclodextrins being water-soluble are able to encapsulated non-polar molecules and thus make these molecules stable in aqueous systems. It is interesting to know which kind of interactions make these guest–host systems stable. Apart from the size considerations that determines the actual fit of the non-polar molecules in the cavities, the delicate balance between the attractive forces between the guest and the host is not well understood. The cavity diameter of β-cyclodextrin is 780 pm and the outer rim diameter is 1530 pm on the polar side of the molecule. Thus this molecule can 'accommodate' molecules that are of the dimensions of ~1000 pm. This leads to interactions that stabilise the whole complex in aqueous solutions. In fact it is possible to calculate the stability constant of this guest–host complex and there are reports that this can be as high as of the order of 10^5 [5].

The biomedical applications of fullerene imply its possible reaction with most essential cellular components like nucleic acids and proteins. It has also been shown that C_{60} fullerene interaction with HIV virus protease inhibited the activity of the latter [6]. The specific role of the C_{60} fullerene in these systems is not very clear. To gain further insight, protein–fullerene interactions were studied using lysozyme as a model protein. Lysozyme is a small, globular protein with 129 amino acid residues. The readily available extensive knowledge about its three dimensional structure both in solution and solid state facilitates our interpretations on the association between fullerene and lysozyme from physicochemical interactions point of view. Lysozyme is a single domain protein with α helical and β sheet structure [7]. The inner core of the protein is hydrophobic. When the interactions between these two molecules were studied by using a mixed solvent system that solubilized both lysozyme and the C_{60} fullerene an adduct was formed between these two that could be easily isolated and was found to be water-soluble giving an amber colored clear solution.

Fluorescence energy transfer study is a versatile technique to study the interactions between any two molecules, provided the emission wavelength of the first molecule overlaps with the excitation wavelength of the second. If the distance between these two molecules is within a certain range (<50 Å), then the emission peak of the first molecule either disappears or shows a decreased intensity with the appearance of the emission peak of the second molecule. The aromatic amino acid groups of lysozyme excited at 280 nm show a peak between 300 and 400 nm corresponding to the emission of aromatic groups (Figure 1a). Surprisingly the lysozyme-C_{60} fullerene adduct did not show the typical emission peak of the lysozyme instead an intense peak was

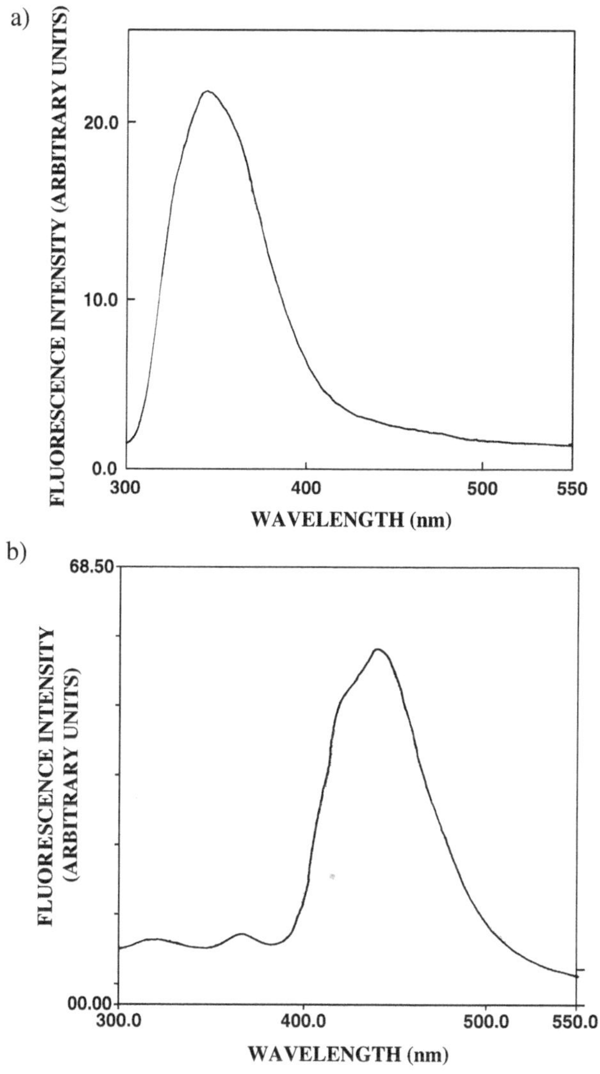

Figure 1. a) Fluorescence emission spectra of lysozyme in aqueous solution. b) Fluorescence emission spectra of the lysozyme–C_{60} fullerene adduct in aqueous media.

observed between 400 and 450 nm resulting from the energy transfer between the excited aromatic groups and the C_{60} fullerene (Figure 1b). This implies that there is an interaction between the two molecules that could be either of the covalent or noncovalent type.

UV–vis spectra of the C_{60} fullerene shows an absorbance maxima at around 330 nm in toluene whereas lysozyme shows an absorbance maximum at 280 nm in water. The spectra of the adduct in water showed the typical absorbance peak of lysozyme and a small peak corresponding to the C_{60} fullerene. Both these peaks were red-shifted by a few nanometers thus confirming the interaction between these two molecules. There was a broadening of the absorbance beyond 400 nm typical of the C_{60} fullerene absorbance in organic solvents. The fluorescence energy transfer studies and UV–vis absorbance studies do not give any indication about the nature of the binding between C_{60} fullerene and lysozyme. To understand the nature of the interaction between the two an aqueous solution of the adduct was shaken with toluene for few minutes and the UV absorbances of the aqueous phase and the organic phase were recorded. The aqueous phase showed the typical absorbance of lysozyme with the disappearance of the absorbance peak corresponding to that of C_{60} fullerene observed in the aqueous solution of the adduct. This suggests that there is no covalent bond formation between the two and the interaction observed in the fluorescence studies is hydrophobic in nature.

Concerted research efforts are underway to use C_{60} fullerene in therapeutic applications. Our studies [8] show that there is a strong possibility for hydrophobic interactions between the various proteins and C_{60} in the biological milieu, when the latter is used for biomedical applications, unless the molecule is delivered effectively at the intended site of action.

References

1. Lehn J.-M., *Supramolecular Chemistry: Concepts and Perspectives*, VCH, Weinheim, 1995.
2. Diedrich F. and Lopez G. M., Supramolecular fullerene chemistry, *Chem. Soc. Rev.* **28** (1999), 263.
3. Desiraju G. R., *Nature* **412** (2001), 397.
4. Freitas R. Jr., *Nanomedicine*, Vol. 1, Landes Bioscience, 1999.
5. Murthy C. N. and Geckeler K. E., *Chem. Commun.* (2001), 1194.
6. Nakamura E., Tokuyam H., Yamago S., Shiraki T. and Sugiura Y., *Bull. Chem. Soc. Jpn.* **69** (1996), 2143.
7. Blake C. C. F., Koenig D. F., Mair G. A., North A. C. T., Phillips D. C. and Sarma V. R., *Nature* **206** (1965), 757.
8. Ratna Prabha C., Patel R. and Murthy C. N., *Fullerenes, Nanotubes and Carbon Nanostructures* **12**(1) (2004), 427.

Hyperfine Interactions (2005) 160:193–197
DOI 10.1007/s10751-005-9164-6

Transport and Magnetic Properties of Eu and Sr Doped Manganite Compound $La_{0.7}Ca_{0.3}MnO_3$

D. S. RANA[1,*], C. M. THAKER[1], K. R. MAVANI[1,3], J. H. MARKNA[1], R. N. PARMAR[1], N. A. SHAH[2], D. G. KUBERKAR[1] and S. K. MALIK[3]
[1]*Department of Physics, Saurashtra University, Rajkot 360 005, India; e-mail: dhanvir@tifr.res.in*
[2]*Department of Electronics, Saurashtra University, Rajkot 360 005, India*
[3]*Tata Institute of Fundamental Research, Colaba, Mumbai 400 005, India*

Abstract. The effect of simultaneous increase in carrier density and size-disorder on the transport and magnetic properties of $La_{0.7}Ca_{0.3}MnO_3$ has been investigated by studying the $(La_{0.7-2x}Eu_x)(Ca_{0.3}Sr_x)MnO_3$ ($0.05 \leq x \leq 0.2$) (LECSMO) compounds. These compounds have been compared with standard $La_{1-x}Ca_xMnO_3$ ($0.3 \leq x \leq 0.5$) (LCMO) in which carrier density alone varies. In LECSMO, the insulator-metal transition temperature (T_p) decreases from 180 K for $x = 0.05$ to 80 K for $x = 0.15$ sample vis-à-vis ~240 K for $x = 0.35$ to ~225 K for $x = 0.45$ in LCMO system. Similarly, the Curie temperature (T_C) in LECSMO, decreases from 205 K for $x = 0.05$ to 75 K for $x = 0.2$ sample against 240 K for $x = 0.35$ to ~220 K for $x = 0.45$ in LCMO system. Also, in LCMO the T_C and T_p are coincident whereas, in LECSMO system, with increasing x there is an increasing disparity between the two. At 5 K, in the metallic region, a large MR (>90%) is observed in $x = 0.15$ sample and is discussed in terms of phase segregation and inter-grain magnetoresistance.

1. Introduction

The application potential of colossal magnetoresistance (CMR), exhibited by ABO_3 type perovskite structured manganites, $(R_{1-x}A_x)MnO_3$ (R = La, Pr, Nd, etc. and A = Ca, Sr, Ba, etc.), has sparked an intense research interest in such compounds [for a review, see [1] and [2]]. The CMR occurs in the vicinity of insulator–metal (I-M) transition temperature (T_p) and paramagnetic to ferromagnetic (PM–FM) transition temperature (T_C). The T_p and T_C are largest when the carrier density is optimal (Mn^{4+} content ~33%) [1, 2] and decrease with decreasing tolerance factor (defined as $t = (\langle r_A \rangle + r_O)/\sqrt{2}(\langle r_B \rangle + \langle r_O \rangle)$ and size-disorder (defined as $\sigma^2 = \Sigma x_i r_i^2 - \langle r_A \rangle^2$) [3–5]. Further, increasing the carrier density beyond optimal amount has a detrimental effect on T_p and T_C [1, 2]. Though the effect of these factors, namely, tolerance factor, size-disorder and carrier density, has been studied individually [1–5], the simultaneous effect of two factors keeping the third constant has not been investigated in detail. In this article, we present the results on electrical and magnetic properties of

* Author for correspondence.

Table I. Size-disorder (σ^2), I-M transition temperature (T_p) and Curie temperature (T_C) for $La_{1-x}Ca_xMnO_3$ ($0.35 \le x \le 0.5$) (LCMO) [6] and $(La_{0.7-2x}Eu_x)(Ca_{0.3}Sr_x)MnO_3$ ($0.05 \le x \le 0.2$) (LECSMO) series

Mn^{4+} content	LCMO			LECSMO		
	σ^2 (Å^2)	T_C (K)	T_p (K)	σ^2 (Å^2)	T_C (K)	T_p (K)
30	0.0003	250	250	0.0003	250	250
35	0.0004	245	245	0.0012	205	180
40	0.0005	240	240	0.0021	175	152
45	0.0006	230	230	0.0030	130	80
50	0.0007	220	220	0.0039	75	–

$(La_{0.7-2x}Eu_x)(Ca_{0.3}Sr_x)MnO_3$ ($0.05 \le x \le 0.2$) (LECSMO) samples. The simultaneous substitution of Eu and Sr has been designed to keep tolerance factor nearly constant at ~0.924 and to monotonically increase the Mn^{4+} content and size-disorder at the A-site. The comparison of physical properties of LECSMO samples may be made with the standard compositions $La_{1-x}Ca_xMnO_3$ ($0.35 \le x \le 0.5$) (LCMO) [6], in which carrier density increases with increasing x, but the size disorder is almost negligible (see below Table I).

2. Experimental

The $(La_{0.7-2x}Eu_x)(Ca_{0.3}Sr_x)MnO_3$ ($0.05 \le x \le 0.2$) samples were synthesized using a standard solid state ceramic preparation technique. The detailed synthesis procedure is reported elsewhere [7]. Powder X-ray diffraction (XRD) patterns were recorded on Siemen's diffractometer using Cu-Kα radiation. Electrical resistivity was measured using standard four-probe configuration in the temperature range of 5–320 K in zero-field and a field of 4 Tesla (PPMS, Quantum Design). Magnetization measurements were performed on a SQUID magnetometer (MPMS, Quantum Design).

3. Results and discussion

From XRD data, the $(La_{0.7-2x}Eu_x)(Ca_{0.3}Sr_x)MnO_3$ ($0.05 \le x \le 0.2$) samples are found to be single-phase compounds crystallizing in a distorted orthorhombic structure (space group-*Pnma*, No. 62). The cell parameters (obtained from Rietveld refinement of the XRD patterns) decrease with increasing x. This may be attributed to the fact that increasing Mn^{4+} content reduces the Jahn–Teller distortion resulting in the structural changes and contraction of cell volume.

Figure 1 shows the electrical resistivity versus temperature plots for all the $(La_{0.7-2x}Eu_x)(Ca_{0.3}Sr_x)MnO_3$ ($0.05 \le x \le 0.2$) samples. It is seen from this figure that, on increasing Eu and Sr content, T_p decreases while the peak resistivity increases. For instance, T_p decreases from 180 K for $x = 0.05$ to 80 K for $x = 0.15$

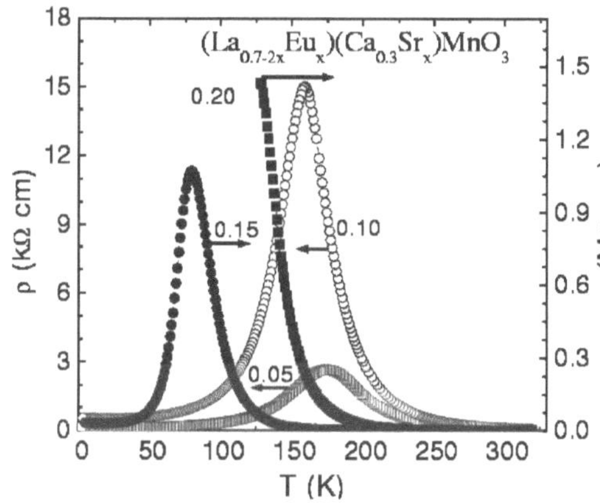

Figure 1. Resistivity *versus* temperature for $(La_{0.7-2x}Eu_x)(Ca_{0.3}Sr_x)MnO_3$ $(0.05 \leq x \leq 0.2)$ samples.

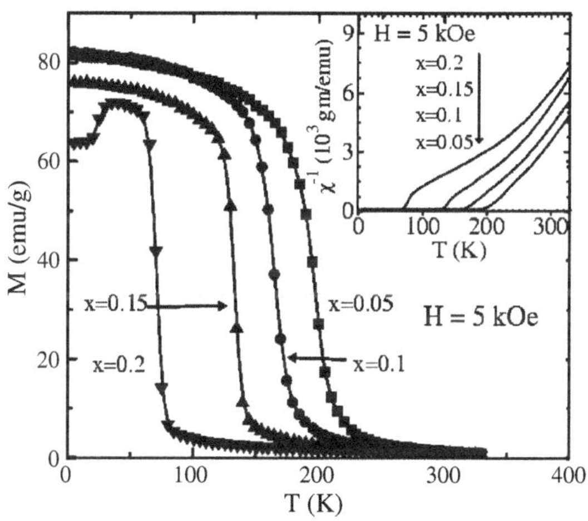

Figure 2. Magnetization *versus* temperature for $(La_{0.7-2x}Eu_x)(Ca_{0.3}Sr_x)MnO_3$ $(0.05 \leq x \leq 0.2)$ samples in a field of 5 kOe. *Inset* shows inverse susceptibility $(1/\chi)$ *versus* temperature for the same samples.

sample. The $x = 0.2$ sample is a charge-ordered sample and hence does not exhibit I-M transition in the absence of applied magnetic field. Figure 2 shows magnetization as a function of temperature in a field of 5 kOe for all the $(La_{0.7-2x}Eu_x)(Ca_{0.3}Sr_x)MnO_3$ $(0.05 \leq x \leq 0.2)$ samples recorded in warming direction after cooling the sample in zero magnetic field. The T_C falls from 205 K for $x = 0.05$ sample to 75 K for $x = 0.2$ sample. Inset in Figure 2 shows the inverse susceptibility versus temperature plots for the same samples.

Table I shows the comparison of σ^2, T_C and T_p of LECSMO and LCMO [6] series. In LECSMO samples, with increasing x, the monotonic fall in both T_p and T_C may be attributed to increase in Mn^{4+} content and size-disorder. The increase in Mn^{4+} content beyond optimal amount increases the Columbic interactions

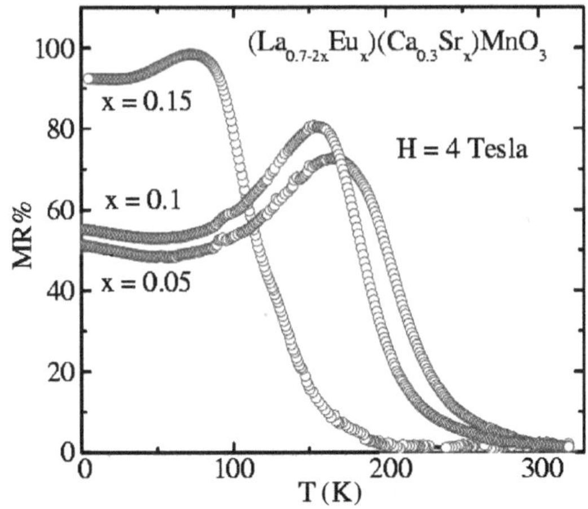

Figure 3. MR% *versus* temperature for $(La_{0.7-2x}Eu_x)(Ca_{0.3}Sr_x)MnO_3$ ($0.05 \leq x \leq 0.2$) samples in a field of 4 Tesla.

while increasing size-disorder results in local structural distortions via random displacement of oxygen ions [5]. Both these factors impede the Zener–Double exchange interactions resulting in a drastic fall in T_p and T_C. It may also be noted that, with increasing x, there is an increasing disparity between T_p and T_C. This may be attributed to the phase segregation near T_C, as has been reported and explained to occur in Gd-doped manganites [8]. Also, a deviation in paramagnetic susceptibility from Curie–Weiss law indicates the presence of magnetic inhomogeneities and hence, the resulting chemical disorder may be responsible for disparity of T_p and T_C. In contrast, the T_C and T_p are coincident and fall less rapidly (250 K for $x = 0.3$ to 220 for $x = 0.5$) for LCMO series [6].

Figure 3 shows magnetoresistance (MR) *versus* temperature plots for $(La_{0.7-2x}Eu_x)(Ca_{0.3}Sr_x)MnO_3$ ($0.05 \leq x \leq 0.15$) samples. The MR has been measured in a magnetic field of 4 Tesla and is calculated using the relation $MR\% = \{(\rho_0 - \rho_H)/\rho_0\} \times 100$. The MR is maximum in the vicinity of I-M transition and increases as T_P decreases with increasing x. Also, the MR increases at low temperature metallic region and reaches ~95% for $x = 0.15$ sample. Such large MR may have its origin both in inter-granular tunneling MR and magnetic disorder at Mn–O–Mn couplings.

4. Conclusions

In summary, we have studied structural, electrical and magnetic properties of $(La_{0.7-2x}Eu_x)(Ca_{0.3}Sr_x)MnO_3$ ($0.05 \leq x \leq 0.2$) compounds to elucidate the simultaneous effect of increasing Mn^{4+} content and size-disorder. The results on this series of samples have been compared with those on standard $La_{1-x}Ca_xMnO_3$ ($0.35 \leq x \leq 0.5$) samples [6]. With simultaneous increase in size-disorder and carrier density, there is a rapid fall in T_C, increasing disparity between T_C and T_p and a large magnetoresistance as compared to the situation in LCMO series where T_C falls less rapidly and T_C and T_p are coincident [6]. This study emphasizes the role of increasing size-disorder on the electrical and magnetic properties, which is almost absent in LCMO series.

Acknowledgements

DGK is thankful to University Grants Commission (UGC), New Delhi, India for financial assistance in the form of a Major Research Project. DSR is thankful to CSIR, India for providing senior research fellowship (SRF).

References

1. Rao C. N. R. and Raveau B. (ed.), *Colossal Magnetoresistance, Charge-ordering and Other Related Properties of Rare-Earth Manganates*, World Scientifics Publishing Company, 1998.
2. Ramirez A. P., *J. Phys., Condens. Matter* **9** (1997), 8171.
3. Goodenough J. B., *Annu. Rev. Sci.* **28** (1998), 1.
4. Hwang H. Y. *et al.*, *Phys. Rev. Lett.* **75** (1995), 914.
5. Rodriguez-Martinez L. M. and Attfield J. P., *Phys. Rev., B* **54** (1996), 15622.
6. Schiffer P. *et al.*, *Phys. Rev. Lett.* **75** (1995), 3336.
7. Rana D. S. *et al.*, *J. Appl. Phys.* **95** (2004), 7097.
8. Sun Y. *et al.*, *Phys. Rev., B* **66** (2002), 94414 and references therein.

Hyperfine Interactions (2005) 160:199–209
DOI 10.1007/s10751-005-9165-5

Oriented Growth of Nanocrystalline Gamma Ferric Oxide in Electrophoretically Deposited Films

TEJASHREE M. BHAVE[1], C. BALASUBRAMANIAN[1], HARSHADA NAGAR[1], SHAILAJA KULKARNI[2], RENU PASRICHA[2], P. P. BAKARE[2], S. K. DATE[1] and S. V. BHORASKAR[1,*]
[1]*Center for Advanced Research in Material Science and Solid State Physics, Department of Physics, University of Pune, Pune 411 007, India; e-mail: svb@physics.unipune.ernet.in*
[2]*National Chemical Laboratory, Dr. Homi Bhabha Road, Pashan, Pune 411 008, India*

Abstract. Films of nanocrystalline γ-Fe_2O_3 were deposited on silicon substrates by using the technique of electrophoretic deposition. The precursor powder was nanocrystalline γ-Fe_2O_3, which was synthesized, using DC arc plasma in the oxygen ambient by vapour–vapour interaction in gas phase condensation; at a stabilized arc current of 40 A. This powder was characterized by X-ray diffraction, Transmission Electron Microscopy, Vibrating Sample Magnetometer and Mössbauer Spectroscopy. An increase in directional coercivity was observed in case of films deposited on silicon substrates, which is dramatically significant. Preferred orientation of almost similar sized nanocrystalline magnetic domains in deposited films is evident from the results of X-ray diffraction and Transmission Electron Microscopy results. The preferred alignment of the nanocrystallites seems to be responsible for the significant changes observed in magnetic properties of films.

1. Introduction

Magnetic thin films consisting of nanocrystalline (10–100 nm) magnetic particles play an important role [1–3] on account of their wide applications as data recording media both in consumer electronics, digital storage media and computational and communicational applications. The role of nanocrystalline magnetic thin film is perhaps most interesting because of the possible magnetic coupling between the different particulates, arranged in the film. It is known that the magnetic properties of small systems are modified in a very interesting way [4–11] because of the intimate coupling of magnetism with elastic strain fields and bond states, the surface spin canting and the thermally activated processes. Apart from their technological applications, the magnetic nanocrystalline films provide basic stimuli for developing better understanding of magnetic phenomena.

* Author for correspondence.

It has been observed, in general, that the magnetization curve (relating the extrinsic behavior of magnetic material) depends on the particle size. Below a certain critical size of particle, the magnetic domains disappear and each particle itself becomes a single domain. With further reduction in size of a particle, the anisotropy barrier is reduced, since it depends on particle volume. Effectively one observes a spontaneous switching of the magnetic moment from one direction to the other if the anisotropy barrier is comparable to the thermal energy of the particle.

This may result into high degree of magnetization and change in coercivity, which can be observed through the hysteresis loop. This effect will be more pronounced if the single domains of nanoparticles are further made to align, in the deposited films, by external forces. In this paper we report the results of such an attempt to align the nanoparticles of gamma ferric oxide.

Maghemite (γ-Fe_2O_3) is one of the most commonly studied [12–17] and used magnetic materials for particulate recording media. Commercially used products consist of particles with sizes typically between 5–7 μm deposited on a 25 μm thick polyester base. Gamma ferric oxide, in bulk form, is a ferromagnetic compound with spinel structure in which iron cations occupy both tetrahedral (A) and octahedral (B) sites. Apart from its applications in recording media, gamma ferric oxide find applications in color imaging, bio-processing, magnetic refrigeration, gas sensors, ferrofluids and several others. [18–24].

There are varieties of techniques for synthesizing magnetic films. They include the methods of Metal Organic Chemical Vapor Deposition (CVD), Pulsed Laser Deposition (PLD), Metal Organic Deposition (MOD), RF Sputtering and Electrophoretic Deposition (EPD). The films prepared with different methods have different properties. For instance, PLD can quickly produce high quality films but is limited for large area (>1 cm^2) where thickness uniformity is an issue. In MOD, deposition is limited to the thickness of about 1 μm or less. MOCVD and RF sputtering include difficulties in optimizing volatile precursors and controlling stoichiometry. In particulate magnetic films the magnetization is created by particles (typically oxides) dispersed through the film binder. In deposited magnetic films the magnetization is created by nanocrystals formed during the deposition process. Thus there is less control over the morphology of the individual magnetic particles, which is actually a prerequisite in particulate recording media [25, 26].

Electrophoretic deposition technique (EPD) was established to synthesize lameller, fiber-reinforced and functionally graded composites. The method can also be used for deposition of oxides and is a cost effective and efficient process to obtain uniform deposit thickness. Present paper reports the magnetic properties of electrophoretically deposited films consisting of nanoparticles of γ-Fe_2O_3. This method [27] involves three processes involving i) charge on the surface of the particles, ii) motion of charged particle in a suspension under dc electric field and iii) deposition on to the oppositely charged substrate. In this method the surface of the particle becomes charged in a solvent because of the ionic adsorption from the solvent due to the difference in the dielectric constants between the particles and the medium. Physical properties of EPD films strongly depend on the processing parameters including the particle size and distribution, the applied field and mixture ratio of solvent and suspension. Electrophoretic method of deposition offers advantages over other methods, like ordered orientation of the particles. This feature is most attractive in ordering the magnetic nanoparticles and it is easy and also cost effective [28]. Film thickness can be tailored over a wide range (10–1,000 nm) and still maintains its usefulness to the desired application. On account of the different zeta potential for different sizes of the particles, the electrophoretic process provides size selective deposition based on the electric field.

The precursor powder of nanocrystalline gamma ferric oxide was synthesized by DC arc plasma assisted gas phase condensation [29] and the crystallinity and morphology were confirmed by X-ray diffraction (XRD) and Transmission Electron Microscopic (TEM) measurements. The paper reports a remarkable improvement in the magnitude of saturation magnetization in the films as compared to the precursor powder as measured by Vibrating Sample Magnetometer (VSM). It also reveals that the substrate plays an important role in deciding the magnitude of saturation magnetization of the film. Silicon as a substrate was seen to enhance the net magnetization. The anisotropic properties of coercivity were also reflected from these measurements.

2. Experimental

Nanoparticles of Iron oxide were synthesized by using transferred DC arc plasma method [29]. Commercially available iron disc (2" in diameter and 1" in thickness) was used as anode. The arc was struck, in oxygen atmosphere, between graphite cathode and the water cooled iron anode by applying a voltage of 22 V. The arc current was maintained at 40 A and the powder formed during the process was allowed to settle on the walls of the chamber, and subsequently it was collected.

Films were prepared from the powder using the technique of electrophoretic deposition. The as synthesized powder was dispersed in a non-conducting (ethyl alcohol) medium. A regulated dc negative voltage of 100 V was applied between silicon substrate and a titanium electrode. The deposition was obtained when titanium was made positive. Magnetic stirrer was used to keep the particles uniformly dispersed in the solution during deposition. Deposition period was approximately 20–30 min. The thickness was estimated from talystep measurements (in which, piezoelectrically driven stylus was moved across a film boundary and the movement of stylus is scanned electronically) and was found to be 25 μm. After deposition the films were transferred to a furnace pre-heated to 380°C and annealed for 1 h in order to enhance the electrical contact with the substrate as well as contact between the nanoparticles.

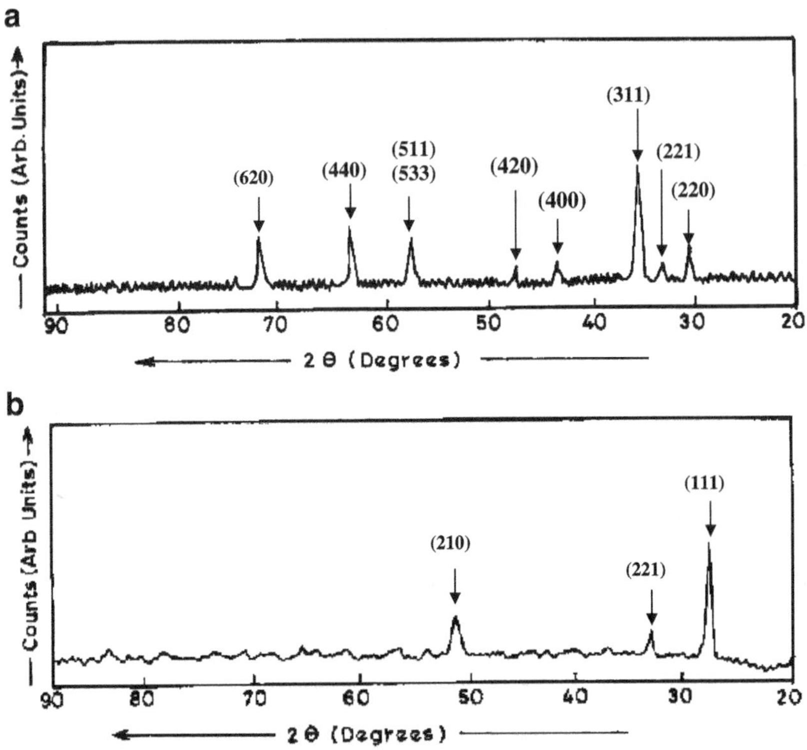

Figure 1. (a) X-ray diffraction pattern of powder. (b) X-ray diffraction pattern of film deposited on silicon substrate.

Figure 2. Scanning Electron Micrograph recorded for EPD film deposited on silicon substrate.

Characterization of the powder as well as deposited films was carried out using X-ray diffraction (XRD), Mössbauer spectroscopy, Transmission Electron Microscopy (TEM) and Vibrating Sample Magnetometer (VSM). A Philips X-ray diffractometer (Model PW 1840) was used with iron target. TEM measurements were carried out using JEOL 1200 EX instrument. Mössbauer spectroscopy was carried out using S-600 Mössbauer spectrometer (Model 1990). VSM analysis was carried out by EG and G PAR 4500 instrument. X-ray Diffraction (XRD) by Philips PW 1710 with Cu – K_α radiation using Ni filter. For recording the TEM of the deposited powder, the films were scraped and the collected powder was dispersed in ethyl alcohol. This dispersion was then deposited on TEM grid. Mössbauer spectroscopy and TEM were used to establish the phase and particle size distribution of iron oxide. Magnetic measurements were carried out with vibrating sample magnetometer (VSM) for both precursor powder as well as the deposited films.

3. Results and discussion

As synthesized powder of iron oxide was brownish orange in colour. The powder was characterized by X-ray diffraction. The X-ray diffraction pattern of the powder showed multiple peaks as shown in Figure 1(a). The '*d*' values of 2.97 Å, 2.78 Å, 2.52 Å, 2.09 Å, 1.88 Å, 1.61 Å, 1.48 Å and 1.32 Å calculated from the Bragg's angles correspond to the cubic γ-Fe_2O_3 phase identified from ASTM data. The line widths in the XRD pattern were used for estimating the average particle size of the powder by the Debye–Scherrer formula and were estimated to be in the range of 25–40 nm. The films deposited by electrophoretic deposition method were also analyzed by grazing angle X-ray diffraction measurements. Figure 1(b) shows the X-ray diffraction pattern for the film deposited on silicon substrate. This spectrum consists of three peaks out of which the one observed at $d = 3.2907$ Å corresponds to Si (111) planes from the substrate and the other two peaks for $d = 2.78$ Å and 1.88 Å are of γ-Fe_2O_3. The reduced number of peaks observed in the film indicates the possibility of preferred orientation. This may be a consequence of a) high crystallinity of substrate used for deposition, which provides an ordered lattice for the nanoparticles to get preferentially oriented, and b) the electrophoretic method of deposition itself may be responsible for orienting the nanocrystallites in a preferred direction. The preferred orientation of nanocrystallites in a particular direction depends on the net dipole moment of the crystallite.

The surface morphology of the film was studied using scanning electron microscope, shown in Figure 2, which again indicates the existence of nanocrystallites in the film. The average size of the particles was observed to be 20 nm.

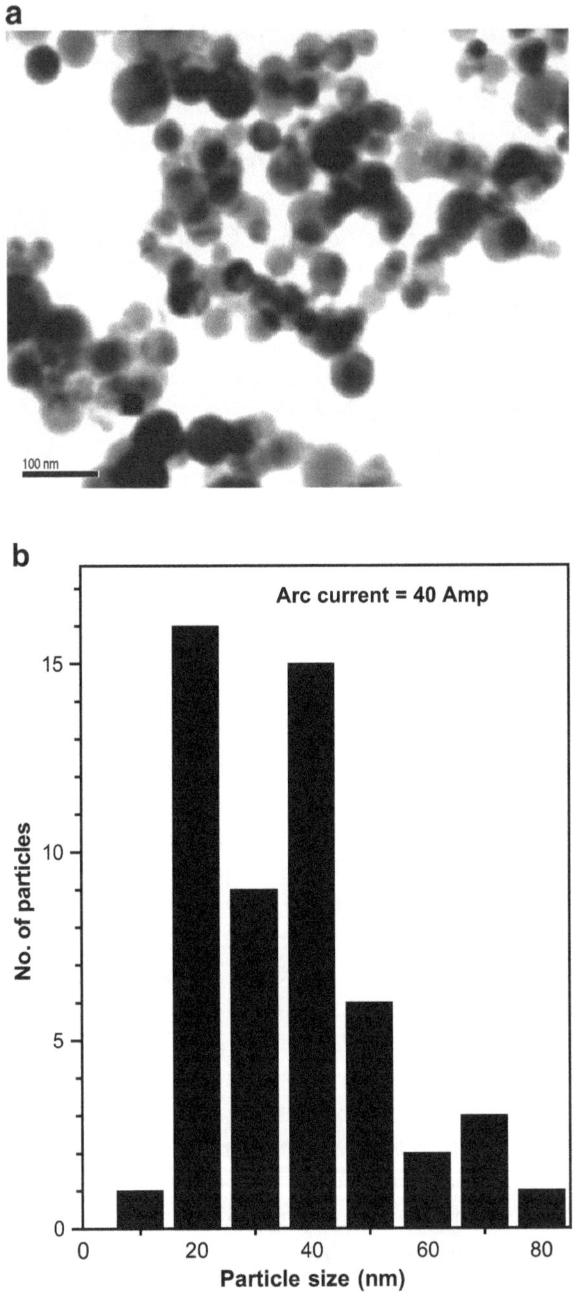

Figure 3. (a) Transmission Electron Micrograph of nanocrystalline γ-Fe_2O_3 powder. (b) A histogram exhibiting a particle size distribution measured from transmission electron micrograph.

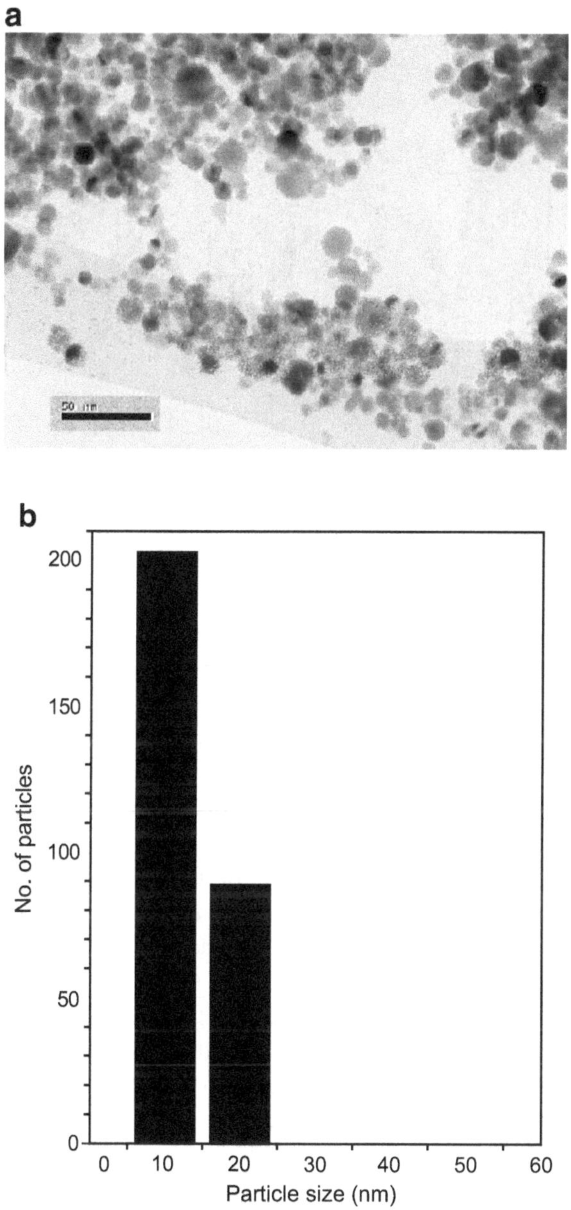

Figure 4. (a) Transmission Electron Micrograph of γ Fe_2O_3 powder collected from EPD film. (b) A histogram showing particle size distribution of γ Fe_2O_3 powder scraped from deposited film.

Figure 3(a) shows a transmission electron micrograph of the powder dispersed on a TEM grid. For recording TEM of the deposited powder, the films were scraped and the collected powder was dispersed in ethyl alcohol. This dispersion was then taken on a TEM grid. Figure 4(a) exhibits a Transmission electron micrograph of the powder deposited using EPD on Si substrate. Figure 3(b) gives

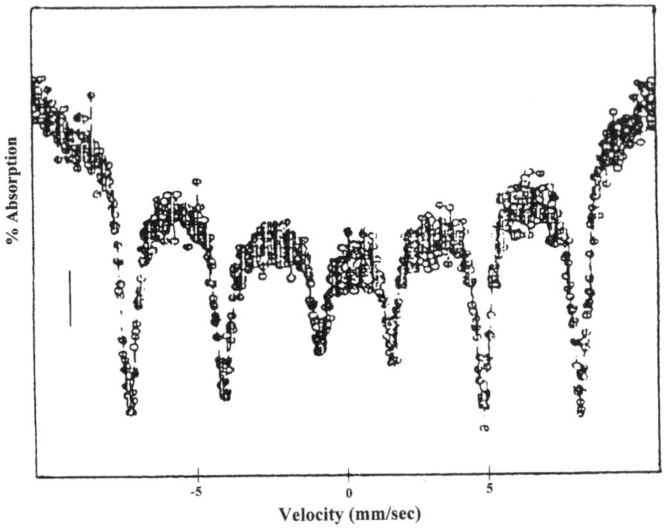

Figure 5. ^{57}Fe Mössbauer spectrum recorded for nanocrystalline γ Fe_2O_3 powder.

a particle size distribution histogram for the precursor powder. The maximum of the histogram lies at 30 nm. TEM micrograph clearly shows the well-dispersed nanoparticles which are almost spherical in shape. Particle size distribution is large and is a typical phenomenon observed in arc plasma method of synthesis. Figure 4(b) shows the particle size distribution histogram for the powder scraped out from the deposited film. Here the particle size distribution seems to have narrowed down along with the reduction in the average particle size (10–20 nm).

^{57}Fe Mössbauer spectra were recorded on the nanocrystalline powder at room temperature (Figure 5), which were analyzed using the standard MOSFIT program. In addition, the spectrum clearly indicates the effects of electron spin relaxation on the total line shape. The broad six-finger pattern is clearly seen to be superimposed upon a non-lorentzian background. The width (FWHM) in the clearly resolved Mössbauer line supports the nanocrystalline nature of the sample [30–32]. The hyperfine interaction parameters namely isomer shift, quadrupole splitting and hyperfine field were found to be 0.36 mm/s, 0.03 mm/s and 493.8 KOe. These values agree with the reported hyperfine interaction data on γ-Fe_2O_3 [30]. Magnetic hysteresis loops for magnetization (M) Vs the magnetic field (H) were recorded for nanocrystalline precursor powder as well as for the electrophoretically deposited films. Figure 6 shows the hysteresis loops for (a) precursor powder and (b) film deposited on crystalline silicon. Coercivity values for the precursor powder and films deposited on crystalline silicon surface are shown in Table I along with the reported magnetic parameters for the bulk γ-Fe_2O_3.

Coercivity value in the parallel direction was found to have increased by three times; however, in the perpendicular direction it was similar to that of the precursor powder. The anisotropy associated with the film geometry evidently

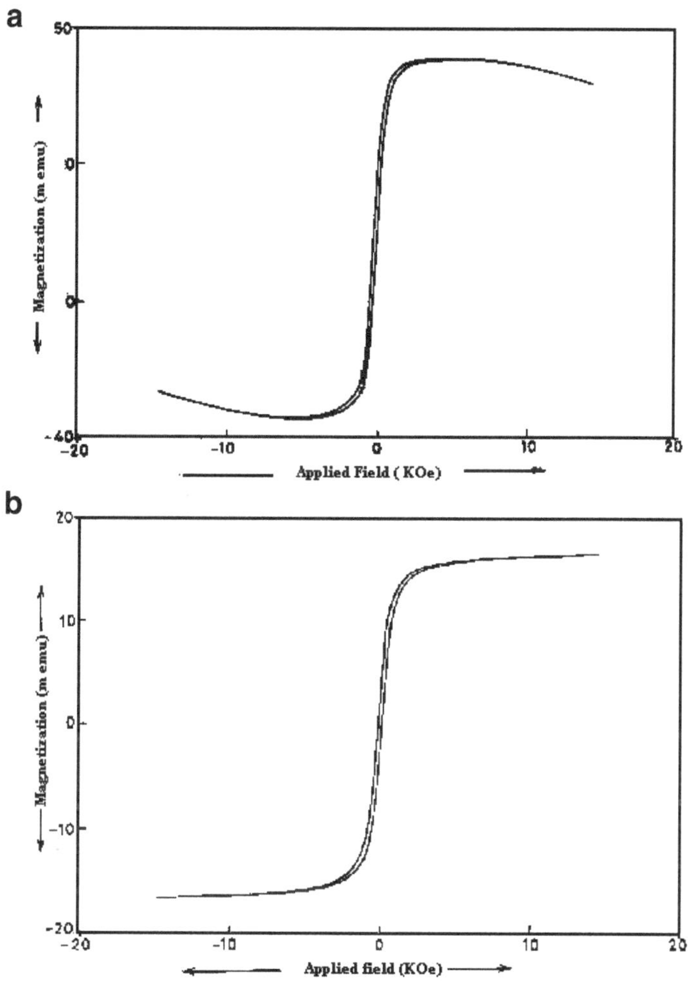

Figure 6. Vibrating sample magnetometer scan measurements for (a) precursor powder, (b) EPD film on Si substrate.

suggests that the magnetic signals originate from the film itself instead of the individual isolated nanoparticles. The values of saturation magnetization were almost same in both the directions for deposited films, however this value was much higher than that obtained for the nanoprecursor. In the Figure 6(a) there is bending of curve. It may be due to non-magnetic mixed phase present or there may be some traces of Fe, so that it may not be going to saturation.

High coercivity associated with anisotropic behavior is an important parameter for the applications of the magnetic material in the recording media. Choice of media influences the way the magnetization is recorded on the disk. Media with particles oriented longitudinally tend to have a much higher remnant magnetization in the longitudinal direction and favor longitudinal recording. Improvements in the performance include improvement in particle alignment and

Table I. Values of magnetization and coercivity for the powder and films of iron oxide as measured by VSM

Sample	Coercivity (Or)		Remnant Magnetization (emu)	
	Parallel	Perpendicular	Parallel	Perpendicular
Precursor powder	170	170	0.0285	0.0285
Powder deposited on silicon substrate	524.7	173.5	0.985	0.02164

morphology. The three-fold increase of the coercivity in the electrophoretically deposited films obviously suggests their unique behavior, which can make them the potential candidate for the recording media.

Moreover, looking at the increase in retentivity in parallel direction that we have obtained, our samples are useful for longitudinal recording purpose. Since the increase in retentivity from powder to films on silicon is approximately five times the films are expected to have very good performance as recording media.

Table I highlights the two parameters a) Coercivity M_{CH} and b) Remnant magnetization M_{RH}. The anisotropy in these two properties points out the possibility of highly oriented domains and their oriented deposition over the silicon substrate on account of its high crystallinity.

4. Conclusion

In conclusion, we have successfully grown films of nanocrystalline maghemite on crystalline silicon which, show a very good increase in saturation magnetization as well as intrinsic coercivity. These films also exhibit anisotropic behavior in coercivity values.

Acknowledgements

One of the authors Tejashree Bhave would like to acknowledge Council for Scientific and Industrial Research for Research Associateship.

References

1. Ichinose N., Ozaki Y. and Kashu S., *Superfine Particle Technology*, Sringer, London, 1992.
2. Gleiter H., *Prog. Mater. Sci.* **33** (1989), 223.
3. Shinde S. R., Kulkarni S. D., Banpurkar A. G., Date S. K. and Ogale S. B., *J. Appl. Phys.* **88** (2000), 3.
4. Pope N. M., Alsop R. C., Chang Y. A. and Sonith A. K. J., *Biomed. Mater. Res.* **28** (1994), 449.
5. Stoner E. C. and Wohlfarth E. P., *Philos. Trans. R. Soc. London Ser. A* **240** (1948), 599.

6. Bauer J., Seeger M., Zern A. and Kronmuller H., *J. Appl. Phys.* **80** (1996), 1667.
7. Weeker J., Schnitzke K., Cerva H. and Grogger W., *Appl. Phys. Lett.* **67** (1995), 563.
8. Yan X. and Xu Y., *J. Appl. Phys.* **79** (1996), 6013.
9. Sanchez R. D., Rivas J., Vazquez-Vazquez C., Lopez-Quintela A., Causa M. T., Tovar M. O. and Seroff S., *Appl. Phys. Lett.* **68** (1996), 134.
10. Schrefl T., Fischer R., Fidler J. and Kronmuller H., *J. Appl. Phys.* **76** (1994), 7053.
11. Hu J., Hu B., Wang K. and Wang Z., *J. Phys., Condens. Matter* **7** (1995), L271.
12. Feltin N. and Pileni M. P., *J. Phys. IV France* (1997), 7.
13. Dormann J. L., Cherkaoui R., Spinu L., Nogues M., Lucari F., D'Orazio F., Fiorani D., Garcia A., Tronc E. and Jolivet J. P., *J. Magn. Magn. Mater.* **187** (1998), L139.
14. Prodan D., Chaneac C., Tronc E., Jolivet J. P., Cherkaour R., Ezzir A., Nogues M. and Dormann J. L., *J. Magn. Magn. Mater.* **203** (1999), 63–65.
15. Kurikka V. P. M. S., Ulman A., Dyal A., Tan X., Yang N.-L., Estournes C., Fournes L., Wattiaux A., White H. and Rafailorich M., *Chem. Mater.* **14** (2002), 1778–1787.
16. Martinez B., O Bradors X., Balcells L., Rouanet A. and Monty C., *Phys. Rev. Lett.* **80** (1998), 1.
17. Bee A., Masssart R. and Neu S., *J. Magn. Magn. Mater* **149** (1995), 6–9.
18. Audram R. G. and Huguenard A. P., U.S. Patent 4302523, 1981.
19. Ziolo R. F., U.S. Patent 4474866, 1984.
20. McMichael R. D., Shull R. D., Swartzendruber L. J., Bennett L. H. and Watson R. E., *J. Magn. Magn. Mater.* **111** (1992), 29.
21. Rosensweig R. E., *Ferrohydrodynamics*, Cambridge, MA, MIT Press, 1985.
22. Takahashi N., Kakuda N., Ueno A., Yamaguchi K. and Fujii T., *J. Appl. Phys.* **28** (1987), 244.
23. Jones H. E., Bissell P. R. and Chantrell R. W., *J. Magn. Magn. Mater* **26** (1990), 21.
24. Chhabra V., Ayyub P., Chattopadhyay S. and Maitra A. N. N., *Mater. Lett.* **26** (1996), 21.
25. Ngo E., Nothwang W., Cole M. and Hubbard C., U.S. Army Research Laboratory, Weapons and Materials Research Directorate, APG, MD 21005.
26. Someswar D., *Bull. Mater. Sci.* **23** (April 2000), 2125–2129.
27. Nelson P. S., Sarkar P. and Datta S., *Am. Ceram. Soc. Bull.* **75** (1996), 48.
28. Datta S., *Bull. Mater. Sci.* **23**(2) (April 2000), 125–129.
29. Madhu Kumar P., Balasubramanian C., Sali N. D., Bhoraskar S. V., Rohatgi V. K. and Badrinarayanan S., *Mater. Sci. Eng., B* **B63** (1999), 215–227.
30. Nikumbh A. K., *J. Mater. Sci.* **25** (1990), 3773–3779.
31. Clark P. E. and Morrish A. H., AIP conference proceedings No. 18, "Magnetism and Magnetic Materials", Bostan, 1937 (American Institute of Physics, New York 1974) p. 1412.
32. Rane K. S., Nikumbh A. K. and Mukhedkar A. J., *J. Mater. Sci.* **16** (1981), 2387.

Hyperfine Interactions (2005) 160:211–217
DOI 10.1007/s10751-005-9166-4

Magnetic and Rheological Characterization of Fe_3O_4 Ferrofluid: Particle Size Effects

KINNARI PAREKH[1,*], R. V. UPADHYAY[2] and R. V. MEHTA[2]
[1]*Physics Department, Faculty of Science, M. S. University of Baroda, Vadodara 390 002, India; e-mail: pkinnarih@rediffmail.com*
[2]*Department of Physics, Bhavnagar University, Bhavnagar 364 002, India; e-mail: rvm@bhavuni.edu*

Abstract. Magnetite particles of different diameters were synthesized by chemical co-precipitation technique and the same are dispersed in dodecane to prepare a magnetic fluid. The results of X-ray diffraction, magnetization measurements, ac susceptibility and viscosity measurements are analyzed and discussed in the text.

Key Words: ac susceptibility, magnetic fluid, size effect.

1. Introduction

During the last few decades much effort has been given to the comprehension of the physical phenomena appearing in magnetic fine particle systems. However, the greatest revolution in this interdisciplinary field has been promoted by the development of new measuring techniques and refinement of synthesis methods, which allow the preparation of the particles at the nanometric scale. At present, it is possible to obtain metallic and oxide particles of a few nanometers in size, either in the form of powder samples, ferrofluids or embedded in a metallic/insulating matrix. One of the main features of the nanomaterials is the fact that their microscopic structure which results from the synthesis method – largely affects the macroscopic properties, giving rise to a wide variety of new phenomena. The remarkable new phenomena observed in nanomaterials arise from the subtle interplay between the intrinsic properties, size distribution of nanoparticle, finite size effects and the interparticle interactions. Finite size effects dominate the magnetic behavior of individual nanoparticles, increasing their relevance as the particle size decreases.

Present work is devoted to tailor the microstructure at the nanometric scale and correlate it to the macroscopic properties. Some of these properties viz. X-ray diffraction, magnetization, ac susceptibility and viscosity measurements are

* Author for correspondence.

discussed here. The sample chosen here is magnetite materials because large no. of papers and applications of ferrofluid contains magnetite as dispersed magnetic materials [1–5].

2. Experimental

2.1. FLUID PREPARATION

The Fe_3O_4 nano-particles were synthesized by co-precipitation technique followed by digestion. After the precipitation, the suspension was kept for digestion at different temperatures and for different time. During this time particles grew and were transformed to crystalline state. The different size of the particles was obtained by application of different digestion temperatures and by changing different base ions for the precipitation. The crystalline nature was confirmed by X-ray diffraction technique. Oleic acid was then added and the mixture was stirred for an hour. The fluid was heated and peptized by adding a small amount of dilute HCl. By magnetic sedimentation the oleic acid coated particles were separated. The suspension containing oleic acid coated particles was repeatedly washed with double distilled water and subsequently washed with acetone to remove the water. Ultimately this acetone wet slurry was dispersed in dodecane and acetone was removed by heating. The fluid thus obtained is stable.

3. Results and discussion

3.1. X-RAY CHARACTERIZATION

The crystal structure of the particles is confirmed using a Bruker D8 Advance powder X-ray diffractometer (Cu Kα radiation, λ = 1.5414 Å). The XRD patterns were analyzed using the Rietveld refinement program [6].

Figure 1 shows the XRD pattern for samples having different particle sizes. The Rietveld refinement of the pattern indicates the formation of a single phase spinel structure. The variation in intensity and linewidth of the XRD pattern reflects the effect of particle size. In case of lowest particle size, it can be seen that there is just a formation of spinel structure and all the peaks are not well developed. But as the particles grow, the width decreases and almost all peaks develop with sharp intensity. Particle size was calculated using Scherrer's formula and is reported in Figure 1.

3.2. MAGNETIZATION MEASUREMENT

The magnetization measurement of the fluid sample was carried out using extraction technique. The instrument was calibrated with a fluid of known magnetization. The response of magnetization measurements for all fluid

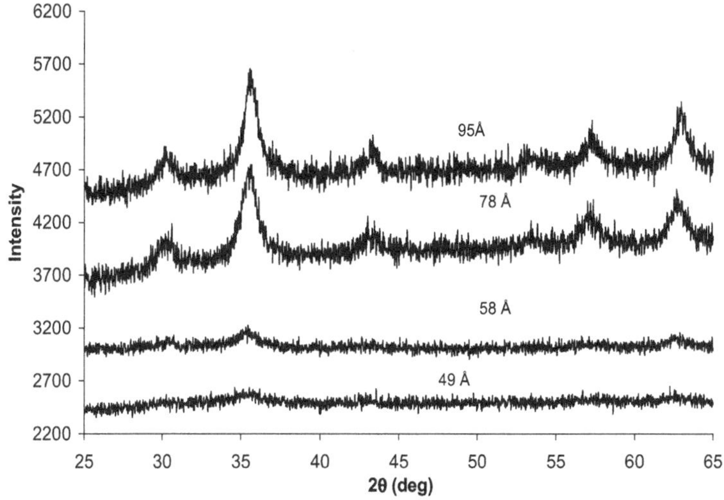

Figure 1. XRD pattern for magnetic particles.

samples is shown in Figure 2. The experimental points were fitted with modified Langevin's theory (line) which includes log-normal particle size distribution. According to this model, the magnetisation M is a function of applied field H and is given by [8].

$$M = \phi M_{\mathrm{d}} \int_0^\infty L(\alpha)P(D)dD \ldots \tag{1}$$

Where, ϕ is particle volume fraction, M_{d} is domain magnetization of the particle and $\alpha = mH/kT$, here m is the magnetic moment of the particle, k is the Boltzmann constant, T is the absolute temperature and $P(D)dD$ is the log-normal distribution function for spherical particles given as

$$P(D)dD = \frac{1}{D\sigma(2\pi)^{1/2}} \exp\left[-\frac{ln(D/D_0)^2}{2\sigma^2}\right] dD \ldots \tag{2}$$

where D is the particle diameter, σ is the standard deviation of ln D and ln D_0 is the mean of ln D. In Figure 2 the solid line is generated using Equation (1). The analysis shows that the domain magnetization of all the samples remain constant though both the particle size and the size distribution changes. The change in particle size is from 80 Å to 105 Å.

3.3. AC SUSCEPTIBILITY MEASUREMENT

The temperature dependent ac susceptibility of both the samples was measured from 77 K to 300 K using an indigenously built ac susceptometer [7]. The data

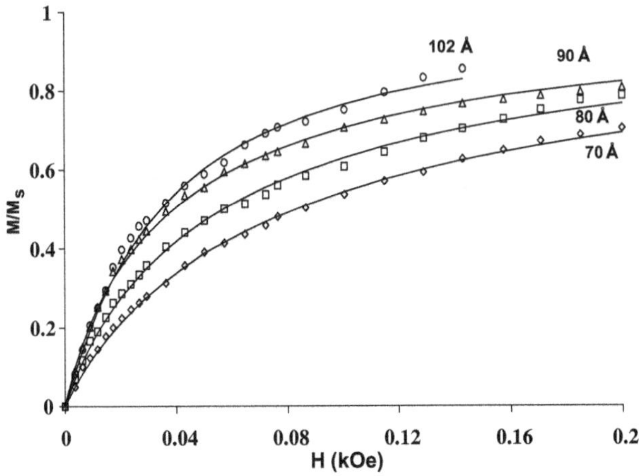

Figure 2. Room temperature magnetization curve.

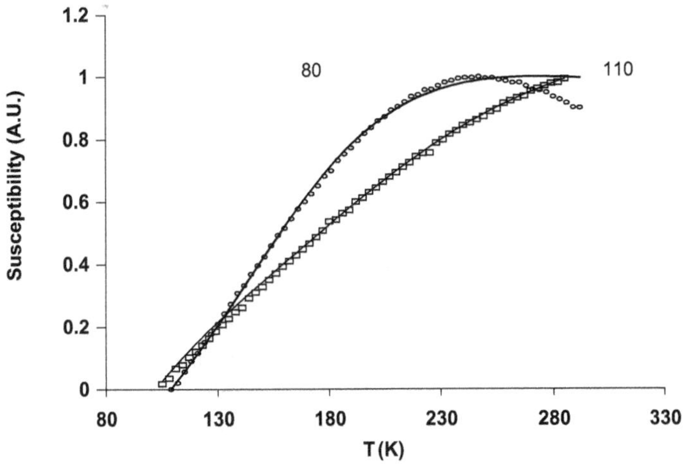

Figure 3. ac susceptibility curve for nano-magnetic particles.

were recorded for the warming cycle. The frequency of measurement for all the samples is kept constant at 211 Hz.

The ac susceptibility measures the reversible part of the magnetization curve. In fine magnetic particle system the peak in susceptibility defines the thermally activated spin reversal which is called as blocking temperature, T_B, of the particles above which all particles are superparamagnetic. Since superparamagnetism depends on the ratio of KV/kT for a given field strength, the peak position will change either by change in the value of the anisotropy constant, K, or by change in the volume, V, of the particle.

Figure 3 shows the temperature dependence of ac susceptibility taken for two different powder samples. The nature of the susceptibility curve can be explained

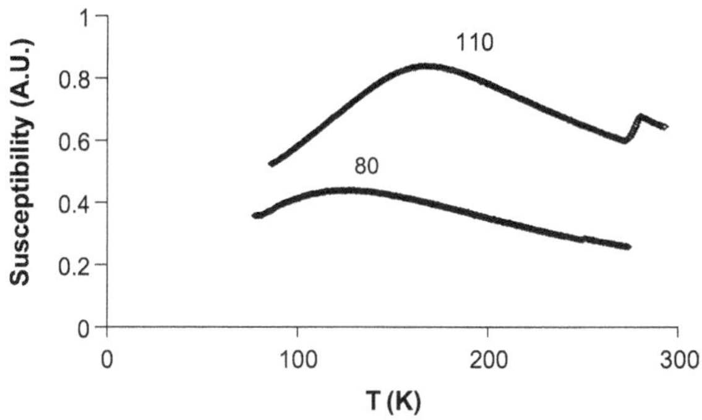

Figure 4. ac susceptibility curve for magnetic fluid samples.

using Neel model [9]. At T_B the relaxation time of thermal activation of grain volume becomes equal to the reciprocal of measuring frequency. According to Neel model each fine particle possess an intrinsic energy barrier and the magnetic vector of the particle can relax over this barrier with a time constant given by $\tau = \tau_0 \exp(\Delta E/kT)$, where τ_0 is a constant of the order of one nano-second. It is evident that the time of measurement is an important factor in determining whether the particle is superparamagnetic or blocked. The susceptibility of a ferrofluid containing monodispersed particles shows a sharp peak at the common blocking temperature of the particles. In present case particles are polydispersed obeying log-normal distribution function; therefore one observes a distribution of blocking temperatures, which will show a broad peak in ac susceptibility. The width of the curve depends on the size distribution. In the low field limit susceptibility can be expressed as [10]

$$\chi(T) = \frac{M_s M_d V_0}{3kT} \exp\left(\frac{3\sigma^2}{2}\right) \int_{-\infty}^{U_p} \frac{1}{\sqrt{2\pi}} \exp\left(\frac{U^2}{2}\right) dU \ldots \tag{3}$$

Where M_s is the saturation magnetisation of the fluid and U_p becomes

$$U_p = \frac{1}{\sigma} \ln\left(\frac{25kT}{KV_0}\right) - 2\sigma \ldots \tag{4}$$

Equation (3) gives the usual Curie form for the susceptibility only when $\sigma \to 0$. In this equation the assumption is made that the domain magnetisation is independent of temperature.

In Figure 3 the line is a theoretical fit to Equation (3) and points are experimental. The values of particle diameter and anisotropy constant are as 80 Å and 110 Å and 4.5×10^5 erg/cm^3 and 5.25×10^5 erg/cm^3, respectively. When the same particles are dispersed in carrier liquid to prepare magnetic fluid

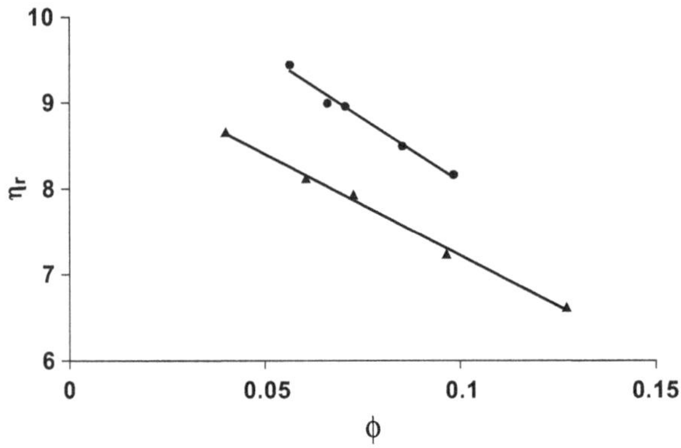

Figure 5. η_r *versus* particle volume fraction ϕ.

the peak temperature will shift to the lower temperature. The same can be seen from Figure 4. Here, the shift in peak temperature is because of reducing the dipole–dipole interaction between the particles. The second peak (near to 300 K) in the Figure 4 is because of the melting point of carrier liquid.

3.4. RHEOLOGICAL MEASUREMENT

The viscosity of the magnetic fluids was studied using a Wells–Brookfield cone/plate digital viscometer model DV-II+. The temperature of the sample cup was maintained up to an accuracy of ±0.01 K with the help of a constant temperature bath.

In absence of magnetic field the rheological behavior of magnetic fluid shows its characteristics like a simple colloid. The shear rate versus shear stress of all the fluids shows a linear behavior, which indicates that the fluids are obeying Newtonian behavior. Further the viscosity of larger size particles is obtained smaller than that of the smaller size particles. This may because of increase of the number density of the smaller size particles. The same can be seen from the Figure 5. Here the term η_r represents the expression $(\eta - \eta_0)/\varphi\eta$. The plot of η_r versus concentration will give a value of hydrodynamic diameter which contains the coating of surfactant layer, δ, on the particle surface [1]. The line in the figure is a plot of Einstein's equation for moderate concentration which is then modified by Rosensweig [1] for the hydrodynamic particle size and the same is given as

$$\frac{\eta - \eta_0}{\varphi\eta} = \frac{5}{2}\left(1 + \frac{\delta}{r}\right)^3 - \left(\frac{\frac{5}{2}\varphi_c - 1}{\varphi_c^2}\right)\left(1 + \frac{\delta}{r}\right)^6 \varphi \ldots \qquad (5)$$

where, φ_c is 0.74, the closest packing for the spherical particles. For the present case the study of two different particle diameters were carried out. From Equation (5) the particle size was obtained from the slope as well as from the intercept. The value obtained from the fit are 63 Å and 71 Å respectively for the 80 Å and 90 Å diameters. The lesser value of the diameter is because of the polydispersity of the sample which is discussed in detail by Kinnari *et al.*, [11].

4. Conclusion

Magnetite particles of different diameters were synthesized by chemical co-precipitation technique. XRD of samples shows the progressive enhancing of the peak intensity with the formation of particles. The width of the peak reduces as the particle size decreases. Magnetization measurements show that the smaller size particles saturates at a higher field compared to the larger size particles. Ac susceptibility of the particles shows the shift in peak temperature with the particle size. This peak will shift again to the lower temperature when the particles are dispersed in the carrier. This is because of reducing the particle–particle interaction in the fluid. Viscosity measurements of the fluid samples confirm the same. For larger size particles the viscosity is less compared to that of the smaller size particles because of reducing the number density of the larger size particles.

Acknowledgements

The work was carried out under DST project No. DST-M-15 sanctioned by Dept. of Science & Technology, Govt. of India, New Delhi. One of the authors (KP) is thankful to DST, New Delhi for providing financial support under FTPYS scheme.

References

1. Rosensweig R. E., *Ferrohydrodynamics*, Cambridge University Press, Cambridge, 1986.
2. Raj K. and Chorney A. F., *Indian J. Eng. Mater. Sci.* **5** (1998), 372.
3. Segal D. L., AERE Harwell, 1980, AERE-R 9976.
4. Jolivet J. P., Chaneac C., Prene P., Vayssieres L. and Tronc E., *J. Phys.* **7** (1997), C1-573.
5. Wu K. T., Kuo P. C., Yao Y. D. and Tsai E. H., *IEEE Trans. Magn.* **37** (2001), 2651.
6. Young R. A., Sakthivel A., Moss T. S. and Paiva-Santos C. O., *User's Guide to Program DBWS-9411 for 'Rietveld Analysis of X-ray and Neutron Powder Diffraction Patterns' with a 'PC' and Various Other Computers*, 1994, p. 21.
7. Jani K., Upadhyay R. V. and Mehta R. V., *Indian J. Eng. Mater. Sci.* **11** (2004), 349.
8. Upadhyay R. V. and Mehta R. V., *Pramana – J. Phys.* **41** (1993), 429.
9. Neel L., *Ann. Geophys.* **5** (1949), 99.
10. Tari A., Popplewell J., Charles S. W., Bunbury P. and Alves K. M., *J. Appl. Phys.* **54** (1983), 3351.
11. Parekh K., Upadhyay R. V. and Mehta R. V., *Indian J. Eng. Mater. Sci.* **5** (1998), 343.

Hyperfine Interactions (2005) 160:219–225
DOI 10.1007/s10751-005-9167-3

Magnetic Behaviour of Nano-Particles of $Ni_{0.8}Cu_{0.2}Fe_2O_4$

S. N. DOLIA* and S. K. JAIN
Department of Physics, University of Rajasthan, Jaipur, India; e-mail: sndolia@uniraj.ernet.in

Abstract. Nano-particles of $Ni_{0.8}Cu_{0.2}Fe_2O_4$ have been synthesized by the co-precipitation method. Langevin function fitting of the superparamagnetic *M–H* curve at 290 K provides a log-normal distribution with median diameter of 30 Å and standard deviation of 0.4. Outside a core of ordered spins, moments in the surface layer are disordered. Magnetization evolves over a long period of time *t* going linearly with log *t*. Magnetic anisotropy which was estimated by fitting the *M*–log *t* curve shows many fold increase over that of bulk particle sample. Major contribution to this enhancement comes form the disordered surface spins.

Key Words: magnetic anisotropy, nano-particles, spinel ferrites, superparamagnetism.

1. Introduction

Particle size has significant effect on the magnetic properties of fine particles. In recent years, magnetic nanoparticles have been a subject of intense research due to their unique magnetic properties [1–6]. Below a critical size, magnetic particles become single domain and exhibit interesting magnetic properties such as superparamagnetism, cluster glass, generally attributed to surface (rather than volume) disorder [7–9]. In nano-particles, change in domain structure causes drastic modifications in magnetic properties such as saturation magnetization and magnetic anisotropy. These unique properties have been made use of in technological application, e.g., information storage [10], ferrofluid technology [11] and magnetocaloric refrigeration [12]. The present work was taken up to study the magnetic behaviour including dynamic behaviour of nanoparticles of $Ni_{0.8}Cu_{0.2}Fe_2O_4$.

2. Experimental details

A ferrofluid of $Ni_{0.8}Cu_{0.2}Fe_2O_4$ in the nano-size range has been prepared using wet chemical process [13] using oleic acid as surfactant and kerosene has been used as the dispersing medium. In order to obtain narrow distribution of particle

* Author for correspondence.

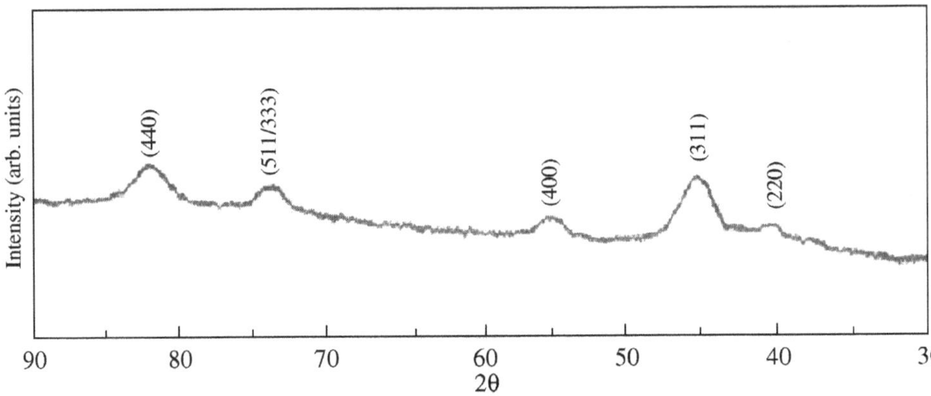

Figure 1. X-ray diffraction of the nanoparticle sample recorded with Fe K_α radiation.

sizes, the fluid was centrifuged at 12,000 rpm. For obtaining dried particles of this ferrite, carrier liquid from the ferrofluid has been removed by repetitive washing with acetone. The powder was then heated at 830°C.

X-ray diffraction (XRD) pattern has been recorded at 290 K, using Fe K_α (1.937 Å) radiation with Mn filter, on a Philips make powder diffractometer model PW 1840. A silicon disc (cubic $a = 5.431$ Å) has been made on the PARC make vibrating sample magnetometer (VSM) model 155. For temperatures down to 20 K, a closed cycle refrigerator cryostat has been used.

3. Results and discussions

Figure 1 shows the indexed XRD pattern recorded at 300 K. The value of lattice parameter is 8.36 Å, which is very close to the value 8.345 Å reported for the bulk sample of $Ni_{0.8}Cu_{0.2}Fe_2O_4$ [14]. This confirms that the sample is formed in single phase cubic spinel structure. Considerably broadened lines in the XRD pattern are indicative of the presence of nano-sized particles. Average crystallite size, estimated with the help of Scherrer equation [15] using the width of 311 reflection is ~30 Å.

Figure 2 show magnetization *versus* magnetic field (*M–H*) curves recorded at 290 K and 20 K. In Figure 2, *M–H* curve recorded at 290 K clearly indicates zero remanence and coercivity which suggest a superparamagnetic behaviour at 290 K and blocking temperature is below 290 K. Figure 3 show hysteresis loops at 20 K recorded after cooling the sample from 290 K in zero field (zfc mode) and after cooling it in a field of 8 kOe (fc mode). This shows two observations (i) even at a magnetic field of 8 kOe, the sample do not indicate sign of magnetic saturation. The saturation magnetization (M_S) has been obtained by extrapolation of M vs. $1/H$ curve to $1/H \to 0$, and it gives a value of ~20 emu/g for M_S, which is much less than 60 emu/g, the value for bulk sample of $Ni_{0.8}Cu_{0.2}Fe_2O_4$. (ii) a large shift of hysteresis loop in fc mode with respect to that in zfc mode. The much reduced

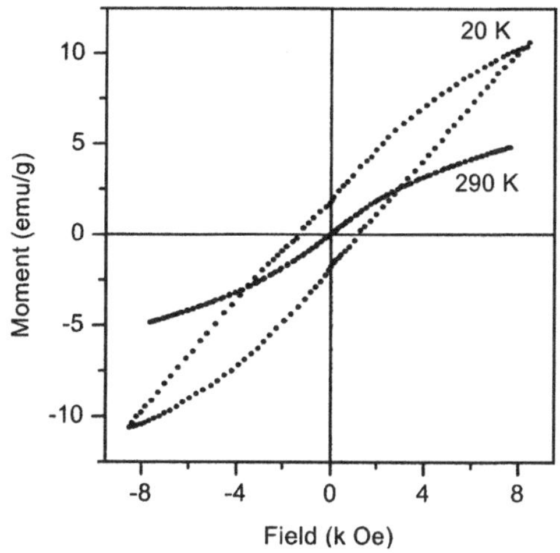

Figure 2. Magnetization-field curves recorded at constant temperatures of 290 K and 20 K.

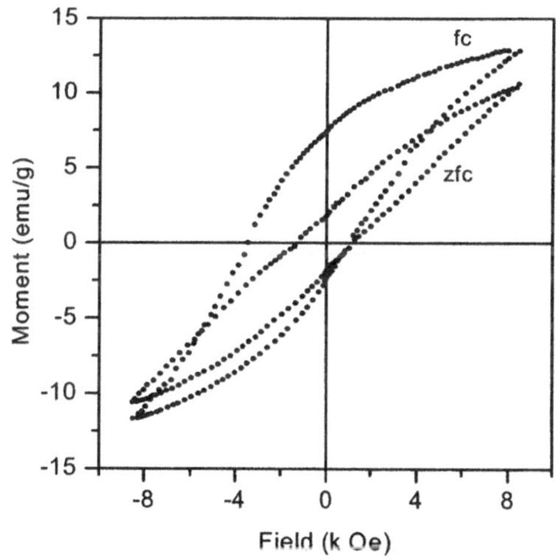

Figure 3. Magnetization-field curves recorded at 20 K after cooling in zfc mode and fc mode (in the field 8 kOe).

M_s and loop shift in the nano-particle sample clearly indicate that outside a core of ordered magnetic moments, those in the surface layer are in a state of frozen disorder [16, 17].

In Figure 4 we show Langevin function fitting on the *M–H* curve at 290 K. Assuming a log-normal distribution for the particle sizes and taking the reported value of M_S = 60 emu/g for bulk particles of $Ni_{0.8}Cu_{0.2}Fe_2O_4$, the Langevin function fitting is obtained with median diameter D_m = 30 Å and standard deviation $\sigma = 0.4$.

Figure 5 shows variation of magnetization *M* as a function of temperature in the range 20 K–290 K in an external field of 45 Oe recorded in zfc and fc modes.

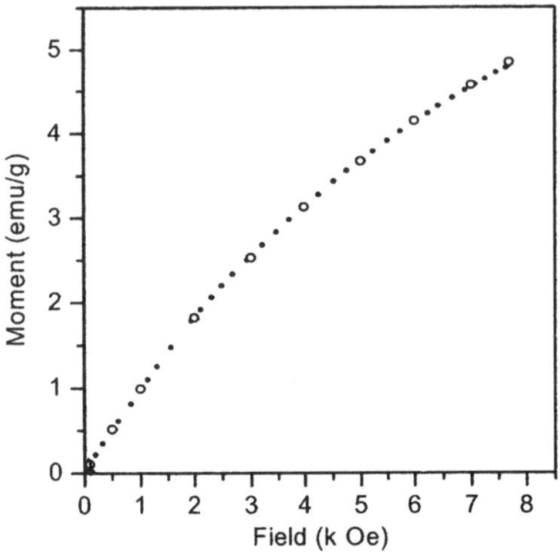

Figure 4. Magnetization-field curves recorded at 290 K along with Langevin function fitting (*open circles*).

In the zfc mode, the sample was cooled in the zero field from 290 K to 20 K and after stabilization of the temperature, a magnetic field of 45 Oe and then measurements were carried out while heating the sample.

In Figure 5, zfc curve exhibits a peak at ~125 K which is suggestive of blocking at 125 K. The blocking temperature (T_B) is the threshold point of thermal activation for the whole nanoparticle sample. When the nanoparticles are cooled in zero magnetic field, the magnetization direction of each nanoparticle align with its easy axis among the nanoparticles, overall susceptibility is almost zero since the applied field is too small to over come the anisotropy alone. Above T_B, magnetic anistropy is overcome by thermal activation and consequently the nanoparticles become superparamagnetic and show paramagnetic properties. Figure 5 also show a divergence between zfc and fc curve, which is a characteristic feature of superparamagnetic behaviour. Such a divergence originates from the anisotropy barrier blocking of the magnetization orientation in the nanoparticles cooled with a zfc process [18].

Further, for studying temporal relaxation of magnetization, we have recorded M as a function of observation time t (in seconds) after zero field cooling of the sample in a field of 500 Oe and at a constant temperature of 20 K. Figure 6 plots the same as a function of log t. Observed near linear variation of M with log t is suggestive of the existence of a distribution of energy barriers. Magnetic anisotropy has been estimated by fitting the observed M–log t curve. Figure 6 depicts the fitting as well.

For arriving at the fitting, following steps have been followed: (i) magnetization m' has been calculated for particles of different sizes, for an observation time t, using the relation $\tau = \tau_o \exp(KV/k_BT)$ [7] where τ_o is assumed to be 10^{-9} s and K is anisotropy constant, V is the volume of the nanoparticle and T is the temperature and (iii) total m for a sample of distributed sizes has been

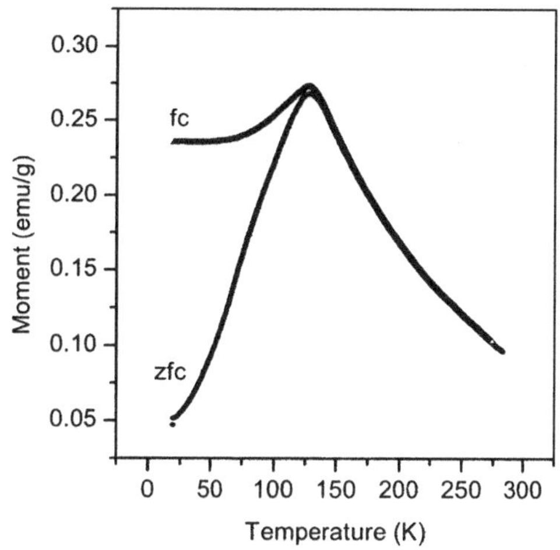

Figure 5. Magnetization-temperature curve recorded in zfc and fc modes (cf. text) in an external magnetic field of 45 Oe.

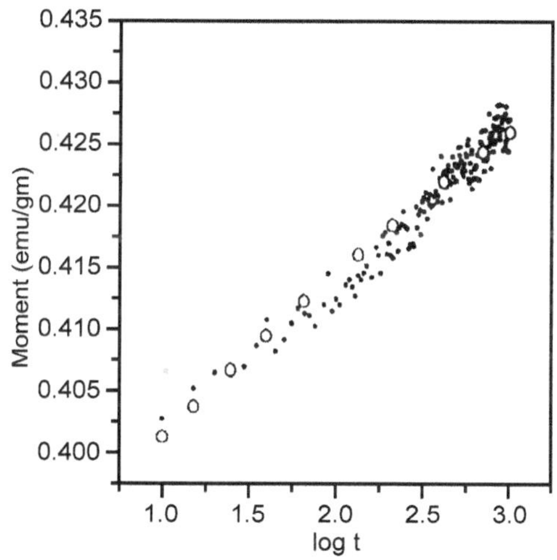

Figure 6. Magnetization M as a function of logarithm of observation time t (in seconds) recorded in zfc mode at the temperature of 20 K. Calculated points are also shown for some observation times by *open circles*.

obtained by summing m' for particles of different sizes weighted with the log-normal distribution. Proceeding in this way m has been computed for different observation times t and then these values have been normalized to the experimental value at some particular t. Fitting of the M–log t curve, using the particle size distribution of $D_m = 30$ Å and $\sigma = 0.4$, provides the value of $K = 820$ kJ/m^3. A comparison with the values reported for bulk particle sample of $NiFe_2O_4$, $K = -10.6$ kJ/m^3 and $K = -6.7$ kJ/m^3 at 90 K and 290 K, respectively [19], shows a huge increase in anisotropy for the nano-particle sample over the bulk particle sample. The above reported values suggest that the observation temperature causes a small change in anisotropy. Thus nano-sizing of the particles results in increase in magnetic anisotropy by several folds. Major contribution to this enhancement comes from the disordered surface spins.

4. Conclusions

Nano-particles of $Ni_{0.8}Cu_{0.2}Fe_2O_4$ have been synthesized by chemical co-precipitation method. The sample shows typical superparamagnetic behaviour with a blocking temperature of 125 K. The dc magnetization measurements show that outside a core of ordered moments, those in the surface layer are disordered. An estimate of magnetic anisotropy by fitting the relaxation curve is ~820 kJ/m^3 which is quite large as compared to a ferrite of a close by composition in its bulk particle state. Disordered spins make a large contribution to the increased anisotropy.

Acknowledgements

The authors thank Prof. A. Krishnamurthy, Prof. B. K. Srivastava and Dr. Subhash Chander for helpful discussions and Ms. Seema Lakhanpal for assistance in sample preparation.

References

1. Morrish A. H. and Haneda K. H., *J. Appl. Phys.* **52** (1981), 2496.
2. Lin D., Numes A. C., Majkrzak C. F. and Berkowitz A. E., *J. Magn. Magn. Mater.* **145** (1995), 343.
3. Jonsson T., Svedlindh P. and Nordblad P., *J. Magn. Magn. Mater.* **140–144** (1995), 401.
4. Chantrell R. W., Coverdale G. N., El-Hilo M. and O'Grady K., *J. Magn. Magn. Mater.* **157** (1996), 250.
5. Dorman J. L. and Fiorani D. (ed.), In: *Magnetic Properties of Fine Particles*, North Holland, Amsterdam, 1992.
6. Dorman J. L., Fiorani D. and Tronc E., In: Prigogine and Rice S. A. (eds.), *Advanced Chemical Physics*, Vol. XCVVII, Wiley, New York, 1997.
7. Been C. P. and Livingston T. D., *J. Appl. Phys.* **30** (1995), 120S.
8. Leslie-Pelecky D. L. and Rieke R. D., *Chem. Mater.* **8** (1996), 1770.
9. Martinez B., Obradors X., Balcells L., Rouanet A. and Monty C., *Phys. Rev. Lett.* **80** (1998), 181.
10. Sun S. and Murray C. B., *J. Appl. Phys.* **85** (1999), 4325.
11. Raj K., Moskowitz R. and Casciari R., *J. Magn. Magn. Mater.* **149** (1995), 174.
12. McMichael R. D., Shull R. D., Swartzendruber L. J. and Bennett L. H., *J. Magn. Magn. Mater.* **111** (1992), 29.
13. Upadhayay R. V. and Mehta R. V., *Pramana-J. Phys.* **41** (1993), 429.

14. Manjura Hoque S., Amanullah Choudhury M. and Islam F., *J. Magn. Magn. Mater.* **251** (2002), 292.
15. Cullity B. D. (ed.), In: *Elements of X-ray Diffraction*, Addition Wesley, Reading, Mass., 1959, p. 132.
16. Chander S., Kumar S., Krishnamurthy A., Srivastava, B. K. and Aswal V. K., *Pramana-J. Phys.* **61**(3) (2003), 617.
17. Kodama R. H. and Beronitz A. E., *Phys. Rev.* **B29** (1999), 632.
18. Rondinone A. J., Samia A. C. S., Zhang Z. J., *J. Phys. Chem.* **B103** (1999), 6876.
19. Chikazumi S. (ed.), In: *Physics of Magnetism*, Robert E. Kreiger Publ. Co., Florida, 1978.

Hyperfine Interactions (2005) 160:227–233
DOI 10.1007/s10751-005-9155-7

Evidence of Clustering in Heusler like Ferromagnetic Alloys

N. LAKSHMI*, RAM KRIPAL SHARMA and K. VENUGOPALAN
Department of Physics, M.L. Sukhadia University, Udaipur 313 001, India;
e-mail: nambakkat@yahoo.com

Abstract. Two Heusler like alloys Fe_2CrAl, and Mn_2CoSn had been prepared and studied using the Mössbauer effect. Both these alloys were ferromagnetic in nature. Their Mössbauer spectra showed the co-existence of a paramagnetic portion along with magnetic hyperfine part upto temperatures well below their Curie temperatures. Low temperature Mössbauer spectra for both these alloys showed a steady variation in the intensity of the paramagnetic portion. X-ray diffraction studies made on these samples ruled out the possibilities of a separate phase. Explanation of the observed phenomena is given by clustering around the magnetic ions present in these samples.

Recently there has been a lot of interest in Heusler alloys because theoretical band structure calculations [1] predict that some of the Heusler alloys are half-metallic ferromagnets. Although Groot *et al.* [2] had predicted the Heusler alloys NiMnSb and PtMnSb to be half-metallic as early as 1983; this property has given rise to a new surge of interest in these systems due to the possibility of use of Heusler alloys as material for spin electronic devices. These systems are attractive because of the considerable change in their magnetic properties that can be made by altering the chemical ordering, substituting for one or more components by other atoms, mechanical treatment etc.

Ternary alloys of stoichiometric composition bearing the general formula X_2YZ are called Heusler alloys. In this class of alloys, X and Y are generally transition elements and Z an sp element. These alloys offer excellent systems for studying magnetic interactions. Many of these alloys can be held at various degrees of order leading to a corresponding change in their magnetic properties. The study of clustering effects is important in both nano as well as bulk materials. It is possible to modify the magnetic properties of such materials by the introduction of proper element to prepare the alloys [3, 4]. In this paper we report the results of Mössbauer studies using Fe-57 and Sn-119 on the Heusler

* Author for correspondence.

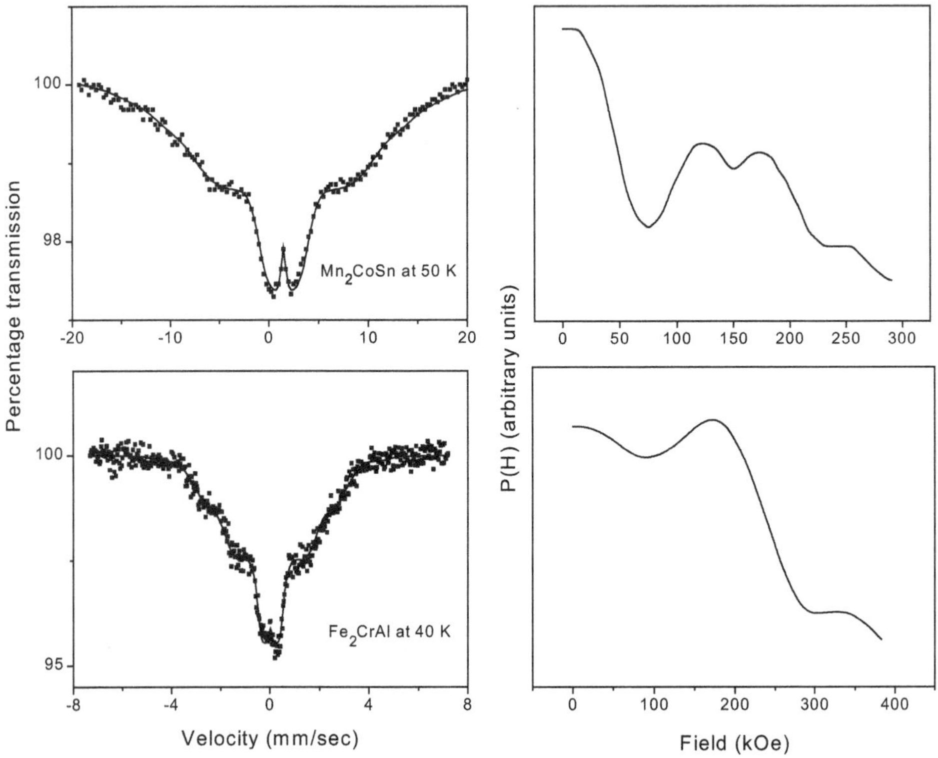

Figure 1. Mössbauer spectra and hyperfine field distributions of Fe_2CrAl at 40 K and Co_2MnSn at 50 K.

like alloys Fe_2CrAl and Mn_2CoSn. The resulting spectra are discussed in terms of clustering around the magnetic atoms in these alloys.

Constituent materials of atleast 99.99% purity (obtained from M/s Spex Inc., U.S.A) were weighed out in stoichiometric proportions for sample preparation. These were then melted in argon atmosphere turning over the melted buttons and re-melting several times to ensure homogeneity. The buttons were then crushed and packed into quartz ampoules evacuated to 10^{-5} torr and kept for annealing. They were annealed at 800°C for 72 h for Fe_2CrAl and 7 days for Mn_2CoSn respectively and allowed to cool down to room temperature in the furnace itself. The sample of Fe_2CrAl used for the study reported here was one that had been further annealed at 200°C for 3 days more. Mn_2CoSn was further annealed at 400°C for a period of 6 days. X-ray studies were made at room temperature using Cu-Kα radiation. Mössbauer spectra for Fe_2CrAl were recorded using a 25 mCi Co-57 (Rh) source over a temperature range of 300 K to 40 K. A 5 mCi Sn-119 in $CaSnO_3$ matrix was used as source for Mn_2CoSn samples. Spectra were recorded for Mn_2CoSn over 670 K to 50 K. Low temperature Mössbauer measurements were carried out using a closed cycle refrigerator. The temper-

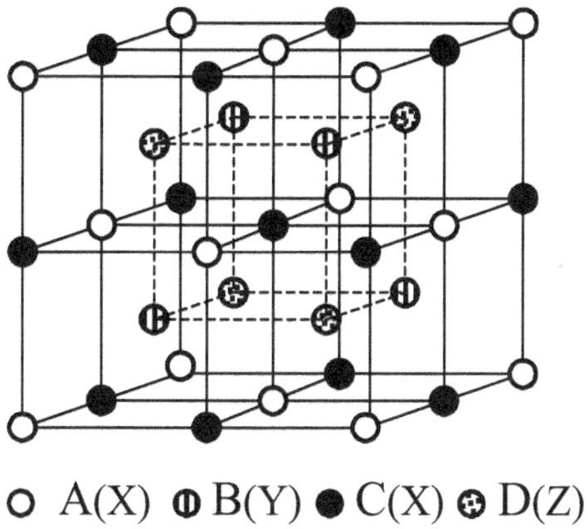

Figure 2. L_{21} structure for X_2YZ type Heusler alloys.

atures were kept steady within 2 K. Figure 1 gives the Mössbauer spectra as well as hyperfine field distributions of Fe_2CrAl at 40 K and of Co_2MnSn at 50 K.

Centroids of the Mössbauer spectra taken for Fe_2CrAl were seen to lie very close to the zero velocity channel. Curie temperature was therefore determined using the zero velocity thermal scan method and was found to be 297 K ± 2 K. Curie temperature for Mn_2CoSn was determined using an a.c. susceptibility equipment and was seen to be 610 ± 5 K. Spectra of both samples well above their T_C showed single lines with no evidence of any quadrupole splitting. Moreover, the centroids of the paramagnetic and magnetic hyperfine portions coincided for both samples ruling out the presence of another chemical phase in these samples.

X-ray diffractograms taken for both alloys showed them to be single phased. The structures were found to conform to the B_2 type. A fully ordered Heusler structure for a material with a general formula X_2YZ is of the L_{21} type. This structure consists of four sets of interpenetrating f.c.c planes A, B, C and D (shown in Figure 2). The A and C sublattices are equivalent and are occupied by X atoms, sublattice B by Y atoms and sublattice D by Z atoms. However, these samples have a B2 type structure. Cr and Fe in the case of Fe_2CrAl and Mn and Co for Mn_2CoSn have, therefore, randomly entered the A[C] and B sites. The lattice parameter $2a_0$ was found to be 5.807 Å and 6.011 Å for Fe_2CrAl and Mn_2CoSn, respectively. Bulk magnetization measurements for both these alloys showed them to be simple ferromagnets. Table I gives the hyperfine field values and integrated intensities (probabilities) of the field components for these samples in the field distributions.

Fe_2CrAl is a part of the series $Fe_{3-x}Cr_xAl$ which had been prepared and studied [5]. At about $x = 0.5$, the Mössbauer spectrum begins to show the presence of a paramagnetic portion co-existing along with the magnetic hyperfine part. The contribution of this paramagnetic part to the spectrum increases with Cr content. Figure 3 gives the probability contribution of the

Table I. Component field values and probabilities of field contributions to the hyperfine distribution in Mössbauer spectra

	Field component	Field (kOe)	P (H) (arbitrary units)
Fe_2CrAl at 40 K	H_{zero} (paramagnetic)	–	0.024
	H_1	176.5	0.021
	H_2	335.0	0.0026
	$H_{average}$	138.0	–
Mn_2CoSn at 50 K	H_{zero} (paramagnetic)	–	0.043
	H_1	120.0	0.033
	H_2	180.0	0.032
	H_3	260.0	0.024
	$H_{average}$	121.5	–

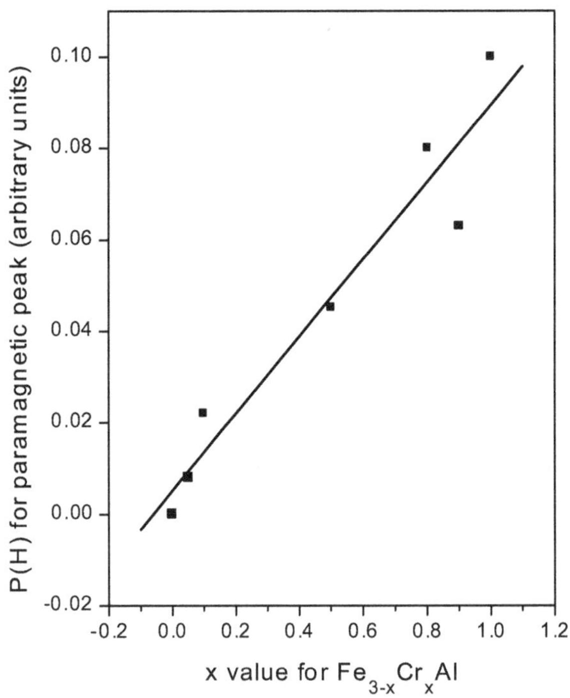

Figure 3. Probability contribution of the paramagnetic portion in the hyperfine field distribution as a function of Cr content in $Fe_{3-x}Cr_xAl$.

paramagnetic portion (integrated peak intensities) in the hyperfine field distribution as a function of Cr content in the $Fe_{3-x}Cr_xAl$ series for Mössbauer spectra at room temperature. It can be seen that there is a nearly linear increase in the paramagnetic contribution to the Mössbauer spectrum with increase in Cr content.

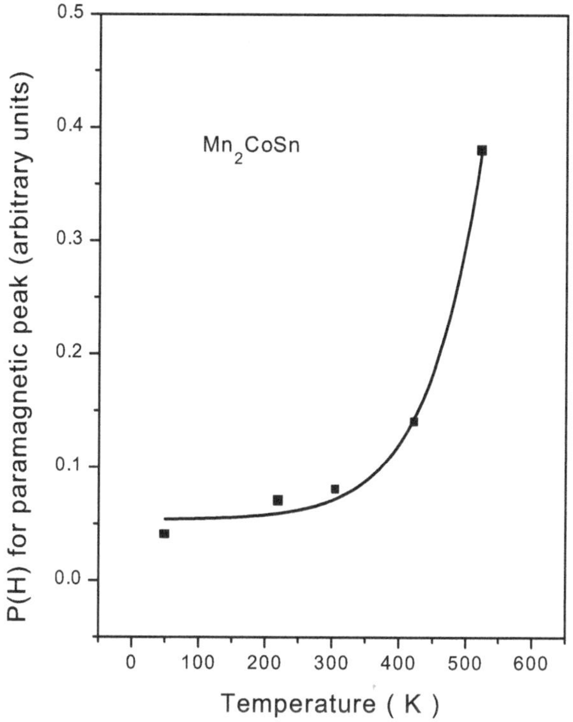

Figure 4. Trend of the zero field probability with temperature in Mn_2CoSn.

Mössbauer spectra were recorded over a range of 40 to 300 K. The paramagnetic portion which had co-existed with the magnetic hyperfine part in the room temperature spectrum was present in all other spectra, down to 40 K. These were fitted using the Window's [6] method of fitting for a distribution in the hyperfine fields. The line widths were constrained to have widths corresponding to α-iron. The distribution curves show one high intensity peak at low field values and a low intensity peak at higher field values. For the spectrum at 40 K, these correspond to 176.5 kOe and 335 kOe. Both these are broadened with $\Delta H/H$ in excess of nearly 30% [7] in all the spectra, pointing to the considerable disorder present. ΔH being the peak width about peak position H in the hyperfine distribution. In addition to these there is a peak corresponding to the zero magnetic field (the paramagnetic part in the spectrum). X-ray studies have ruled out the presence of any other chemical phases in this alloy. The average hyperfine field for this spectrum is 138 kOe. The value of the average hyperfine field is much lower than the weighted averages of both these field values because of the contribution of the paramagnetic part. From Table I it can be seen that the probabilities (integrated intensities) for the zero field components (paramagnetic part) and the 176.5 kOe peak in the distribution are nearly equal.

The coexistence of paramagnetic peak well below the Curie temperature points to the wide spread in the relaxation time in the system. Such a relaxation phenomenon is also seen in Fe_3Al system [8]. This has been ascribed to the fluctuations that occur in the number of next neighbour atoms leading to an appreciable dispersion of the relaxation time. The effect of addition of Cr into disordered Fe_3Al resulted in the preferential population of Cr atoms into the sites that would have been normally occupied by Fe in an ordered system [5]. Thus the coexistence of paramagnetic peak in the Fe_2CrAl hyperfine spectrum can therefore be attributed to the clustering of Cr atoms.

A study of the series $Fe_{3-x}V_xAl$ [9] showed that despite good chemical ordering, the Mössbauer spectra showed the coexistence of paramagnetic as well as magnetic structures with an increase in the paramagnetic contribution with increase in V content. The authors have explained the observed behaviour in terms of presence of magnetic clusters. Similarly Ritcey *et al.* [10] had observed that for a sample of $Co_2CrAl_{0.98}Sn_{0.02}$, the Mössbauer spectrum at tin site showed a six line Lorentzian along with a central singlet. They had also reported that the singlet peak was found to increase substantially in amplitude upon annealing. No theory was put forward to explain this phenomenon.

From the appearance of the paramagnetic portion at a particular concentration of Cr along with the continuous increase with increase in Cr concentration, as also the persistence of this paramagnetic part over a very large temperature range, well below T_C, indicates the presence of Cr clusters which screen some of the Fe atom from neighbouring Fe atoms, which are the only magnetic atoms in these alloys which give rise to the hyperfine fields.

In this sample, the magnetic moment is carried by the Co atom [11]. As in the case of Fe_2CrAl, it was seen that this sample too had formed the (disordered) B2 structure. However, it is single phased. Sn-119 Mössbauer spectra were recorded over a range of 50 K to 670 K. In all these spectra co-existence of a magnetic hyperfine component with a central paramagnetic peak was observed at all temperatures [12]. The spectra were fitted using the Window method with widths constrained to those of natural Sn. The field distributions had three distinct field components along with a zero probability component. For the spectrum at 50 K, the field values corresponded to 120 kOe, 180 kOe and 260 kOe (Table I). In this case too $\Delta H/H$ values were large, approximately 40%. The average field value was 121.5 kOe. The paramagnetic contribution is quite large (Table I) which lead to the lowering of the average hyperfine field value in Mn_2CoSn.

Figure 4 gives the trend of the zero field probability (from the paramagnetic contribution) with temperature. In this alloy, Co is the magnetic moment carrying ion, and so a fluctuation in the number of next neighbour atoms of Sn which senses the hyperfine field, would lead to a spread in the relaxation times. Thus the coexistence of paramagnetic peak in Mn_2CoSn can be understood to be due to clustering of Mn atoms around some of the Sn atoms. Another such disordered alloy, Mn_2NiSn was prepared and studied in an identical manner [12]. Interestingly, this alloy behaved like an ordinary ferromagnet and did not show the presence of any paramagnetic peak co-existing with the magnetic hyperfine spectrum, indicating that the disorder in this case was confined to a nearly equal distribution of Ni and Mn between the A[C] and B sites.

Thus in conclusion we see that presence of Cr clusters around Fe in the case of Fe_2CrAl and of Mn around Co in Mn_2CoSn in both cases gave rise to the co-existence of a paramagnetic portion with the magnetic hyperfine part in their Mössbauer spectra. From the values of the probability of the zero field contribution to the hyperfine field distributions in these alloys, it can be concluded that in both these alloys about half of the magnetic ions are shielded by clustering of the non-magnetic atoms leading to a change in the properties at a microscopic level. Such granularity is not visible at the macroscopic level since as already observed, magnetization measurements showed these to behave like ordinary ferromagnets.

Acknowledgements

This work has been supported by the UGC-DRS and COSIST schemes of the Department of Physics, M.L. Sukhadia University, Udaipur. One of us (Ram Kripal) acknowledges financial support by CSIR.

References

1. Nanda B. R. K. and Dasgupta I., *J. Phys., Condens Matter* **15** (2003), 7307.
2. de Groot R. A., Mueller F. M., van Engen P. G. and Buschow K. H. J., *Phys. Rev. Lett.* **50** (1983), 2024.
3. Jay J. P., Wójcik M. and Panissod P., *Z. Phys., B* **101** (1996), 471.
4. Alvarado-Leyva P. G. and Dorantes-Dávila J., *Superf. Vacío* **13** (2001), 117.
5. Lakshmi N., Venugopalan K. and Varma J., *Phys. Rev., B* **47** (1993), 14054.
6. Window B., *J. Phys., E* **4** (1971), 401.
7. Lakshmi N., Venugopalan K. and Varma J., *Pramana J. Phys.* **59** (2002), 531.
8. Czer L., Dezsi I., Keszthelyi L., Ostanevich L. and Pal L., *Phys. Lett.* **19** (1965), 55.
9. Tuszynski M., Zarek W. and Popiel E. S., *Hyperfine Interact.* **59** (1990), 369.
10. Ritcey S. P. and Dunlap R. A., *J. Appl. Phys.* **55** (1984), 2051.
11. Surikov V. V., Zhordochkin V. N. and Yu Astakhova T. *Hyperfine Interact.* **59** (1990), 469.
12. Lakshmi N., Pandey A. and Venugopalan K., *Bull. Mater. Sci.* **25** (2002), 309.

Hyperfine Interactions (2005) 160:235–239
DOI 10.1007/s10751-005-9156-6

Influence of 50 MeV Lithium Ion Irradiation on Hyperfine Interactions of $Cr_{0.5}$ $Li_{0.5}$ Fe_2O_4

S. G. PARMAR[1], RAVI KUMAR[2] and R. G. KULKARNI[3,*]
[1]*Department of Physics, Saurashtra University, Rajkot 360005, India*
[2]*Nuclear Science Centre, New Delhi 110067, India*
[3]*Physics Department, Shivaji University, Kolhapur 416004, India;*
e-mail: rgkrajkot@hotmail.com

Abstract. Spinel oxide $Cr_{0.5}$ $Li_{0.5}$ Fe_2O_4 has been irradiated at Nuclear Science Centre, New Delhi, by 50 MeV lithium ions of fluence $5*10^{13}$ ions/cm^2 and irradiation effect on hyperfine interactions has been investigated by Mossbauer spectroscopy. The Mossbauer spectrum of irradiated sample shows no paramagnetic doublet contribution and the hyperfine fields corresponding to the Fe^{3+} in the octahedral (B) and the tetrahedral (A) sites are very well separated. That is the observed superimposed A and B sites in unirradiated sample are split into separate lines after Li irradiation. Further an increase of the intensity of the lines (2)–(5) with respect to (1)–(6) signals an orientation of the hyperfine magnetic field towards a direction perpendicular to the ion path due to the irradiation induced strain by the latent tracks. The computer simulation of Mossbauer spectra indicated that the irradiated Fe^{3+}-site occupancy of the A-site hyperfine field increased from 43% to 55% whereas the B-site hyperfine field decreased from 57% to 45% compared to unirradiated sample.

1. Introduction

Swift heavy ion irradiation (SHI) on magnetic oxides causes micro-structural defects and disorder, which affects magnetic properties. In materials like $Y_3Fe_5O_{12}$, $BaFe_{12}O_{19}$ and $NiFe_2O_4$, the magnetic properties are very sensitive to the irradiation-induced disorder, which results in a decrease of the saturation magnetization [1, 2]. Mossbauer spectrometry experiments have verified [3] that a paramagnetic phase is induced. The damage induced in magnetic insulators by high-energy heavy ion irradiation has resulted in the formation of latent tracks. Recently, such latent tracks have been observed by high-resolution electron microscopy (HREM) in magnetic oxides, a garnet $Y_3Fe_5O_{12}$ and a magnetoplombite $BaFe_{12}O_{19}$ [1, 2]. The induced chemical and physical modifications by Xenon ions (3.1 GeV) in magnetic oxides with spinel structure MFe_2O_4, with $M = Mg^{2+}$, Ni^{2+}, Fe^{2+} and Zn^{2+} have been studied [4] by magnetization mea-

* Author for correspondence.

surements, Mossbauer spectroscopy and HREM. These results [4] indicate that continuous trails of damage, called discontinuous latent tracks, have been observed in $NiFe_2O_4$ and $MgFe_2O_4$. Conversely electron microscopy has shown that the spinel $ZnFe_2O_4$ and the magnetite Fe_3O_4 compounds exhibit no discontinuous tracks, but extended spherical defects aligned along the ion wakes [4]. The study of the heavy ion irradiation induced change in the magnetic properties of Fe_3O_4 [5] has led to the hypothesis that magnetic domain walls could be pinned by the extended defects.

The most of the high-energy heavy ion irradiation studies reported above [1–5] on magnetic oxides are in energy range of GeV. So far no irradiation studies with light ions like lithium, oxygen etc in the hundred MeV range on spinel ferrites are reported. Moreover, substituted lithium ferrites of the type $Li_{0.5}Fe_{2.5-x}M_xO_4$ with M = V, Cr, and Rh [6] and M = Ga [7] have become attractive for microwave applications as replacement for garnets and other spinel ferrites. Motivated by the above observations, it is decided to study the effect of 50 MeV Li ion irradiation on hyperfine interactions of $Cr_{0.5}$ $Li_{0.5}$ Fe_2O_4 by Mossbauer sectroscopy.

2. Experimental

The samples of Cr^{3+} and Li^{1+} co-substituted spinel ferrite composition $Cr_{0.5}$ $Li_{0.5}$ Fe_2O_4 were prepared by the standard ceramic method using the oxides of Fe_2O_3, Cr_2O_3 and Li_2O_3, all 99.9% pure supplied by E. Mereck. Thin pellets of polycrystalline sample of $Cr_{0.5}$ $Li_{0.5}$ Fe_2O_4 having diameter 0.5 mm was irradiated at the Nuclear Science Centre, New Delhi by 50 MeV lithium ions of fluence $5{*}10^{13}$ ions/cm^2 using 15 UD pelletron facility. The X-ray diffraction (XRD) patterns of the sample before and after Li ion irradiation were obtained using FeK_α radiation on the Philips X-ray diffractrometer (PM9920) for checking any changes in structure. The Mossbauer spectra of the sample before and after irradiation were recorded in the transmission geometry with a constant acceleration transducer and a 512 multichannel analyser operating in time mode. A gamma source of ^{57}Co(Pd) of 10 mCi was used.

3. Results and discussion

The X-ray diffractograms of both irradiated and unirradiated $Cr_{0.5}$ $Li_{0.5}$ Fe_2O_4 samples showed sharp lines corresponding to single-phase cubic spinel. The observed XRD peaks were modelled by modified by Gaussian functions and refined unit cell parameters calculated using the least squares program are 8.2554 A^0 and 8.2405 A^0 for irradiated and unirradiated samples, respectively. The lattice parameter of the irradiated sample (8.2554 A^0) is slightly larger than the unirradiated one (8.2405 A^0) suggesting irradiation has expanded the lattice.

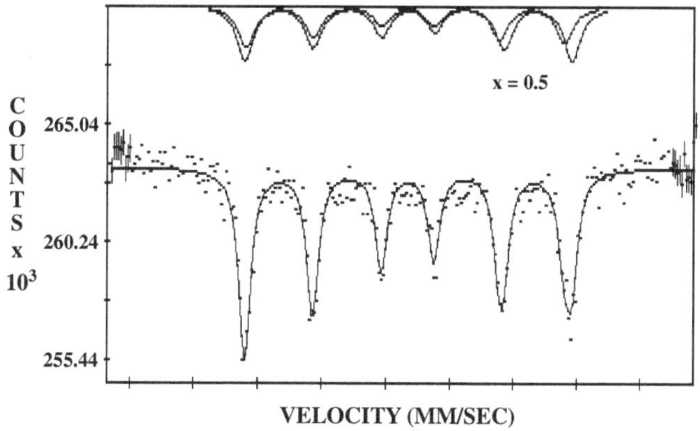

Figure 1. Mössbauer spectrum of $Cr_{0.5}$ $Li_{0.5}$ Fe_2O_4 (a) unirradiated and (b) irradiated sample.

The room temperature (300 K) Mossbauer spectra of the unirradiated and irradiated samples of $Cr_{0.5}$ $Li_{0.5}$ Fe_2O_4 are shown in Figure 1. The Mossbauer spectrum (Figure 1a) of unirradiated sample shows a superposition of two Zeeman sextets corresponding to the magnetically ordered Fe^{3+} ions at the tetrahedral (A) and octahedral (B) sites. The Mossbauer spectrum (Figure 1b) of Li ion irradiated sample exhibits the damage creation in two ways:

(1) No paramagnetic doublet contribution appears in the centre of spectrum, indicating that within the Mossbauer experimental errors (+/− 0.3%) the paramagnetic phase induced by lithium ion irradiation is nearly absent.

(2) The Zeeman patterns corresponding to the Fe^{3+} in octahedral (B) and tetrahedral (A) sites hyperfine fields are well separated. That is the observed superimposed A and B sites in unirradiated sample (Figure 1a) are split into separate hyperfine fields after Li ion irradiation. Further an increase of the intensity of the lines (2)–(5) with respect to (1)–(6) signals an orientation of the hyperfine magnetic field towards a direction perpendicular to the ion path due to irradiation induced strain by the latent tracks.

These both phenomena [(1) and (2)] have been observed first in magnetic Fe_3O_4 below T_N [4] and it was shown that the hyperfine magnetic field seems to rotate towards a direction perpendicular to the ion path possibly due to the strain induced by the latent tracks.

The Mossbauer spectra (Figure 1) were least squares fitted using the MOSFIT program developed at University of Cincinnati (U.S.A). The computer simulation of Mossbauer spectra allows the determination of the hyperfine field distribution of Fe^{3+} ions in the A and B sites assuming an identical Lamb–Mossbauer for A and B sites. The solid lines through the data points in Figure 1(a,b) are results of computer fits using two magnetic sextets assuming equal line widths for A and B sites. The results are presented in Table I for both irradiated and unirradiated samples.

Table I. Mossbauer results for $Cr_{0.5}$ $Li_{0.5}$ Fe_2O_4

Sample	Fe^{3+}-site occupancy		A-site occupancy	B-site occupancy	Mossbauer	
	$I_A(\%)$	$I_B(\%)$	$(Li^{1+})^A$	$(Cr^{3+}Li^{1+})^B$	I_B/I_A	(Neel) $n_B^N(\mu_B)$
Irradiated	55.1(15)	44.9(15)	0.0	(0.5 + 0.)	0.815	1.5
Unirradiated	43.0(10)	57.0(10)	0.15	(0.5 + 0.35)	1.33	3.0

It is evident from Table I that the irradiated Fe^{3+} site occupancy for the A-site hyperfine field increased from 43% to 55% whereas the B-site hyperfine field decreased from 57% to 45% compared to unirradiated sample. According to Neel's two sublattice model of ferrimagnetism, the moment per formula unit in μ_B, n_B^N, is expressed as

$$n_B^N(x) = M_B(x) - M_A(x)....(1),$$

where M_B and M_A are the B and A sublattice magnetic moments. On the basis of site occupancies listed in Table I, an approximate cation distribution for irradiated $Cr_{0.5}$ $Li_{0.5}$ Fe_2O_4 sample can be written as $[Fe_{0.85}Li_{0.15}]^A$ $[Cr_{0.5}Li_{1.15}]^B$ 04. Therefore Neel's moment n_B^N for irradiated sample is

$$n_B^N(irrad) = \left[0.5\mu(Cr^{3+}) + 0.15(Li^{1+}) + 1.15\mu(Fe^{3+}\right]^B - \left[0.85\mu(Fe^{3+}) + 0.15\mu(Li^{4+})\right]^A.$$

Where $\mu(Fe^{3+})$, $\mu(Cr^{3+})$ and $\mu(Li^{1+})$ are ionic moments with respective values $5\mu_B$, $3\mu_B$ and $O\mu_B$. The calculated n_B^N values for irradiated and unirradiated samples are listed in Table I. The magnetic moment obtained for unirradiated sample from Mossbauer data using Neel's model is $1.5\mu_B$ which is nearly two times lower than the irradiated sample ($3.0\mu_B$). Mossbauer spectrum of $Cr_{0.5}$ $Li_{0.5}$ Fe_2O_4 sample suggests an irradiation induced anisotropy of the hyperfine magnetic field leading to the hypothesis that magnetic domain walls could be pinned by the extended defects. Similar behaviour has been observed in Fe_3O_4 [5] due to the change induced by heavy ion irradiation in magnetic properties.

Acknowledgements

The authors are thankful to NSC, New Delhi for providing experimental facilities. One of the author (SGP) is thankful to UGC for providing financial aid in the form of Teacher Fellowship.

References

1. Studer F., Groult D., Nguyen N. and Toulemonde M., *Nucl. Instrum. Methods* **B19/20** (1987), 856.
2. Fuchs G., Studer F., Balanzat E., Groult D., Jousset J. C. and Raveau B., *Nucl. Instrum. Methods* **B-12** (1985), 471.
3. Toulemonde M., Fuchs G., Studer F., Nguyen N. and Groult D., *Phys. Rev.* **B35** (1987), 6560.
4. Studer F., Pascard H., Groult D., Houpert C., Nguyen N. and Toulemonde M., *Nucl. Instrum. Methods* **B32** (1988), 389.
5. Meillon S., Studer F., Hervieu M. and Pascard H., *Nucl. Instrum. Methods* **B107** (1996), 363.
6. Blasse G., *Philips Rept. Suppl.* **3** (1964), 1.
7. Al-Dallal S., Khan M. N. and Ahmed A., *J. Mater. Sci.* **25** (1990), 407.

Hyperfine Interactions (2005) 160:241–246
DOI 10.1007/s10751-005-9157-5

Concentration Dependence of Room Temperature Magnetic Ordering in Dilute Fe: $Ge_{1-x}Te_x$ Alloy

D. R. S. SOMAYAJULU*, NARENDRA PATEL, MUKESH CHAWDA, MITESH SARKAR and K. C. SEBASTIAN
Department of Physics, M.S. University of Baroda, Baroda, India;
e-mail: drs_somayajulu@yahoo.com

Abstract. Hyperfine interaction techniques like Mossbauer spectroscopy are very sensitive tools to study the local probe interactions in dilute magnetic semiconductors. We report here a Mossbauer study on the concentration dependence in $Fe_{0.008}Ge_{1-x}Te_x$ for $x = 0$, 0.008, 0.016, 0.03 and 0.05. At room temperature magnetic interactions were observed for all concentrations of Te and the population of magnetic site was found to increase gradually with the Te concentration. A constant magnetic hyperfine field of 136 KOe was found. A quadrupole doublet due to the $FeTe_2$ compound phase was also seen.

Key Words: dilute magnetic semiconductors, magnetic ordering, Mossbauer spectroscopy.

1. Introduction

Dilute Magnetic Semiconductors prepared by doping transition metals impurities into nonmagnetic semiconductors have recently become one of the key research areas in the field of SPINTRONICS [1]. Various efforts have been made to raise the Curie temperature up to room temperature by achieving greater understanding of the magnetic interaction and transport properties of the materials. Recently in magnetically doped ZnO [2], GaN [3] and TiO_2 [4] the Curie temperature was found to be 280, 250 and 400 K, respectively. In ternary compounds like $(Cd_{0.8}Mn_{0.2})GeP_2$ [5], $(Zn_{0.8}Mn_{0.2})SnAs_2$ [6] and $(Zn_{0.94}Mn_{0.06})GeP_2$ [7], ferromagnetic ordering was observed at 320, 329 and 312 K, respectively.

However in Group IV semiconductors (Ge) with a band gap of about 0.65 eV (at 300 K), the dilute quantity of Mn, Cr and Fe substitution showed values of T_c of 284 [8], 126 [9] and 233 K [9], respectively. The GeTe based DMS thin films showed magnetic ordering at 100 K [10]. Thus RT ferromagnetic ordering in Ge based alloys is yet to be achieved. Also it becomes important to study the type of interactions at the Fe site experienced in producing the magnetic order. To study such local probe–impurity interactions, microscopic hyperfine interaction techniques are the best tools compared to bulk solid state techniques.

* Author for correspondence.

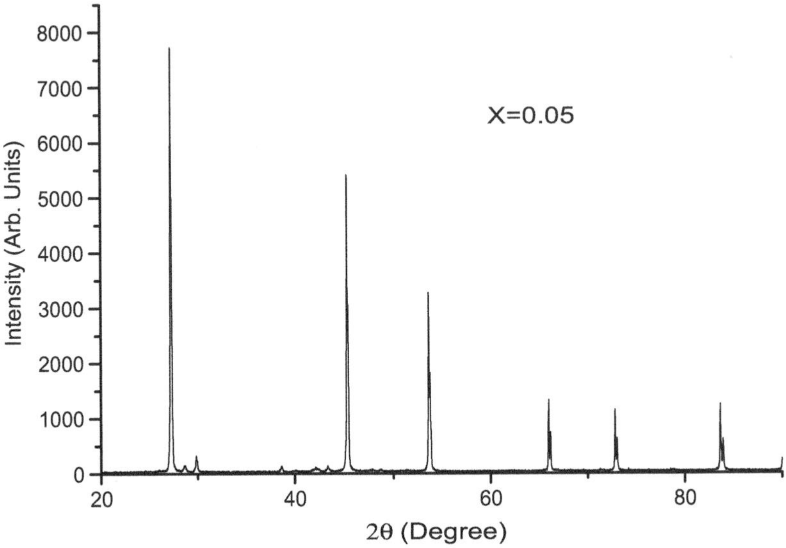

Figure 1. XRD spectra of $Fe_{0.008}Ge_{0.942}Te_{0.05}$.

Here we report the room temperature hyperfine magnetic field at the Fe site in dilute Fe: $Ge_{1-x}Te_x$ alloy. The Te concentration was varied from $0.008 \leq x \leq 0.05$, at an Fe concentration of 0.8 at %.

2. Experimental

High-purity germanium (99.99%), enriched Fe^{57} and tellurium (99.99%) were taken to make the $Fe_{0.008}Ge_{1-x}Te_x$ alloys, where x = 0, 0.008, 0.016, 0.03 and 0.05. The samples were enclosed in small quartz ampoules, which were then evacuated to 10^{-5} Torr and sealed properly. After encapsulation, the sealed ampoules were shaken to mix the contents thoroughly and then loaded into a furnace. The ampoules were heated to 1050°C for 24 h and then slowly cooled to room temperature. The samples prepared in this way were in the form of globules, which were powdered finely by grinding. Approximately 100 mg of each sample was then spread uniformly over a 1.2 cm^2 area to make the Mossbauer absorbers. To check the uniform distribution of Fe and Te in the host matrix, the globules prepared were broken into different parts and Mossbauer spectra were recorded for each. The results were identical within the experimental errors. The spectra were recorded using a Mossbauer constant acceleration spectrometer whose line width is 0.28 mm/s. The source used was ^{57}Co in a Rh matrix.

3. Results and discussion

XRD spectra were obtained with the Cu-Kα radiation (λ = 1.54 Å). Figure 1 shows the XRD spectrum of a sample with x = 0.05 and reveals shifts in the peak

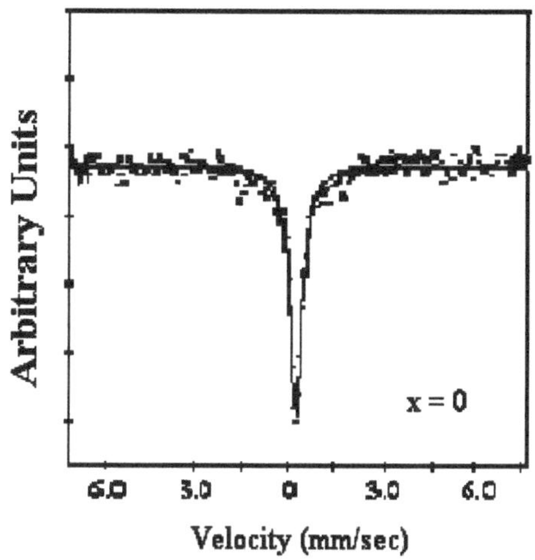

Figure 2. Mossbauer spectra of the sample $Fe_{0.008}Ge_{0.992}$ at room temperature.

positions of Ge with respect to its Ge standard 2θ values. The lattice constant determined from the position of the (111) peak is 5.069 and 5.039 Å for $Fe_{0.008}Ge_{0.942}Te_{0.05}$ and Ge, respectively. The change in lattice constant with the addition of the Te impurity in Ge argues strongly in favor that the Te ions are incorporated in substitutional sites of the Ge host lattice. The Fe concentrations being very small the XRD spectra are not sensitive to it.

Figures 2 and 3 show typical Mossbauer spectra of all the samples. The spectra recorded were least-square fitted using the standard Meerwal program and the parameters evaluated are listed in the Table I. Here the quoted values of the isomer shift (IS) are with respect to natural iron α-Fe. Figure 2 illustrates the singlet state for $x = 0$. The isomer shift evaluated is 0.31 mm/s, which agrees quite well with the earlier reported value of $FeGe_2$ phase [11]. While the $FeGe_2$ phase is antiferromagnetic below room temperature, it shows a paramagnetic behavior at 300 K and appears to be singlet in ^{57}Fe Mossbauer spectra. Thus it can be concluded that there is a formation of the $FeGe_2$ compound phase in $Fe_{0.008}Ge_{0.992}$ ($x = 0.008$) alloy.

Four concentrations of Te were introduced into the $Fe_{0.008}Ge_{1-x}Te_x$ alloy. They are $x = 0.008$, 0.016, 0.03 and 0.05. In all the samples (see in Figure 3), we observed a magnetic hyperfine (HMF) site A, the quadrupole (QS) site B and the singlet site C. This indicates that Fe is located at three different environments in the matrix. The calculated hyperfine interaction parameters for sites A, B and C remain constant for all concentrations of Te. Table I shows that the singlet site C has an isomer shift of IS = 0.29 ± 0.02 mm/s which indicates that the $FeGe_2$ compound phase is forming. However the population of site C decreases with increasing concentration of Te, as shown in Figure 4.

Site B parameters are evaluated as QS = 0.51 ± 0.02 mm/s and IS = 0.30 ± 0.02 mm/s, respectively, which matches with the earlier reported values for

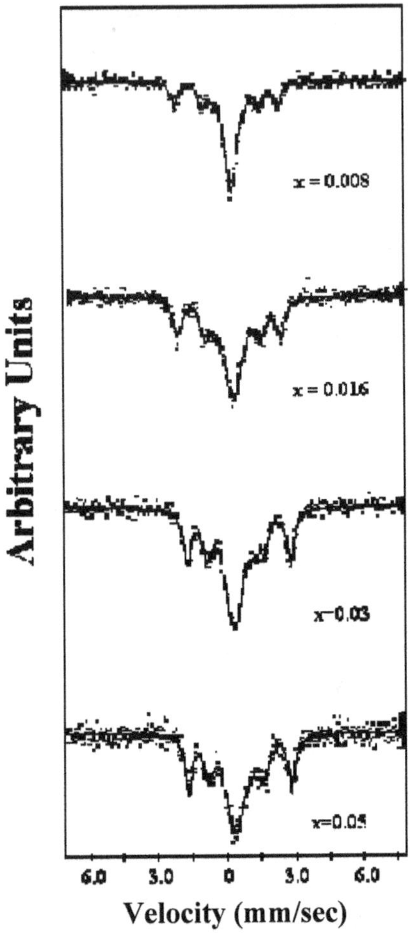

Figure 3. Mossbauer spectra of the samples $Fe_{0.008}Ge_{1-x}Te_x$ ($x = 0.008, 0.016, 0.03, 0.05$) at room temperature.

Table I. Mossbauer Parameters of Fe: Ge_{1-x} Te_x alloys. x=0.008, 0.016, 0.03, 0.05

Composition of samples	Magnetic field (KOe)	Quadrupole splitting (mm/s)	Isomer shift (mm/s)			Fraction (%)		
$Fe_{0.008}Ge_{1-x}Te_x$	A	B	A	B	C	A	B	C
$x = 0$	–	–	–	–	0.31(2)	–	–	100
$x = 0.008$	136(1)	0.51(2)	0.37(2)	0.26(2)	0.29(2)	43(2)	43(1)	13(0.5)
$x = 0.016$	136(1)	0.49(2)	0.35(2)	0.26(2)	0.25(2)	53(2)	31(1)	14(0.5)
$x = 0.03$	135(1)	0.51(2)	0.35(2)	0.27(2)	0.27(2)	63(2)	29(1)	7(0.5)
$x = 0.05$	136(1)	0.50(2)	0.37(2)	0.30(2)	0.25(2)	78(2)	21(1)	2(0.5)

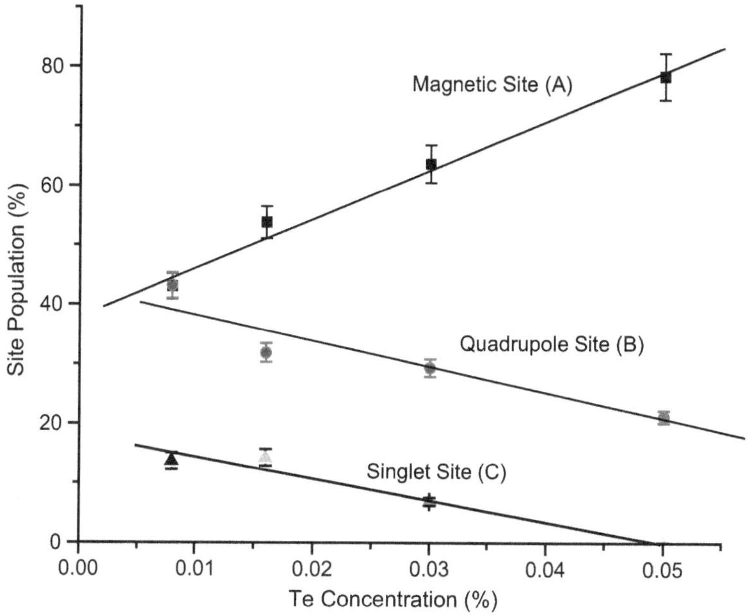

Figure 4. Plot of Site (*Magnetic*, *Quadrupole* and *Singlet*) population (%) vs. concentration of Te.

$FeTe_2$ [12] at RT. The quadrupole splitting remains constant for all Te concentrations, but the population of site B decreases with increasing concentration of Te.

The site A is magnetic with a hyperfine field of 136 ± 1 KOe at all concentrations of Te. There is a gradual increase in population of site A with increasing Te concentration as shown in Figure 4. It can also be seen that the population of site A increases at the cost of site B and C. Fe is an acceptor and Te is a donor in the Ge matrix. Because of the difference in electronegativity, Te is expected to be strongly attracted towards Fe and therefore forming donor–acceptor pairs and consequently breaking covalent bonds between Fe and Ge. This may be responsible for forming Fe local moments. On the other hand doping with acceptor impurities like indium (In) in the Fe–Ge system does not show any hyperfine magnetic field [13]. Furthermore with the increase in the donor Te concentration conduction electron densities are expected to increase in the system. These conduction electrons can bring about the RKKY type of interactions to polarize the local moments of Fe and give rise to the observed magnetic order in the $Fe_{0.008}Ge_{1-x}Te_x$ alloys. The decrease in the population of the quadrupole site B with simultaneous increase in the population of A site as the Te concentration increases, indicates that in Ge the donor (Te)-acceptor (Fe) pair formation is more probable than forming the $FeTe_2$ (B site) compound phase. Thus it appears that a ternary alloy of Ge–Fe–Te is responsible for the magnetic site in the system.

References

1. Ohno H., *Science* **281** (1998), 951.
2. Ueda K., Tabata H. and Kawai T., *Appl. Phys. Lett.* **79** (2001), 988.
3. Reed M. L., El-Masry N. A., Stadelmaier H. H., Ritums M. K., Reed M. J., Parker C. A., Roberts J. C. and Bedair S. M., *Appl. Phys. Lett* **79** (2001), 3473.
4. Matsumoto Y. *et al.*, *Science* **291** (2001), 854.
5. Medvedkin G. A., Ishibashi T., Nishi T., Hayata K., Hasegawa Y. and Sato K., *Jpn. J. Appl. Phys.* Part 2 **39** (2000), L949.
6. Choi S., Cha G. B., Hong S. C., Cho S., Kim Y., Ketterson J. B., Jeong S. Y. and Yi G. C., *Solid State Commun.* **122** (2002), 165.
7. Cho S. *et al.*, *Phys. Rev. Lett.* **88** (2002), 257203.
8. Sze S. M. and Irvin J. C., *Solid State Electron.* **11** (1968), 599.
9. Choi S., Hong S. C., Cho S., Kim Y., Ketterson J., Jung C.-U., Rhie K., Kim B.-J. and Kim, Y. C., *J. Appl. Phys.* **93** (2003), 7670.
10. Fukuma Y., Asada H., Miyashita J., Nishimura N. and Koyanagi T., *J. Appl. Phys* **93** (2003), 7667.
11. Delima J. C., Silva J., Grandi T., Sartorelli M., Silva M., Filho A., Sanchez D. and Baggio-Saitovitch E., *Hyperfine Interact.* **136** (2001), 45–56.
12. Ward J. B. and Howard D. G., *J.Appl. Phys.* **47** (1976), 389.
13. Patel N., Ph. D. Thesis, 2005, M S University, Baroda (Unpublished).

Hyperfine Interactions (2005) 160:247–252
DOI 10.1007/s10751-005-9158-4

Magnetic Ordering in $Fe_{0.008}Ge_{1-x}D_x$ (D = As, Bi)

NARENDRA PATEL[1,*], MUKESH CHAWDA[1], MITESH SARKAR[1], K. C. SEBASTIAN[1], D. R. S. SOMAYAJULU[1] and AJAY GUPTA[2]
[1]*Department of Physics, M. S. University of Baroda, Baroda, India;*
e-mail: narendra_msu@yahoo.co.in
[2]*Inter University Consortium, Indore, India*

Abstract. The formation of local moments and the effect of charge carriers in dilute magnetic semiconductors can be well understood using local probe techniques like Mossbauer Spectroscopy. We report here on Mossbauer studies in the systems $Fe_{0.008}Ge_{1-x}D_x$ (D = As, Bi), $Fe_{0.008}Ge_{1-x}In_x$, and $Fe_{0.008}Ge_{1-x}Sn_x$. At room temperature magnetic interactions were observed for donor (D) impurities at the Fe site in the $Fe_{0.008}Ge$ system. No such magnetic ordering was observed for acceptor (In) or neutral (Sn) impurities.

Key Words: dilute magnetic semiconductors, magnetic ordering, Mossbauer spectroscopy.

1. Introduction

The theory of carrier-induced ferromagnetism in Dilute Magnetic Semiconductors (DMS) [1, 2] has effectively explained the mechanism to raise the Curie temperature to room temperature and above. The above-room-temperature ferro-magnetism in Mn-doped III–V-group semiconductors ZnO [3], GaN [4] and GaP [5] is well understood through this mechanism, where a long-range interaction between Mn^{2+} ions occurs via mediated polarised itinerant carriers. Such carrier-induced magnetic ordering was also observed in other IV–VI compound systems such as GeTe doped with Cr, Mn or Fe [6], but the Curie temperatures were much below room temperature. The itinerant carriers (especially holes) responsible for the magnetic interactions in all these materials are brought by controlled doping of acceptor impurities of transition metals like Mn or Fe in nonmagnetic lattice sites of the compound semiconductors. Thus hole-mediated ferromagnetism is now well known, but the electron-mediated ferromagnetism (i.e., in the n-type material) is yet to be achieved at room temperature.

Here we investigate the effect of dilute donor impurities (As, Bi) in $Fe_{0.008}Ge$ alloys. In such dilute systems the local probe hyperfine interaction studies are

* Author for correspondence.

Table I. Mossbauer Parameters of Magnetic Hyperfine Field (HMF), Quadrupole Splitting (QS) and Isomer Shift (IS) for $Fe_{0.008}Ge_{0.982}$, $Fe_{0.008}Ge_{1-x}D_x$, where (D = As, Bi, In, Sn)

Composition of samples	Magnetic hyperfine field (kOe)	Quadrupole splitting (mm/s)	Isomer shift (mm/s)			Area under the curve (%)		
	A	B	A	B	Singlet C	A	B	Singlet C
$Fe_{0.008}Ge_{0.982}$	–	–	–	–	0.31 (2)	–	–	100 (1)
$Fe_{0.008}Ge_{1-x}As_x$ $x = 0.01$	121.2 (1.5)	–	0.22 (3)	–	0.29 (2)	49 (2)	–	51 (1)
$Fe_{0.008}Ge_{1-x}As_x$ $x = 0.03$	–	1.63 (2)	–	0.27 (1)	–	–	100 (1)	–
$Fe_{0.008}Ge_{1-x}As_x$ $x = 0.05$	–	1.63 (2)	–	0.27 (1)	–	–	100 (1)	–
$Fe_{0.008}Ge_{1-x}Bi_x$ $x = 0.01$	129.6 (1.2)	–	0.21 (8)	–	0.29 (8)	46 (2)	–	54 (1)
$Fe_{0.008}Ge_{1-x}Bi_x$ $x = 0.03$	131.4 (2.0)	–	0.27 (1)	–	0.27 (7)	51 (2)	–	48 (1)
$Fe_{0.008}Ge_{1-x}Bi_x$ $X = 0.05$	129 (1)	–	0.22 (4)	–	0.29 (1)	67 (2)	–	33 (1)
$Fe_{0.008}Ge_{1-x}In_x$ $X = 0.05$	–	–	–	–	0.31 (1)	–	–	100 (1)
$Fe_{0.008}Ge_{1-x}Sn_x$ $X = 0.05$	–	–	–	–	0.31 (1)	–	–	100 (1)

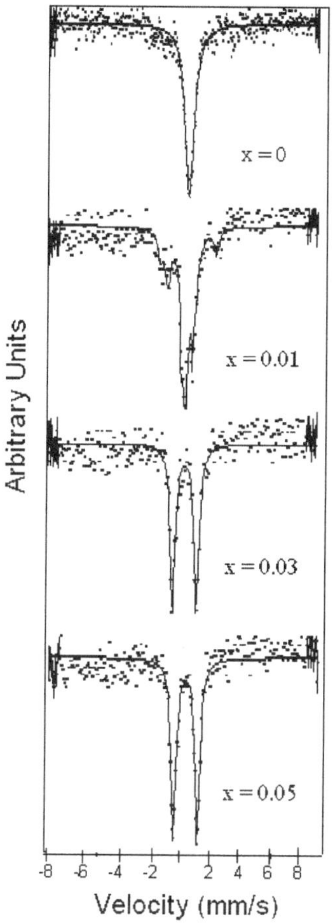

Figure 1. Mossbauer spectra of $Fe_{0.008}Ge_{1-x}As_x$ for x = 0, 0.01, 0.03 and 0.05.

expected to give appropriate information. Hence the present study was done using the Mossbauer Spectroscopic technique at room temperature.

2. Experimental

$Fe_{0.008}Ge_{1-x}D_x$ alloys were made by doping pentavalent impurities As and Bi at concentrations x = 0.01, 0.03 and 0.05 at.%. Here the Fe concentration being low, enriched ^{57}Fe powder was used to make these alloys. High purity (>5 N) Ge, ^{57}Fe and high purity (99.99%) of As or Bi were taken in the required proportions in the quartz ampoules. The ampoules were then sealed under a vacuum better than 10^{-5} Torr. The sealed ampoules were shaken well to mix the contents thoroughly, heated to 1050°C for 24 h and then slowly cooled to room temperature. The samples so prepared were in the form of globules, which were powdered finely to make the Mossbauer absorbers.

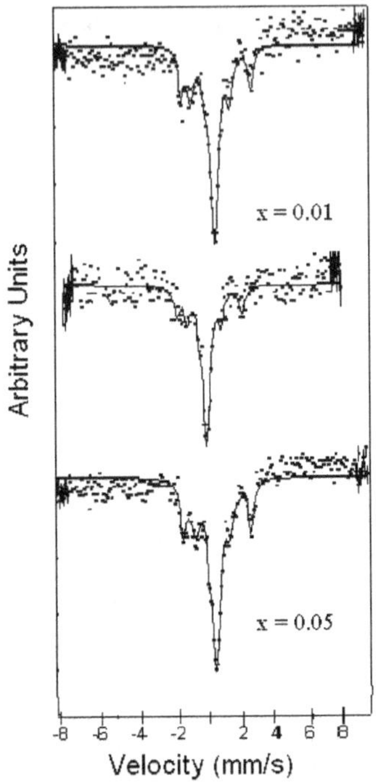

Figure 2. Mossbauer spectra of $Fe_{0.008}Ge_{1-x}Bi_x$ for $x = 0.01$, 0.03 and 0.05.

A similar procedure was used to make $Fe_{0.008}Ge_{1-x}In_x$ and $Fe_{0.008}Ge_{1-x}Sn_x$ alloys. The Mossbauer spectra of all the samples were recorded using a constant acceleration spectrometer, whose line width was 0.28 mm/s. The source used was Co^{57} in a Rh matrix. All the spectra were least-square fitted using the Meerwal program; the deduced Mossbauer parameters, magnetic hyperfine field, quadrupole splitting (QS) and isomer shift (IS), are listed in Table I. The quoted values of the isomer shift are with respect to natural iron.

3. Results and discussion

XRD spectra for all the systems were obtained with the Cu-Kα radiation (λ = 1.54 Å). The spectra reveal the shift in peaks of Ge with respect to its standard 2θ values. The small change in lattice constant with the addition of donor impurities in Ge argues strongly that the donor ions are incorporated in substitutional sites of Ge host lattice. The Fe concentrations being very small, the XRD spectra are not sensitive to it. However, to check the uniform dispersion of Fe in the sample, absorbers were made from different parts of the same globule. The Mossbauer spectra of the different parts of a globule were found to be identical. The lines being reasonably sharp Fe can be expected to be in a defined unique position.

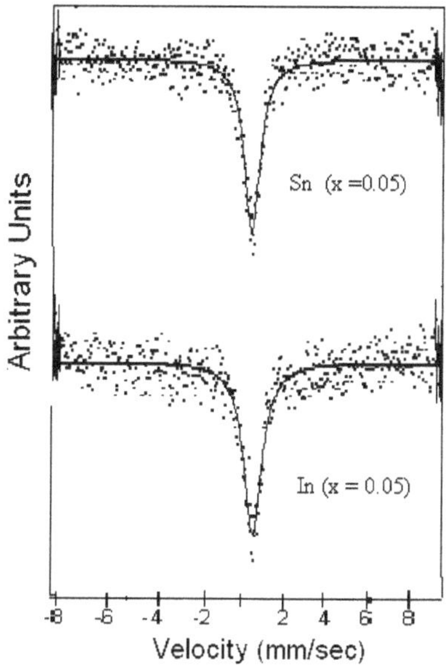

Figure 3. Mossbauer spectra of $Fe_{0.008}Ge_{1-x}In_x$ and $Fe_{0.008}Ge_{1-x}Sn_x$.

The Mossbauer spectra of $Fe_{0.008}Ge$ and $Fe_{008}Ge_{1-x}D_x$(D = As, Bi) for x = 0.01, 0.03 and 0.05 are shown in Figures 1 and 2. The $Fe_{0.008}$ Ge (x = 0) sample showed only a single line and the parameter IS= 0.31 mm/s compares with the reported value of the $FeGe_2$ compound [8] as shown in Figure 1.

In the case of As at x = 0.01 (Figure 1), one magnetic site (site A) and a singlet site (site C) were observed. The magnetic hyperfine field (MHF) was evaluated to be 121 kOe with IS = 0.22 mm/s. The isomer shift of the site C, IS= 0.29 mm/s, is same as that for the $FeGe_2$ compound. At the higher concentrations of x = 0.03 and 0.05, only a quadrupole site (site B) with QS = 1.63 mm/s and IS = 0.27 mm/s was observed. These Mossbauer parameters coincide with the reported [7] values of the $FeAs_2$ compound. Thus when the As-concentration exceeds x = 0.01, Fe seems to have a high probability of forming the compound phase $FeAs_2$.

Figure 2 shows the Mossbauer spectra for $Fe_{0.008}Ge_{1-x}Bi_x$ for x = 0.01, 0.03 and 0.05. For all three concentrations a magnetic sextet (site A) and a singlet (site C) could be fitted to the spectra. Best fits gave for the site A a MHF of 129 kOe. The IS value obtained for site C is found to be same as that for the $FeGe_2$ compound.

Fe acts as an acceptor [9] and As and Bi are known to be donors in Ge. The electro-negativity of Fe, Ge, As and Bi are 1.83, 2.01, 2.18 and 2.01, respectively. Thus As has a strong affinity towards Fe to form compounds and, indeed, for higher concentrations $FeAs_2$-compound formation was observed (site B). At the lower As-concentration (x = 0.01), donor (As)-acceptor (Fe) pairs (DAP) are likely to be formed, even the formation of a possible complex of the type Ge–Fe–As–Ge. In such a complex there can be sp–d exchange interactions between

Fe and As, which may be responsible for producing the magnetic moment on Fe. The line widths of all the lines did not exceed 0.31 mm/s thereby suggesting unique Fe sites. At higher As-concentrations, compound formation seems to dominate over DAP formation, but for Bi the DAP formation seems to be likely at higher concentrations. We have not found any information on magnetic properties of the Fe–Bi compounds in the literature. For all concentrations of Bi a magnetic site was observed. The population of this site gradually increased as Bi concentration is increased with a corresponding decrease in the $FeGe_2$ component. Hence it can be concluded that with Bi in the system the formation of Donor (Bi)-Acceptor (Fe) pairs is highly probable and through sp–d exchange mechanism a local magnetic moment may be produced on the Fe atoms. The increase in the population of site A may indicate that with increasing Bi concentration more charge carriers (electrons in this case) are brought into the system and a carrier mediated polarization of the local moments may be taking place.

To test the assumption of the formation of DAPs in Ge, In (acceptor) and Sn (neutral) impurities were doped along with Fe. The Mossbauer spectra of the systems $Fe_{0.008}Ge_{0.942}In_{0.05}$ and $Fe_{0.008}Ge_{0.942}Sn_{0.05}$ (shown in Figure 3) did not show any magnetic interactions. Only singlet sites (C) were observed in both cases, which correspond to $FeGe_2$. This indicates that if a donor is present in the system along with Fe, there is a likelihood of forming DAPs in Ge. Hence it seems reasonable to expect that the donor-acceptor pairs are responsible for the local moment formation. This suggests that the ternary alloy Ge–Fe–D–Ge where D is a donor, is likely to produce magnetic ordering in Ge.

Acknowledgements

The authors would like to thank Dr. Raghvendra Reddy and Dr. Suresh Bhardwaj for their kind support in carrying out experimental work. The authors also like to thank the Director, Inter University Consortium, Indore, for extending us to use the experimental facilities.

References

1. Ohno H. *et al.*, *Phys. Rev. Lett.* **68** (1992), 2664.
2. Story T., Galazka R. R., Frankel R. B. and Wolff P. A., *Phys. Rev. Lett.* **56** (1986), 777.
3. Ueda H., Tabata H. and Hawai T., *Appl. Phys. Lett.* **79** (2001), 988.
4. Reed M. L. *et al.*, *Appl. Phys. Lett.* **79** (2001), 3473.
5. Pearton S. J. *et al.*, *Phys. Status Solidi, A* **195** (2003), 222.
6. Fukuma Y., Asada H., Miyashita J., Nishimura N. and Koyanagi T., *J. Appl. Phys.* **93** (2003), 7667.
7. Ioffe P. A., Tsemekhman K., Parshukova L. N. and Boblovskii A. G., Russ. *J. Inorg. Chem.* **30** (1985), 1566.
8. Delima J. C., Silva J., Grandi T., Sartorelli M., Silva M., Filho A., Sanchez D. and Baggio-Saitovitch E., *Hyperfine Interact.* **136** (2001), 45.
9. Sze S. M. and Irvin J. C., *Solid-State Electron.* **11** (1968), 599.

Hyperfine Interactions (2005) 160:253–260
DOI 10.1007/s10751-005-9159-3

^{151}Eu Mössbauer Spectroscopy in $La_{0.48}Eu_{0.29}Ca_{0.33}MnO_3$

T. GOVARDHAN REDDY[1,*], P. YADAGIRI REDDY[2],
V. RAGHAVENDRA REDDY[2], AJAY GUPTA[2] and K. RAMA REDDY[1]
[1]*Department of Physics, Osmania University, Hyderabad 500 007, India;*
e-mail: govardhanreddyturpu@yahoo.com
[2]*UGC-DAE Consortium for Scientific Research, Khandwa Road, Indore 452 017, India*

Abstract. Electrical conductivity with and without magnetic field, d.c. magnetization and ^{151}Eu Mössbauer studies were carried out in $La_{0.48}Eu_{0.29}Ca_{0.33}MnO_3$ perovskite manganite system. An insulating ground state is found throughout the temperature range with charge ordered (CO) state emerging at $T_{CO} \sim 140$ K, where as an external magnetic field of 6 T induces metal–insulator transition at ~120 K. D.C. magnetization measurements show the antiferromagnetic (AFM) transition occurring at $T_N \approx 48$ K. The temperature dependent ^{151}Eu Mössbauer measurements showed that the substituted Eu replaces La^{3+} in the 3+ charge state and a small magnetic moment gets induced at the Eu nucleus at low temperatures. The anomalous variation of the *f*-factor with temperature occurring around T_N and T_{Co} corroborates the occurrence of antiferromagnetic (AFM) and charge ordering (CO) transition, respectively.

1. Introduction

Several physical properties of doped manganites, $R_{1-x}A_xMnO_3$ (R being rare earth ion and A divalent cation), have created a flurry of research activity to understand the interplay of charge, lattice and spin dynamics [1–3]. The prototype compound of this family $LaMnO_3$ is an antiferromagnetic insulator characterised by a strong structural distortion interpreted in terms of co-operative Jahn–Teller (J–T) effect. Upon doping with a divalent cation (Ca, Sr) at La site, various interesting phenomenon like charge ordering (CO), orbital ordering (OO), antiferromagnetism (AFM), canted structures and ferromagnetism (FM) were observed [4–6]. The J–T phonons play a crucial role in understanding the resistance behaviour with temperature in these systems [7]. Spectroscopic studies based on neutron [8, 9], EXAFS [10], μSR [11] reveal that the role of the J–T phonons is vital in controlling the electrical transport. The charge carriers coupled to spins in the system produce simultaneous magnetic transition also [12].

* Author for correspondence.

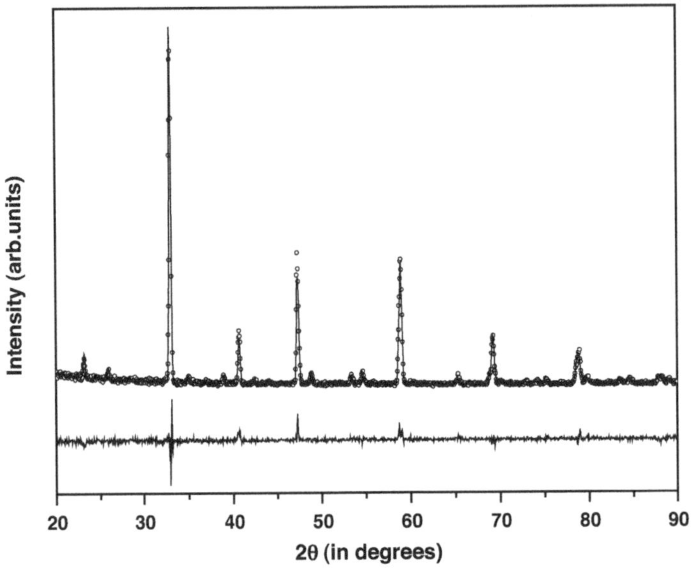

Figure 1. The XRD pattern of $La_{0.38}Eu_{0.29}Ca_{0.33}MnO_3$ and Rietveld refined pattern along with the difference in both.

Widely studied CMR system $La_{0.67}Ca_{0.33}MnO_3$ shows a ferromagnetic (FM) transition at ~250 K [1]. Partial replacement of La by Eu in this system is expected to reveal the role of spin and lattice polarons around the metal–insulator transition as the existence of small polarons in the insulating paramagnetic phase was questioned based on ^{57}Fe Mössbauer studies [13]. In our recent studies, it was confirmed through Eu substituted system that small- and spin-polarons do coexist around the metal–insulator transition [14]. In this context an attempt to understand the role of the J–T distortions in the $La_{0.38}Eu_{0.29}Ca_{0.33}MnO_3$ system is made through ^{151}Eu Mössbauer studies along with electrical and magnetic transport measurements. Temperature variation of *f*-factor and quadrupole coupling constant is interpreted in terms of the role of the J–T distortions in the CO and AFM phase transitions.

2. Experimental

Samples with composition of $La_{0.67-x}Eu_xCa_{0.33}MnO_3$ (x = 0.17, 0.21, 0.25 and 0.29) were prepared through the standard solid state route using stoichiometric quantities of La_2O_3, Eu_2O_3, $CaCO_3$ and MnO_2 with nominal purities higher than 99.9%, after standardising the procedure for consistency of results. In this report the results on the sample with $x = 0.29$ is being presented, as other concentrations of Eu yielded normal CMR system with huge resistivity cusp [14]. The phase purity and crystal structure of the synthesised samples were examined by the step scanned powder diffraction pattern recorded at 300 K using Siemens D5000

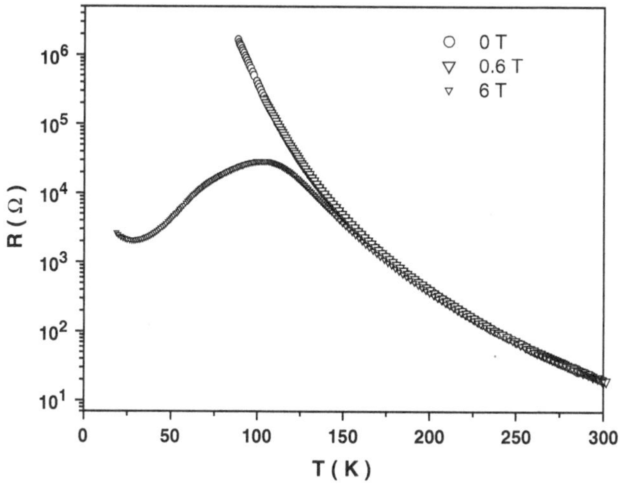

Figure 2. The temperature dependence of the electrical resistivity of $La_{0.38}Eu_{0.29}Ca_{0.33}MnO_3$ at different external magnetic fields 0 T, 0.6 T and 6 T.

diffractometer with the Cu Kα radiation. The electrical resistance measurements of the samples were carried out as a function of temperature and magnetic field by four-probe technique in the superconducting magnet (OXFORD-make). ZFC and FC dc magnetization measurements were performed at 100 Oe field with a DC extraction magnetometer using physical properties measurements systems (PPMS, Quantum design). The field was applied in the plane of the sample. The ^{151}Eu Mössbauer measurements were carried out in the transmission geometry. The Mössbauer source was SmF_3 (^{151}Sm, ~60 mCi). Mössbauer spectra were recorded at different temperatures in the temperature range of 18–300 K, using the Closed Cycle Refrigerator (Jani's-make).

3. Results and discussion

Figure 1 gives the powder X-ray diffraction pattern of the sample $La_{0.38}Eu_{0.29}Ca_{0.33}MnO_3$ along with the Rietveld refined pattern [15] and difference of both. The Rietveld refinement of diffraction pattern is done, by assuming the orthorhombic structure with Pbnm space group. The evaluated lattice parameters are $a = 54.245$ nm, $b = 54.542$ nm and $c = 76.680$ nm. The decrease in the lattice parameters due to partial substitution of Eu in the parent compound $La_{0.67}Ca_{0.33}MnO_3$ [16] at La site is also observed. This is due to the increased distortions caused by the substitution of the lesser ionic radii element at La in the system, which leads to the suppression of the double exchange interaction and makes the ground state to be an insulator. The temperature dependence of the electrical resistance is shown in Figure 2 at different fields 0 T, 0.6 T and 6 T. The system is an insulator at 0 T and 0.6 T magnetic fields. The temperature

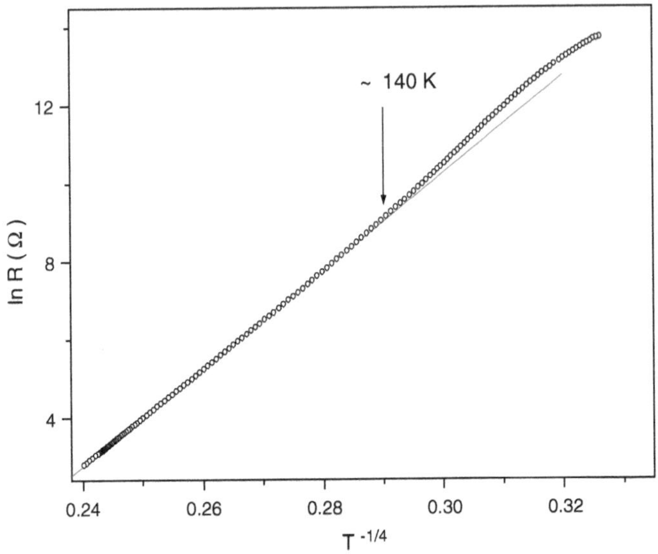

Figure 3. The Variable Range Hopping model fitting to the zero field R–T data.

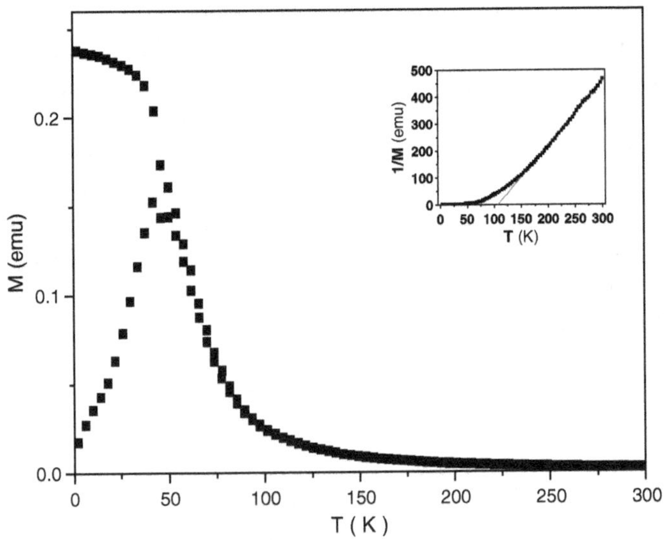

Figure 4. The ZFC and FC magnetization behaviour with temperature of $La_{0.38}Eu_{0.29}Ca_{0.33}MnO_3$, inset shows the deviation from Curie–Wiess behaviour around ~150 K.

variation of zero field resistance follows the Variable Range Hopping (VRH) model [17] down to ~150 K, as shown in Figure 3. The zero field resistivity variation with temperature shows a small but discernible change in the slope around 140 K, which corresponds to the onset of charge ordering [5]. Application of external magnetic field drastically modifies such an insulating charge-ordered state and induces a metal–insulator transition as observed in Figure 2.

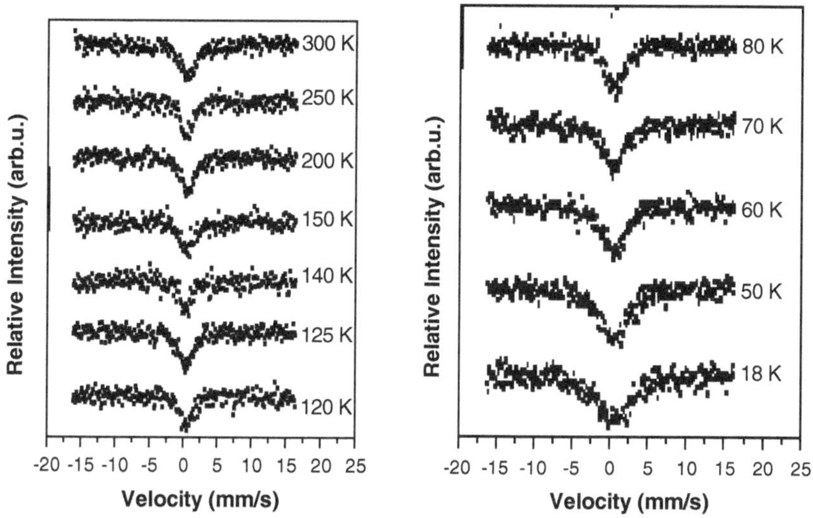

Figure 5. The ^{151}Eu Mössbauer spectra of $La_{0.38}Eu_{0.29}Ca_{0.33}MnO_3$ at various temperatures from 300 K to 18 K.

The ZFC and FC magnetization measurements shown in Figure 4 indicate the AFM character of the sample below $T_N \approx 48$ K. At high temperatures χ follows Curie–Wiess behaviour with a deviation in the $T < 150$ K region (Figure 4 inset) [4]. A long range antiferromagnetic ordering occurs with stabilisation of the charge and orbital ordering due to the co-operative J–T distortions at ~48 K. The temperature region $T_N < T < T_{co}$ is of mixed tendencies and the co-operative J–T distortions increase stabilising the charge ordered AFM state.

The Mössbauer spectra of ^{151}Eu shown in Figure 5 were recorded at several temperatures. The isomer shift is found to be $\delta = 0.5$ mm/s (with reference to Eu_2O_3), which clearly shows that Eu is stabilised in the 3+ state [18] occupying the La^{3+} rare earth site.

In the temperature region above $\sim T_N$ the Mössbauer spectra were analysed for quadrupole splitting. The temperature variation of quadrupole coupling constant is shown in Figure 6. It is seen that minimum in the quadrupole interaction occurs around $T \approx T_{CO}$ and increase of quadrupole interaction below T_{co} is attributed to the onset of co-operative J–T distortion. Our results are also consistent with that of the charge ordering observed through the X-ray scattering [5]. Analysis of Mössbauer spectra for the evaluations of magnetic field in the AFM region gives the induced field at Eu site as 3.5 T at 18 K.

The *f*-factor, defined as the ratio of area under Mössbauer line to the background, shown in Figure 7 has drastic variations around T_{CO} which may be attributed to the change in the crystal structure parameters due to charge ordering [5]. Debye behaviour when fitted to the experimental *f*-factor values without considering the points around the phase transitions gave Debye temperature θ_D

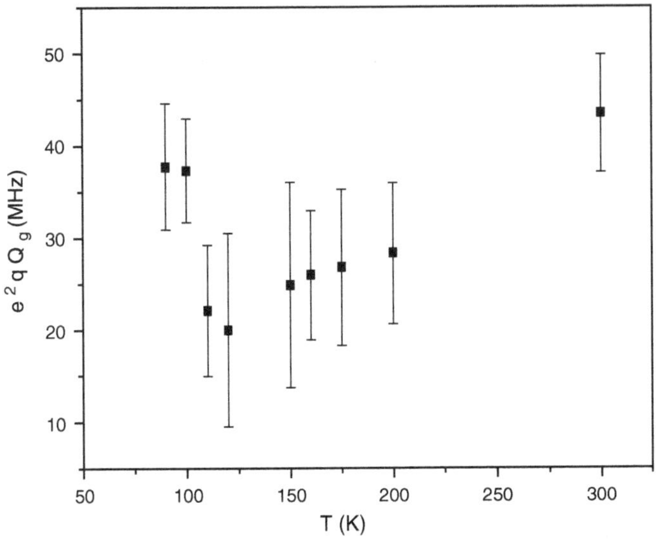

Figure 6. Temperature variation of the quadrupole coupling constant above T_N.

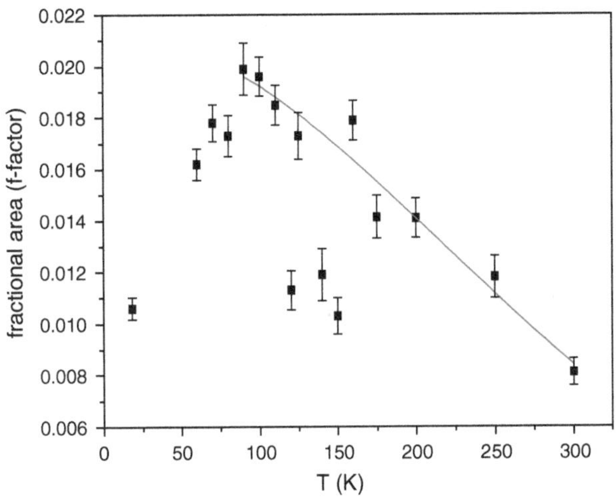

Figure 7. The temperature dependence of the Mössbauer *f*-factor (fractional area under the curve) of $La_{0.38}Eu_{0.29}Ca_{0.33}MnO_3$, the *solid line* depicts fit to the Debye behaviour.

~265 K ± 30 K which is typical of manganite systems [10]. It is obvious to expect non-Debye behaviour in *f*-factor at phase transition as the mean square displacement of Eu ion varies drastically due to structural readjustments at phase transition boundaries [7–10]. A similar trend is also observed at AFM transition at lower temperatures, as shown in Figure 7.

4. Conclusions

The partial substitution of Eu for La in the $La_{0.67}Ca_{0.33}MnO_3$ system suppresses DE interaction and beyond a critical composition of Eu, the system becomes insulator with the onset of charge ordering followed by AFM ordering. The transferred hyperfine field at Eu nuclei is found to be 3.5 T in the AFM region around 18 K. The anomalous variation of the *f*-factor with temperature manifests the structural variation at charge ordering T_{CO} and long range AFM ordering effecting the *f*-factor similar to the superconducting transitions [19].

Acknowledgements

One of the authors Mr. T. G. Reddy thanks the Council of Scientific and Industrial Research (CSIR), India for financial support through Senior Research Fellowship. Authors from Osmania University thank Department of Science and Technology, India for research support vide project No. SP/S2/M-67/96. We thank Dr. Mukul Gupta of Paul Scherrer Institute, Switzerland for providing the magnetization data on our samples. We also thank Dr. Rajeev Rawat of UGC-DAE CSR, Indore for helping us in getting the in-field electrical resistance measurements.

References

1. Tokura Y. (ed.), *Colossal Magnetoresistive Oxides*, Gordon and Breach Science Publishers, 2000.
2. Salamon M. B. and Jaime M., *Rev. Mod. Phys.* **73** (2001), 583.
3. Nagaev E. L., *Phys. Rep.* **346** (2001), 387.
4. Aluraj A., Biswas A., Raychaudhuri A. K., Rao C. N. R., Woodward P. M., Voft T., Cox D. E. and Cheetam A. K., *Phys. Rev., B* **57** (1997), R8115.
5. Cox D. E., Radaelli P. G., Marezio M. and Cheong S.-W., *Phys. Rev., B* **57** (1997), 3305.
6. Tomioka Y., Asamitsu A., Kuwahara H., Moritomo Y. and Tokura Y., *Phys. Rev., B* **53** (1995), R1689.
7. Millis A. J., Littlewood P. and Shraiman B., *Phys. Rev. Lett.* **74** (1995), 5144.
8. Radaelli P. G., Marezio M., Hwang H. Y., Cheong S.-W. and Batlogg B., *Phys. Rev., B* **54** (1996), 8992.
9. Dai P., Zhang J., Mook H. A., Liou S.-H., Dowben P. A. and Plummer E. W., *Phys. Rev., B* **54** (1996), R3694.
10. Meneghini C., Cimino R., Pascarelli S., Mobili S., Raghu C. and Sarma D. D., *Phys. Rev., B* **56** (1997), 3520.
11. Teressa J. M. D., Ibarra M. R., Algarabel P., Morellon L., Garcia-Landa B., Marquina C., Ritter C., Maignan A., Martin C., Raveau B., Kurbakov A. and Trounov V., *Phys. Rev., B* **65** (2002), 100403R.
12. Zener C., *Phys. Rev.* **82** (1951), 403.
13. Chechersky V., Nath A., Isaac I., Franck J. P., Ghosh K. and Greene R. L., *Phys. Rev., B* **63** (2001), 052411.
14. Reddy T. G., Reddy P. Y., Reddy V. R., Gupta A. and Reddy K. R., *Solid State Commun.* **133** (2005), 77.

15. Young R. A. (ed.), *The Rietveld Method*, Oxford University Press, New York, 1993.
16. Balsco J. *et al.*, *Phys. Rev., B* **55** (1997), 8905.
17. Mott N. F. (ed.), *Metal–Insulator Transition*, 2nd edn. Taylor & Francis, p. 67.
18. Greenwood N. N. and Gibb T. C. (eds.), *Mössbauer Spectroscopy*, Chapman and Hall Ltd., London, 1971.
19. Jia Y. Q., Jin M. Z. and Liu X. W., *Phys. Status Solidi, B* **182** (1994), 177.

Author Index to Volume 160 (2005)

Ashtaputre S., 81
Asokan K., 181
Avasthi D. K., 95

Bakare P. P., 199
Balasubramanian C., 199
Bangal M., 81
Bertschat H. H., 3
Bhave T. M., 199
Bhoraskar S. V., 199
Butz T., 27

Chawda M., 241, 247
Chiou J. W., 181

Date S. K., 199
Dederichs P. H., 57
Dietrich M., 3, 17
Dogra A., 143
Dolia S. N., 143, 219

Esquinazi P., 27
Ethiraj A., 81

Gosavi S. W., 81
Gupta A., 123, 143, 157, 247, 165, 253
Gupta M., 157
Gupta R., 39, 107

Han K.-H., 27
Hebalkar N., 81
Höhne R., 27

Jain S. K., 219
Jan J. C., 181
Joshi P. B., 173

Katayama-Yoshida H., 57
Kavita S., 157
Knobel M., 143
Kuberkar D. G., 193
Kulkarni R. G., 235
Kulkarni S. K., 81
Kulkarni S., 199
Kumar D., 165
Kumar R., 143, 235, 181
Kurup V., 173

Lakshmi N., 227
Lieb K. P., 39, 107
Lieb K.-P., 1

Malik S. K., 193
Manzhur Y., 3
Marathe G. R., 173
Marathe S., 81
Markna J. H., 193
Mavani K. R., 193
Mehta R. V., 211
Müller G. A., 39, 107
Murthy C. N., 189

Nagar H., 199

Parekh K., 211
Parmar R. N., 193
Parmar S. G., 235
Pasricha R., 199
Patel N., 241, 247
Pong W. F., 181
Potzger K., 3
Prandolini M. J., 3
Pratap A., 173

Rana D. S., 193
Rao K. V. R., 181
Ray S., 67
Reddy K. R., 253

Reddy P. Y., 253
Reddy T. G., 253
Reddy V. R., 157, 253

Samokhvalov V., 17
Sarkar M., 241, 247
Sarma D. D., 67
Sato K., 57
Schaaf P., 39, 107
Schneider F., 17
Sebastian K. C., 241, 247
Setzer A., 27
Shah N. A., 193
Sharma R. K., 227
Sharma S. K., 143
Singh M., 143
Siva Kumar V. V., 143
Somayajulu D. R. S., 1, 241, 247
Spemann D., 27

Thaker C. M., 193
The Isolde Collaboration, 17
Tsai H. M., 181
Tsai M.-H., 181

Unterricker S., 17
Upadhyay R. V., 211
Urban J., 81

Venugopalan K., 227

Weber A., 3

Zeitz W.-D., 3
Zhang K., 39

FSC
www.fsc.org
MIX
Papier | Fördert
gute Waldnutzung
FSC® C083411

Kolibri 360 GmbH
Ferdinand-Jühlke-Straße 7
99095 Erfurt, Deutschland
produktsicherheit@kolibri360.de